Theory of Randomized Search Heuristics

Foundations and Recent Developments

SERIES ON THEORETICAL COMPUTER SCIENCE

ISSN: 1793-849X

This book series provides an up-to-date and reliable source of information from fundamental knowledge to emerging research areas related to theoretic computer science. The target audiences cover researchers, engineers, scientists, students and professionals.

Vol. 1: Theory of Randomized Search Heuristics: Foundations and Recent Developments
edited by Anne Auger and Benjamin Doerr

Series on Vol. 1
Theoretical Computer Science

Theory of Randomized Search Heuristics

Foundations and Recent Developments

Editors

Anne Auger
INRIA, France

Benjamin Doerr
Max-Planck-Institut für Informatik, Germany

NEW JERSEY • LONDON • SINGAPORE • BEIJING • SHANGHAI • HONG KONG • TAIPEI • CHENNAI

Published by

World Scientific Publishing Co. Pte. Ltd.
5 Toh Tuck Link, Singapore 596224
USA office: 27 Warren Street, Suite 401-402, Hackensack, NJ 07601
UK office: 57 Shelton Street, Covent Garden, London WC2H 9HE

British Library Cataloguing-in-Publication Data
A catalogue record for this book is available from the British Library.

Series on Theoretical Computer Science — Vol. 1
THEORY OF RANDOMIZED SEARCH HEURISTICS
Foundations and Recent Developments

ISBN-13 978-981-4282-66-6
ISBN-10 981-4282-66-9

Printed in Singapore by World Scientific Printers.

To Ingo Wegener

Preface

Randomized search heuristics such as evolutionary algorithms, evolution strategies, genetic algorithms, simulated annealing or ant colony and particle swarm optimization are recognized as powerful optimization algorithms and play an important role in modern algorithmics nowadays. While their success in many application fields is undoubted, we have little theoretical understanding why and when these methods work well.

The theory of randomized search heuristics tries to answer these fundamental questions. In spite of being a young field, substantial advances have been made in the last years. In this book, the first one covering theoretical aspects of different randomized search heuristics, we aim at both giving an introduction to this field and presenting recent progress. By collecting important results spread over many different conference or journal papers so far, this book shall also serve as a reference for future research work.

The different chapters cover randomized search heuristics for discrete/combinatorial and continuous search spaces. Though single objective optimization is mainly addressed, one chapter is fully devoted to multi-objective optimization. Another chapter is dedicated to the famous No-Free-Lunch theorem.

We editors are deeply indebted to the contributors of this book. They did an outstanding job making their expertise easily accessible to a broad audience. We are thankful to the reviewers of the different chapters for their constructive careful reviews. Our deepest thanks also go to Liang Quan, Gregory Lee, and all other staff of World Scientific Publishing Co. for their help and patience. Finally, we are very happy that Ming-Yang Kao, associate editor of the book series on Theoretical Computer Science, suggested to write this book and provided the contact to the publisher.

This book is dedicated to Ingo Wegener, who much too early passed away on November 26, 2008. Without his pioneering work this field would have never gotten the maturity it now has.

We hope that the reader enjoys this book. We are grateful for any comments and any notifications of (hopefully only very few) errors. We plan to set up a web-site to communicate them. Needless to say, any comment also helps improving a new edition of the book.

September 2010 *Anne Auger and Benjamin Doerr*

Contents

Chapter 1

Analyzing Randomized Search Heuristics: Tools from Probability Theory

Benjamin Doerr
Max-Planck-Institut für Informatik,
Campus E1 4,
66123 Saarbrücken,
Germany
doerr@mpi-inf.mpg.de

In this chapter, we collect a few probabilistic tools that are useful for analyzing randomized search heuristics. This includes elementary material like Markov, Chebyshev and Chernoff bounds, but also lesser known topics like dealing with sums of random variables that are only close to being independent or a strong lower bound for the time needed by the coupon collector process. Such results, while also of general interest, seem to be particularly useful in the analysis of randomized search heuristics.

Contents

1.1. Introduction

Rigorous analyses of randomized search heuristics, naturally, need some understanding of probability theory. However, unlike one would expect at first, most of them are not very complicated. As the reader will see in this book, the art of analyzing randomized search heuristics consists of finding a clever way to apply one of a small collection of simple tools much more often than of having to prove deep theorems in probability theory. This, in fact, is one of the beauties of the theory of randomized search heuristics.

In this chapter, we collect many of these tools. As will be visible, some of them are very basic and could be proven in a few lines. Those which rely on deeper methods, however, often have an easy to use variant that usually is sufficient in most applications. An example for this is the famous Azuma martingale inequality. In its most common use, the bounded differences method, you will not find any martingales.

This chapter aims at serving three purposes.

- For the reader not yet experienced with this field, we aim to provide the necessary material to understand most of the proofs in the following chapters. To this aim, we start at a very elementary level in the first sections. We would also like to highlight three simple results, the linearity of expectation fact (Lemma 1.4), the waiting time argument (Theorem 1.6) and a simple version of the Chernoff bound (Corollary 1.10), that already suffice to prove many interesting results.
- The reader more familiar with the theory of randomized search heuristics might like to find an easily accessible collection of the tools he frequently uses. To this aim, we state many results in different versions. Clearly, often one easily implies the other, but we feel that saving us in future from looking for the right version of, e.g., the Chernoff bound, either by searching in different sources or by again and again deriving the version we need from the one we find in our favorite reference, is a worthwhile aim.
- Finally, we shall also present some results that are lesser known in the community, partially because they recently appeared in papers on theory of classical algorithms, partially because they adapt classical results to the needs of people working on theory of randomized search heuristics. Examples include a lower bound for the coupon collector process with weakly exponential failure probabil-

ity (Theorem 1.24), Chernoff bounds for negatively correlated random variables (Theorem 1.16) and other ways to deal with slightly dependent random variables (Lemmas 1.18 to 1.20).

1.2. Prerequisites

We shall assume that the reader has some basic understanding of the concepts of *probability spaces*, *events* and *random variables.* As usual in probability theory and very convenient in analysis of algorithms, we shall almost never explicitly state the probability space we are working in. Hence an intuitive understanding of the notion of a random variable should be enough to follow the remainder of this book. Having said this, here is the one statement that is easier expressed in terms of events rather than random variables.

Lemma 1.1 (Union bound). *Let $E_1, \ldots, E_n$ be arbitrary events in some probability space. Then*

$$\Pr\Big(\bigcup_{i=1}^{n} E_i\Big) \leq \sum_{i=1}^{n} \Pr(E_i).$$

While very elementary and often far from a tight estimate, the union bound is often sufficient to prove that with high probability none of a set of bad events does happen. Of course, each of the bad events has to be shown to be sufficiently rare.

Almost all random variables in this book will be *discrete*, that is, they take a finite number of values only (with non-zero probability). As a simple example, consider the random experiment of independently rolling two distinguishable dice. Let X_1 denote the outcome of the first roll, that is, the number between 1 and 6 which the first die displays. Likewise, let X_2 denote the outcome of the second roll. These are already two random variables. We formalize the statement that with probability $\frac{1}{6}$ the first die shows a one by saying $\Pr(X_1 = 1) = \frac{1}{6}$. Also, the probability that both dice show the same number is $\Pr(X_1 = X_2) = \frac{1}{6}$. The *complementary event* that they show different numbers, naturally has a probability of $\Pr(X_1 \neq X_2) = 1 - \Pr(X_1 = X_2) = \frac{5}{6}$.

We can add random variables (defined over the same probability space), e.g., $X := X_1 + X_2$ is the sum of the numbers shown by the two dice, and we can multiply a random variable by a number, e.g., $X := 2X_1$ is twice the number shown by the first die.

The most common type of random variable we shall encounter in this book is an extremely simple one called *binary random variable* or *Bernoulli random variable*. It takes the values 0 and 1 only. In consequence, the probability distribution of a binary random variable X is fully described by its probability $\Pr(X = 1)$ of being one, since $\Pr(X = 0) = 1 - \Pr(X = 1)$.

Binary random variable usually show up as indicator variables for some random event. For example, if the random experiment is a simple roll of a fair (six-sided) die, we may define a random variable X by setting $X = 1$, if the die shows a 'six', and $X = 0$ otherwise. We say that X is the indicator random variable for the event "die shows a 'six' ".

Indicator random variables are useful for counting. If we roll a die n times and $X_1, \ldots, X_n$ are the indicator random variables for the events that the corresponding roll showed a 'six', then $\sum_{i=1}^{n} X_i$ is a random variable describing the number of times we saw a 'six' in these n rolls. In general, a random variable X that is the sum of n binary random variables being one all with equal probability p, is called a *binomial random variable* (with success probability p). We have $\Pr(X = k) = \binom{n}{k} p^k (1 - p)^{n-k}$ for all $k \in \{0, \ldots, n\}$.

A different question is how long we have to wait until we roll a 'six'. Assume that we have an infinite sequence of die rolls and $X_1, X_2, \ldots$ are the indicator variables for the event that the corresponding roll showed a 'six'. Then we are interested in the random variable $Y = \min\{k \in \mathbb{N} \mid X_k = 1\}$. Again for the general case of all X_i being one with probability p, this random variable Y is called *geometric random variable* (with success probability p). We have $\Pr(Y = k) = (1 - p)^{k-1} p$ for all $k \in \mathbb{N}$.

Before continuing with probabilistic tools, let us mention two simple, but highly useful estimates.

Lemma 1.2. *For all $x \in \mathbb{R}$,*

$$1 + x \leq e^x. \tag{1.1}$$

While valid for all $x \in \mathbb{R}$, naturally this estimate is strongest for x close to zero. This is demonstrated by the following estimate.

Lemma 1.3. *For all $n \in \mathbb{N}$,*

$$(1 - \tfrac{1}{n})^n \leq \tfrac{1}{e} \leq (1 - \tfrac{1}{n})^{n-1}. \tag{1.2}$$

Note that the left inequality is just a special case of the previous lemma.

1.3. Linearity of Expectation

The *expectation* (or mean) of a random variable X taking values in some set $\Omega \subseteq \mathbb{R}$ is defined by $E(X) = \sum_{\omega \in \Omega} \omega \Pr(X = \omega)$, where we shall always assume that the sum exists and is finite. As a trivial example, we immediately see that if X is a binary random variable, then $E(X) = \Pr(X = 1)$.

An elementary, but very useful property is that expectation is linear.

Lemma 1.4. *Let $X_1, \ldots, X_n$ be arbitrary random variables and $a_1, \ldots, a_n \in \mathbb{R}$. Then*

$$E\Big(\sum_{i=1}^{n} a_i X_i\Big) = \sum_{i=1}^{n} a_i E(X_i).$$

This fact is usually exploited by writing a complicated random variable as sum of simpler ones and then deriving its expectation from the expectation of the simple random variables. For example, let X be a binomial random variable with success probability p, that is, we have $\Pr(X = k) = \binom{n}{k} p^k (1-p)^{n-k}$. Then, as seen above, X is the sum of independent binary random variables $X_1, \ldots, X_n$, each satisfying $\Pr(X_i = 1) = p$. In consequence, $E(X) = E(X_1 + \ldots + X_n) = E(X_1) + \ldots + E(X_n) = np$. Note that we did not need that the X_i are independent. We just proved the following.

Lemma 1.5. *Let X be a binomial random variable describing n trials with success probability p. Then $E(X) = pn$.*

Computing the expectation of a geometric random variable is slightly more difficult. Assume that X is a geometric random variable with success probability p. Intuitively, we feel that the time needed to see a success is $1/p$. This intuition is guided by the fact that after $1/p$ repetitions of the underlying binary random experiment, the expected number of successes is exactly one.

Note that this does *not* prove our claim. Fortunately the claim is still true, and follows either from standard results in Markov chain theory, or some elementary, though non-trivial, calculations. It is one of the results most often used in the analysis of randomized search heuristics.

Theorem 1.6 (Waiting time argument). *Let X be a geometric random variable with success probability p. Then $E(X) = 1/p$.*

Proof. By definition, $E(X) = \sum_{i=1}^{\infty} i(1-p)^{i-1}p$. Note that $\sum_{i=1}^{\infty}(1-p)^{i-1}p = \sum_{i=1}^{\infty} \Pr(X=i) = \Pr(X \in \mathbb{N}) = 1$ by definition of the geometric random variable. We thus compute

$$\begin{aligned} E(X) &= \sum_{i=1}^{\infty} i(1-p)^{i-1}p \\ &= \sum_{i=1}^{\infty}(1-p)^{i-1}p + (1-p)\sum_{i=1}^{\infty}(i-1)(1-p)^{i-2}p \\ &= 1 + (1-p)\sum_{i=2}^{\infty}(i-1)(1-p)^{i-2}p \\ &= 1 + (1-p)\sum_{i=1}^{\infty} i(1-p)^{i-1}p = 1 + (1-p)E(X). \end{aligned}$$

Solving this equation shows that $E(X) = 1/p$. □

1.4. Deviations from the Mean

Often, we are interested not so much in the expectation of some random variable, e.g., the run-time of an algorithm, but we would like to have a bound that holds with high probability. Since computing the expectation is so easy, a successful approach is to first compute the expectation and then bound the probability that the random variable exceeds this expectation by a certain amount. Inequalities serving this purpose are called *tail inequalities* or *large deviation inequalities.*

1.4.1. *Markov Inequality*

A simple one-line proof shows the most general deviation bound, valid for *all* non-negative random variables.

Lemma 1.7 (Markov inequality). *Let X be a non-negative random variable. Then for all $\lambda \geq 1$,*

$$\Pr(X \geq \lambda E(X)) \leq \tfrac{1}{\lambda}.$$

Proof. We have $E(X) = \sum_{\omega} \omega \Pr(X = \omega) \geq \sum_{\omega \geq \mu} \mu \Pr(\omega) = \mu \Pr(X \geq \mu)$ for all $\mu \geq 0$. Setting $\mu = \lambda E(X)$ proves the claim. □

1.4.2. *Chebyshev Inequality*

For completeness, we quickly state a second elementary inequality, although it seems that is seldom used in the theory of randomized search heuristics. The *variance* of a discrete random variable X is $\text{Var}(X) = E((X - E(X))^2) = E(X^2) - E(X)^2$. Just by definition, it is a measure of how well X is concentrated around its mean. Applying Markov's inequality to the random variable $(X - E(X))^2$ easily yields the following.

Lemma 1.8 (Chebyshev inequality). *For all* $\lambda > 0$,

$$\Pr(|X - E(X)| \geq \lambda\sqrt{\text{Var}(X)}) \leq \tfrac{1}{\lambda^2}.$$

Note that Chebyshev's inequality automatically yields two-sided bounds (that is, for both cases that the random variable is larger and smaller than its expectation), as opposed to Markov's inequality (giving just a bound for exceeding the expectation) and the Chernoff bounds in the following section (giving different bounds for both sides). There is a one-sided version of the Chebyshev inequality attributed to Cantelli, replacing the $\frac{1}{\lambda^2}$ by $\frac{1}{\lambda^2+1}$, a gain not adding a lot in most applications.

1.4.3. *Chernoff Bounds for Sums of Independent Random Variables*

Above, we computed that the expectation of a binomial random variable is np. For example, if we flip a fair coin n times, we expect to see 'tails' $n/2$ times. However, we also feel that the actual outcome should be relatively close to this expectation if n is sufficiently large. That this is true in a very strong sense, can be computed from the probability distribution of this random variable. However, deeper methods show such results for much more general settings. We shall omit most of the proofs and refer the interested reader to, e.g., the standard text book by Alon and Spencer (2008). A condensed, but self-contained proof was given by Hagerup and Rüb (1990).

In the general setting, we assume that our random variable of interest is the sum of independent random variables, not necessarily having the same distribution. The bounds presented below are all known under names like *Chernoff* or *Hoeffding* bounds, referring to the seminal papers by Chernoff (1952) and Hoeffding (1963). Since the first bounds of this type were actually proven by Bernstein (1924), the name Bernstein inequality would be more appropriate. We shall not be that precise and instead use the most common name *Chernoff inequalities* for all such bounds.

For the readers' convenience, we shall give several versions of these bounds.

Theorem 1.9. *Let $X_1, \ldots, X_n$ be independent random variables taking values in $[0,1]$. Let $X = \sum_{i=1}^n X_i$.*

(a) Let $\delta \in [0,1]$. Then $\Pr(X \le (1-\delta)E(X)) \le \left((\frac{1}{1-\delta})^{1-\delta} e^{-\delta}\right)^{E(X)}$.
(b) Let $\delta \ge 0$. Then $\Pr(X \ge (1+\delta)E(X)) \le (\frac{e^\delta}{(1+\delta)^{1+\delta}})^{E(X)}$.

Corollary 1.10. *Let $X_1, \ldots, X_n$ be independent random variables taking values in $[0,1]$. Let $X = \sum_{i=1}^n X_i$.*

(a) Let $\delta \in [0,1]$. Then $\Pr(X \le (1-\delta)E(X)) \le \exp(-\delta^2 E(X)/2)$.
(b) Let $\delta \ge 0$. Then $\Pr(X \ge (1+\delta)E(X)) \le (\frac{e}{1+\delta})^{(1+\delta)E(X)}$.
(c) Let $\delta \in [0,1]$. Then $\Pr(X \ge (1+\delta)E(X)) \le \exp(-\delta^2 E(X)/3)$.
(d) For all $d \ge 6E(X)$, $\Pr(X \ge d) \le 2^{-d}$.

Part (a) and (b) follow from estimating the corresponding expressions in (a) and (b) of Theorem 1.9. Part (c) and (d) are derived from bound (b) in Theorem 1.9, and are often more convenient to work with. The different constants 1/2 and 1/3 in (a) and (c) cannot be avoided. Ignoring the constants, (a) and (c) tell us that the probability that such a random variable deviates from its expectation by a small constant factor, is exponentially small in the expectation. In consequence, we get inverse polynomial failure probabilities if the expectation is at least logarithmic, and exponentially small ones, if the expectation is $\Theta(n)$, e.g., if the X_i are one with constant probability.

We also see that we obtain useful bounds for $\delta = o(1)$. More precisely, from $\delta \gg 1/\sqrt{E(X)}$ on, the bounds become non-trivial, and for $\delta \gg \log(n)/\sqrt{E(X)}$ inverse-polynomial.

If $E(X) = \Theta(n)$, it is often more convenient to use a Chernoff bound for additive deviations. The following version (which is Theorem 2 in the seminal paper by Hoeffding (1963)) in addition does not require the random variables not be non-negative.

Theorem 1.11. *Let $X_1, \ldots, X_n$ be independent random variables. Assume that each X_i takes values in a real interval of length c_i only. Let $X =$*

$\sum_{i=1}^{n} X_i$. *Then for all* $\lambda > 0$,

$$\Pr(X \geq E(X) + \lambda) \leq \exp\big(-2\lambda^2 / \sum_i c_i^2\big),$$

$$\Pr(X \leq E(X) - \lambda) \leq \exp\big(-2\lambda^2 / \sum_i c_i^2\big).$$

Using similar arguments as those that prove Theorem 1.11, Hoeffding (1963) also proves a bound that takes into account the variance of the random variables.

Theorem 1.12. *Let* $X_1, \ldots, X_n$ *be independent random variables. Let* b *be such that* $X_i \leq E(X_i) + b$ *for all* $i = 1, \ldots, n$. *Let* $X = \sum_{i=1}^{n} X_i$. *Let* $\sigma^2 = \frac{1}{n}\sum_{i=1}^{n} \mathrm{Var}(X_i)$. *Then*

$$\Pr(X \geq E(X) + \lambda) \leq \left(\left(1 + \frac{b\lambda}{n\sigma^2}\right)^{-\left(1+\frac{bt}{n\sigma^2}\right)\frac{\sigma^2}{b^2+\sigma^2}} \left(1 - \frac{\lambda}{nb}\right)^{-\left(1-\frac{\lambda}{nb}\right)\frac{b^2}{b^2+\sigma^2}}\right)^n$$

$$\leq \exp\left(-\frac{\lambda}{b}\left(\left(1 + \frac{n\sigma^2}{b\lambda}\right)\ln\left(1 + \frac{b\lambda}{n\sigma^2}\right) - 1\right)\right).$$

As discussed in Hoeffding (1963), the second bound is the same as inequality (8b) in Bennett (1962). Using the abbreviations $r = \lambda/b$ and $s = b\lambda/(n\sigma^2)$, the bound of Bennett (1962) becomes $\exp(-r((1 + 1/s)\ln(1+s) - 1))$. This is stronger than the bound of $\exp(-rs/(2+\frac{2}{3}s)) = \exp(-\lambda^2/(2n\sigma^2(1 + b\lambda/(3n\sigma^2))))$ due to Bernstein (1924) and the bound of $\exp(-r \arcsin(s/2)/2)$ due to Prokhorov (1956).

We end this subsection with a real gem from Hoeffding's paper, which, in spite of its usefulness, seems rarely known. It states that the last two theorems also hold, simultaneously, for all partial sums of the sequence $X_1, \ldots, X_n$.

Theorem 1.13. *In the settings of Theorem 1.11 or 1.12, let* $Y_k := \sum_{i=1}^{k} X_i$. *Let* $X = \max\{Y_k \mid 1 \leq k \leq n\}$. *Then the bounds of Theorem 1.11 and 1.12 also hold for* X *defined like this.*

1.4.4. *Chernoff Bounds for Geometric Random Variables*

If X is a sum of independent geometric random variables $X_1, \ldots, X_n$, we would still think that X displays some concentration behavior similar to that guaranteed by Chernoff bounds. It is true that the X_i are unbounded, but the probability $\Pr(X_i = k)$ decays exponentially with k. Therefore, we would feel that an X_i taking a large valued should be a sufficiently rare event. In fact, the following is true.

Theorem 1.14. *Let $p \in]0,1[$. Let $X_1, \ldots, X_n$ be independent geometric random variables with* $\Pr(X_i = j) = (1-p)^{j-1}p$ *for all $j \in \mathbb{N}$ and let $X := \sum_{i=1}^n X_i$.*

Then for all $\delta > 0$,

$$\Pr(X \geq (1+\delta)E(X)) \leq \exp\left(-\frac{\delta^2}{2}\frac{(n-1)}{1+\delta}\right).$$

Proof. Let $Y_1, Y_2, \ldots$ be an infinite sequence of independent, identically distributed binary random variables such that Y_i is one with probability $\Pr(Y_i = 1) = p$. Note that the random variable "smallest j such that $Y_j = 1$" has the same distribution as each X_i. In consequence, X has the same distribution as "smallest j such that exactly n of the variables $Y_1, \ldots, Y_j$ are one". In particular, $\Pr(X \geq j) = \Pr(\sum_{i=1}^{j-1} Y_i \leq n-1)$ for all $j \in \mathbb{N}$. This manipulation reduces our problem to the analysis of independent binary variables and enables us to use the classical Chernoff bounds.

By Theorem 1.6, the expected value of each X_i is $E(X_i) = \frac{1}{p}$. Hence $E(X) = \frac{n}{p}$. Let $Y := \sum_{i=1}^{\lceil(1+\delta)E(X)-1\rceil} Y_i$. By the above,

$$\Pr(X \geq (1+\delta)E(X)) = \Pr(Y \leq n-1).$$

The expected value of Y is bounded by

$$E(Y) = \lceil(1+\delta)E(X) - 1\rceil p \geq (1+\delta)n - p > (1+\delta)(n-1).$$

Now let $\delta' := 1 - \frac{n-1}{E(Y)}$. Then $0 < \delta' \leq 1$ and $\Pr(Y \leq n-1) = \Pr(Y \leq (1-\delta')E(Y))$. Hence we can apply Corollary 1.10 to get

$$\begin{aligned}
\Pr(X \geq (1+\delta)E(X)) &= \Pr(Y \leq (1-\delta')E(Y)) \\
&\leq \exp\left(-\frac{1}{2}E(Y)\left(1 - \frac{n-1}{E(Y)}\right)^2\right) \\
&\leq \exp\left(-\frac{1}{2}E(Y)\left(1 - \frac{1}{1+\delta}\right)^2\right) \\
&\leq \exp\left(-\frac{1}{2}(n-1)(1+\delta)\left(\frac{\delta}{1+\delta}\right)^2\right).
\end{aligned}$$

□

1.4.5. *Chernoff-type Bounds for Independent Variables with Limited Influence*

Often, the random variable of interest can be expressed as sum of independent random variables. Then, as seen above, Chernoff bounds give excellent estimates on the deviation from the mean. Sometimes, the random variable we are interested in is also determined by the outcomes of many independent random variables, however, not as simply as in the case of a sum. Nevertheless, if each of the independent random variables only has a limited influence on the outcome, then bounds similar to those of Theorem 1.11 can be proven.

Theorem 1.15 (Azuma's inequality). *Let $X_1, \dots, X_n$ be independent random variables taking values in the sets $\Omega_1, \dots, \Omega_n$, respectively. Let $\Omega := \Omega_1 \times \dots \times \Omega_n$. Let $f : \Omega \to \mathbb{R}$ and $c_1, \dots, c_n > 0$ be such that for all $i \in \{1, \dots, n\}$, $\omega, \bar{\omega} \in \Omega$ we have that if for all $j \neq i$, $\omega_j = \bar{\omega}_j$, then $|f(\omega) - f(\bar{\omega})| \leq c_i$. Let $X = f(X_1, \dots, X_n)$. Then for all $\lambda > 0$,*

$$\Pr(X \geq E(X) + \lambda) \leq \exp\big(-2\lambda^2 / \sum_i c_i^2\big),$$

$$\Pr(X \leq E(X) + \lambda) \leq \exp\big(-2\lambda^2 / \sum_i c_i^2\big).$$

Theorem 1.15 is usually known under the name Azuma's inequality or *method of bounded differences.* It is a special case of a similar bound for martigales due to Azuma (1967), which, however, already appeared in Hoeffding (1963). Again, we will stick to the most common name and not care whether it is the most appropriate one.

The version of Azuma's inequality given above is due to McDiarmid (1998), while most authors state a slightly weaker bound of $\exp(-\lambda^2/2\sum_i c_i^2)$.

1.4.6. *Dealing with Non-independent Random Variables*

All bounds of Chernoff type presented so far build on a large number of *independent* random variables. Often, in particular, in the analysis of randomized search heuristics, this is too much to ask for. In this subsection, we present two ways to still obtain useful bounds.

The first approach uses negative correlation. Let $X_1, \dots, X_n$ binary random variables. We call them *negatively correlated,* if for all $I \subseteq \{1, \dots, n\}$

the following holds.

$$\Pr(\forall i \in I : X_i = 0) \le \prod_{i \in I} \Pr(X_i = 0),$$

$$\Pr(\forall i \in I : X_i = 1) \le \prod_{i \in I} \Pr(X_i = 1).$$

In simple words, we require that the event that a set of variables is all zero or all one, is at most as likely as in the case of independent X_i. It feels that this condition should make life rather easier than in the independent case, and exactly this is what Panconesi and Srinivasan (1997) were able to prove.

Theorem 1.16 (Chernoff bounds, negative correlation). *Let $X_1, \ldots, X_n$ be negatively correlated binary random variables. Let $a_1, \ldots, a_n \in [0,1]$ and $X = \sum_{i=1}^n a_i X_i$. Then X satisfies the Chernoff bounds given in Theorem 1.9 and Corollary 1.10.*

Here is an example of how to apply Theorem 1.16, namely to derive Chernoff bounds for *hypergeometric* distributions. Say we choose randomly n elements from a given N-element set *without replacement.* For a given m-element subset of full set, we wonder how many of its elements we have chosen. This random variable is called hypergeometrically distributed with parameters N, n and m.

More formally, let S be any N-element set. Let $T \subseteq S$ have exactly m elements. Let U be a subset of S chosen uniformly among all n-element subsets of S. Then $X = |U \cap T|$ is a random variable with hypergeometric distribution (with parameters N, n and m).

It is easy to see that $E(X) = |U||T|/|S| = mn/N$: Enumerate $T = \{t_1, \ldots, t_m\}$ in an arbitrary manner (before choosing U). For $i = 1, \ldots, m$, let X_i be the binary random variable that is one if and only if $t_i \in U$. Clearly, $\Pr(X_i = 1) = |U|/|S| = n/N$. Since $X = \sum_{i=1}^m X_i$, we have $E(X) = mn/N$ by linearity of expectation (Lemma 1.4).

It is also obvious that the X_i are not independent. If $n < m$ and $X_1 = \ldots = X_n = 1$, then necessarily we have $X_i = 0$ for $i > n$. Fortunately, however, these dependencies are of the negative correlation type. This is intuitively clear, but also straight-forward to prove. Let $I \subseteq \{1, \ldots, m\}$. Let $W = \{t_i \mid i \in I\}$ and $w = |W| = |I|$. Then $\Pr(\forall i \in I : X_i = 1) = \Pr(W \subseteq U)$. Since U is uniformly chosen, it suffices to count the number of U that contain W (these are $\binom{|S \setminus W|}{|U \setminus W|}$) and compare them with the total number of possible U. Hence $\Pr(W \subseteq U) = \binom{N-w}{n-w}/\binom{N}{n} = \frac{n\cdot\ldots\cdot(n-w+1)}{N\cdot\ldots\cdot(N-w+1)} < (n/N)^w = \prod_{i\in I}\Pr(X_i = 1)$. A similar argument shows that the X_i also satisfy the first part of the definition of negative correlation.

Theorem 1.17. *If X is a random variable with hypergeometric distribution, then is satisfies the Chernoff bounds given in Theorem 1.9 and Corollary 1.10.*

A second situation often encountered is that a sequence of random variables is not independent, but each member of the sequence has a good chance of having a desired property conditional on each possible outcome of its predecessors. More formally, we have the following.

Lemma 1.18 (Chernoff bound, lower tail, moderate independence). *Let $X_1, \ldots, X_n$ be arbitrary binary random variables. Let $X_1^*, \ldots, X_n^*$ be binary random variables that are mutually independent and such that for all i, X_i^* is independent of $X_1, \ldots, X_{i-1}$. Assume that for all i and all $x_1, \ldots, x_{i-1} \in \{0,1\}$,*

$$\Pr(X_i = 1 | X_1 = x_1, \ldots, X_{i-1} = x_{i-1}) \geq \Pr(X_i^* = 1).$$

Then for all $k \geq 0$, we have

$$\Pr\Big(\sum_{i=1}^{n} X_i < k\Big) \leq \Pr\Big(\sum_{i=1}^{n} X_i^* < k\Big),$$

and the latter term can be bounded by Chernoff bounds for independent random variables.

Lemma 1.19 (Chernoff bound, upper tail, moderate independence). *Let $X_1, \ldots, X_n$ be arbitrary binary random variables. Let $X_1^*, \ldots, X_n^*$ be binary random variables that are mutually independent and such that for all i, X_i^* is independent of $X_1, \ldots, X_{i-1}$. Assume that for all i and all $x_1, \ldots, x_{i-1} \in \{0,1\}$,*

$$\Pr(X_i = 1 | X_1 = x_1, \ldots, X_{i-1} = x_{i-1}) \leq \Pr(X_i^* = 1).$$

Then for all $k \geq 0$, we have

$$\Pr\Big(\sum_{i=1}^{n} X_i > k\Big) \leq \Pr\Big(\sum_{i=1}^{n} X_i^* > k\Big),$$

and the latter term can be bounded by Chernoff bounds for independent random variables.

Both Lemmas are simple corollaries from the following general result, which itself might be useful in the analysis of randomized search heuristics when encountering random variables taking more values than just zero and one.

Lemma 1.20 (Dealing with moderate dependencies). *Let $X_1, \ldots, X_n$ be arbitrary integral random variables. Let $X_1^*, \ldots, X_n^*$ be random variables that are mutually independent and such that for all i, X_i^* is independent of $X_1, \ldots, X_{i-1}$. Assume that for all i and all $x_1, \ldots, x_{i-1} \in \mathbb{Z}$, we have $\Pr(X_i \geq m | X_1 = x_1, \ldots, X_{i-1} = x_{i-1}) \leq \Pr(X_i^* \geq m)$ for all $m \in \mathbb{Z}$, that is, X_i^* dominates $(X_i | X_1 = x_1, \ldots, X_{i-1} = x_{i-1})$. Then for all $k \in \mathbb{Z}$, we have*

$$\Pr\Big(\sum_{i=1}^{n} X_i \geq k\Big) \leq \Pr\Big(\sum_{i=1}^{n} X_i^* \geq k\Big).$$

Proof. Define $P_j := \Pr(\sum_{i=1}^{j} X_i + \sum_{i=j+1}^{n} X_i^* \geq k)$ for $j \in \{0, \ldots, n\}$ and $\mathcal{X}_k^n := \{(x_1, \ldots, x_n) \in \mathbb{Z}^n | \sum_{i=1}^{n} x_i = k\}$. Then

$$\begin{aligned}
P_{j+1} &= \Pr\left(\sum_{i=1}^{j+1} X_i + \sum_{i=j+2}^{n} X_i^* \geq k\right) \\
&= \sum_{m \in \mathbb{Z}} \Pr\left(\sum_{i=1}^{j} X_i + \sum_{i=j+2}^{n} X_i^* = k - m \wedge X_{j+1} \geq m\right) \\
&= \sum_{m \in \mathbb{Z}} \sum_{(x_1, \ldots, x_j, x_{j+2}, \ldots, x_n) \in \mathcal{X}_{k-m}^{n-1}} \Pr\left(X_1 = x_1, \ldots, X_j = x_j\right) \cdot \\
&\qquad \Pr\left(X_{j+1} \geq m | X_1 = x_1, \ldots, X_j = x_j\right) \cdot \prod_{i=j+2}^{n} \Pr\left(X_i^* = x_i\right) \\
&\leq \sum_{m \in \mathbb{Z}} \Pr\left(\sum_{i=1}^{j} X_i + \sum_{i=j+2}^{n} X_i^* = k - m\right) \cdot \Pr\left(X_{j+1}^* \geq m\right) \\
&= \Pr\left(\sum_{i=1}^{j} X_i + \sum_{i=j+1}^{n} X_i^* \geq k\right) \\
&= P_j.
\end{aligned}$$

Thus, we have

$$\Pr\Big(\sum_{i=1}^{n} X_i \geq k\Big) = P_n \leq P_{n-1} \leq \cdots \leq P_1 \leq P_0 = \Pr\Big(\sum_{i=1}^{n} X_i^* \geq k\Big). \qquad \square$$

Lemma 1.19 follows directly by noting that its assumptions imply the domination assumption in Lemma 1.20. To prove Lemma 1.18, note that the assumptions there imply that $-X_i^*$ dominates $-(X_i|X_1 = x_1, \ldots, X_{i-1} = x_{i-1})$ for all $x_1, \ldots, x_{i-1}$. Hence applying Lemma 1.20 to the negated variables proves the claim.

1.5. Coupon Collector

The coupon collector problem is the following simple question. Assume that there are n types of coupons available. Whenever you buy a certain product, you get one coupon, with its type chosen uniformly at random from the n types. How long does it take until you have a coupon of each type?

Besides an interesting problem to think about, understanding its solution also pays off when analyzing randomized search heuristics, because very similar situations are encountered here. Since often the setting is similar, but not exactly as in the coupon collector problem, it helps a lot not just to know the coupon collector theorem, but also its proof. We start with a simple version.

Theorem 1.21 (Coupon collector, expectation). *The expected time to collect all n coupons is nH_n, where $H_n := \sum_{i=1}^{n} \frac{1}{i}$ is the nth harmonic number. Since $\ln(n) < H_n < 1 + \ln(n)$, the coupon collector needs an expected time of $(1 + o(1))n\ln(n)$.*

Proof. Given that we already have i different coupons for some $i \in \{0, \ldots, n-1\}$, the probability that the next coupon is one that we do not already have, is $(n-i)/n$. By the waiting time argument (Theorem 1.6), we see that the time T_i needed to obtain a new coupon given that we have exactly i different ones, satisfies $E(T_i) = n/(n-i)$. Clearly, the total time T needed to obtain all coupons is $\sum_{i=0}^{n-1} T_i$. Hence, by linearity of expectation (Lemma 1.4), $E(T) = \sum_{i=1}^{n-1} E(T_i) = nH_n$.

The estimates on H_n follow from the fact that $\int_1^{n+1} \frac{1}{x}dx$ is a lower bound for H_n, and likewise, $1 + \int_1^{n} \frac{1}{x}dx$ is an upper bound. □

As discussed before, the expectation is only one aspect of a random variable, and often we would be happy to also know that the random variable is close to its expectation with good probability. The always applicable Markov inequality tells us that the probability to need more than cnH_n rounds, is bounded by $1/c$.

Chebyshev's inequality (Lemma 1.8) can be used to prove the following two-sided bound.

Lemma 1.22. *Let T denote the time needed to complete the coupon collector process with n types of coupons. Then* $\Pr(|T - nH_n| \geq cn) \leq 2/c^2$.

Stronger bounds for the upper tail can be derived very elementary. Note that the probability that a particular coupon is missed for t rounds, is $(1 - \frac{1}{n})^t$. By a union bound argument (see Lemma 1.1), the probability that there is a coupon not obtained within t rounds, and hence equivalently, that $T > t$, satisfies $\Pr(T > t) \leq n(1 - \frac{1}{n})^t$. Using the simple estimate of Lemma 1.2 (or 1.3), we compute $\Pr(T \geq cn\ln(n)) \leq n\exp(-c\ln n) = n^{-c+1}$.

Theorem 1.23. *Let T denote the time needed to collect all n coupons. Then* $\Pr(T \geq (1+\varepsilon)n\ln(n)) \leq n^{-\varepsilon}$.

Note that we used the variable c to parameterize the deviations in the estimates above, because often a constant c is a useful choice. However, since there are no asymptotics involved, c may as well depend on n. The strongest results in terms of the asymptotics, to be found, e.g., in the book by Motwani and Raghavan (1995), is that for any real *constant* c, we have

$$\lim_{n\to\infty} \Pr(n\ln(n) - cn \leq T \leq n\ln(n) + cn) = e^{-e^{-c}} - e^{-e^{c}}.$$

For the analysis of randomized search heuristics, however, the non-asymptotic results given previously might be more useful.

Surprisingly, there seem to be no good lower bounds for the coupon collector time published. In the following, we present a simple solution to this problem.

Theorem 1.24. *Let T denote the time needed to collect all n coupons. Then for all $\varepsilon > 0$,* $\Pr(T < (1-\varepsilon)(n-1)\ln(n)) \leq \exp(-n^{\varepsilon})$.

Proof. Let $t = (1-\varepsilon)(n-1)\ln(n)$. For $i = 1, \ldots, n$, let X_i be the indicator random variable for the event that a coupon of type i is obtained within t rounds. Note first that $\Pr(X_i = 1) = 1-(1-\frac{1}{n})^t \leq 1-\exp(-(1-\varepsilon)\ln(n)) = 1 - n^{-1+\varepsilon}$, where the estimate follows from Lemma 1.3.

Let $I \subset \{1, \ldots, n\}$, $j \in \{1, \ldots, n\} \setminus I$. By the law of total probability, we have

$$\begin{aligned} \Pr(\forall i \in I : X_i = 1) = {} & \Pr(\forall i \in I : X_i = 1 | X_j = 1) \cdot \Pr(X_j = 1) + \\ & \Pr(\forall i \in I : X_i = 1 | X_j = 0) \cdot \Pr(X_j = 0). \end{aligned} \tag{1.3}$$

We have $\Pr(\forall i \in I : X_i = 1 | X_j = 0) \geq \Pr(\forall i \in I : X_i = 1)$—the left part is the probability to obtain all coupons with types in I if in each round a random coupon out of $n-1$ different ones (including I) is drawn. This is clearly easier than achieving the same with n coupons, which is the right hand side.

From (1.3) we conclude that $\Pr(\forall i \in I : X_i = 1 | X_j = 1) \leq \Pr(\forall i \in I : X_i = 1)$. Equivalently, we have $\Pr(\forall i \in I \cup \{j\} : X_i = 1) \leq \Pr(\forall i \in I : X_i = 1) \cdot \Pr(X_j = 1)$. By induction, we conclude

$$\begin{aligned}\Pr(\forall i \in \{1, \ldots, n\} : X_i = 1) &\leq \prod_{i=1}^{n} \Pr(X_i = 1) \\ &\leq (1 - n^{-1+\varepsilon})^n \\ &\leq \exp(-n^{\varepsilon})\end{aligned}$$

by Lemma 1.2. □

1.6. Further Tools

There are a number of results that, to the best of our knowledge, have not been used in the analysis of randomized search heuristics, but where we see a good chance that they might be useful. Therefore, we briefly describe them in this section.

1.6.1. *Talagrand's Inequality*

Consider the setting of Azuma's inequality (Theorem 1.15), that is, we have a function f defined over a product of n probability spaces such that changing a single component of the input to f has only a limited influence on the function value. Assume for simplicity that all these influences c_i are at most one.

One weakness of Azuma's inequality is that does give useful bounds only for deviations λ of size $\Omega(\sqrt{n})$. This is different to most Chernoff type bounds, where it often suffices that $\lambda = \Omega(\sqrt{(E(X))})$. The bound $\Pr(X \leq (1-\delta)E(X) \leq \exp(-\delta^2 E(X)/2)$ for example can be rewritten as $\Pr(X \leq E(X) - \lambda) \leq \exp(-\lambda^2/(2E(X)))$.

Talagrand's inequality is a way to overcome this short-coming of Azuma's inequality. It comes, however, with a certain prize. Still in the notation of Azuma's theorem, let m denote a median of $X = f(X_1, \ldots, X_n)$. Hence m is a number such that $\Pr(X \leq m) \geq 1/2$ and $\Pr(X \geq m) \geq 1/2$. Assume further that f has certificates of size s for exceeding m. That means, that for all ω with $f(\omega) \geq m$, there are s components of ω such that any ω' which agrees with ω on these components, also satisfies $f(\omega') \geq m$. This sounds more obscure than it actually is. If each $\Omega_i = \{0,1\}$ and f is simply the sum of the components, then a certificate for having sum greater than m are just m components that are one.

Given that f is as described above, then a simple version of Talagrand's bound ensures

$$\Pr(X \leq m - \lambda) \leq 2\exp(-t^2/(4s)).$$

Hence if the certificate size can be chosen of order $E(X)$, we obtain bounds of order comparable to Chernoff bounds, apart from the issue that we also need to show that the median is close to the expectation. This usually is true if the random variable is sufficiently concentrated (which we hope for anyway).

Nevertheless, it is clear that we pay with a significant technical overhead for the hope for a stronger bound. Talagrand's bound first appeared in Talagrand (1995). It is also covered in the textbooks by Alon and Spencer (2008) and Janson *et al.* (2000). Our presentation is greatly inspired by the unpublished lecture notes by Jirka Matoušek and Jan Vondrák, accessible from the first author's homepage.

1.6.2. *Kim-Vu Inequality*

Both Azuma's and Talagrand's inequality require that each variable has only a bounded influence on the function value. One might imagine situations where this is not fulfilled, but still a strong concentration behavior is observed. Such settings might in particular occur if some variables may have a large influence on the function value, but only with very small probability. An example met in this text (though solved with different methods) are sums of independent, geometrically distributed random variables.

In such situations, we might hope that a suitable condition on the typical influence of each variable does suffice. Indeed, this is what the recent inequality due to Kim and Vu (2000) does. The result, however, is quite technical. Therefore, we omit further details and point the interested reader to the original paper by Kim and Vu (2000) or the textbook by Alon and Spencer (2008).

1.7. Conclusion

In this chapter, we collected a number of tools that found application in the analysis of randomized search heuristics. The collection shows a slight dichotomy. On the one hand, a number of standard arguments from the classical theory of randomized algorithms can be used just as simple as there. On the other hand, the particular nature of randomized search heuristics asks for special tools. This can be because the objective of the analysis is different, e.g., we rarely care for constant factors, but do care a lot about guarantees that hold with high probability. This motivates the use of a different coupon collector theorem. Another reason is the type of random experiments that show up when running a randomized search heuristic. Here we often observe an abundance of dependencies. Hence methods that allow to use classical estimates also in such dependent environments are highly needed. This is different from the classical theory of randomized algorithms. Here, the researcher would design the algorithms in a way that both the algorithm is efficient and that avoids, as far as possible, such difficulties in the analysis.

While the theory community in the randomized search heuristics field made great progress in the last ten year to find mathematical tools suitable for the analysis of such heuristics, it remains a challenge to go further. When analyzing a particular problem, how often do we stumble upon the situation that we 'know' what is the answer, but it seems very hard to prove that, simply because the setting is minimally different from a classical setting? This is when not only solving the particular problem is asked for, but in addition to develop tools that are robust against such slight modifications, which cause all the trouble (and much of the fun as well).

References

Alon, N. and Spencer, J. H. (2008). *The Probabilistic Method*, 3rd edn. (Wiley-Interscience, New York).

Azuma, K. (1967). Weighted sums of certain dependent variables, *Tohoku Math. Journal* **3**, pp. 357–367.

Bennett, G. (1962). Probability inequalities for the sum of independent random variables, *Journal of the American Statistical Association* **57**, pp. 33–45.

Bernstein, S. (1924). On a modification of Chebyshevs inequality and of the error formula of Laplace, *Ann. Sci. Inst. Sav. Ukraine, Sect. Math. 1* **4**, pp. 38–49.

Chernoff, H. (1952). A measure of asymptotic efficiency for tests of a hypothesis based on the sum of observations, *Ann. Math. Statistics* **23**, pp. 493–507.

Hagerup, T. and Rüb, C. (1990). A guided tour of Chernoff bounds, *Inf. Process. Lett.* **33**, pp. 305–308.

Hoeffding, W. (1963). Probability inequalities for sums of bounded random variables, *J. Amer. Statist. Assoc.* **58**, pp. 13–30.

Janson, S., Łuczak, T. and Ruciński, A. (2000). *Random Graphs* (Wiley-Interscience, New York).

Kim, J. H. and Vu, V. H. (2000). Concentration of multivariate polynomials and its applications, *Combinatorica* **20**, pp. 417–434.

McDiarmid, C. (1998). Concentration, in *Probabilistic Methods for Algorithmic Discrete Mathematics, Algorithms Combin.*, Vol. 16 (Springer, Berlin), pp. 195–248.

Motwani, R. and Raghavan, P. (1995). *Randomized Algorithms.* (Cambridge University Press).

Panconesi, A. and Srinivasan, A. (1997). Randomized distributed edge coloring via an extension of the Chernoff–Hoeffding bounds, *SIAM J. Comput.* **26**, pp. 350–368.

Prokhorov, Y. V. (1956). Convergence of random processes and limit theorems in probability theory, *Theory of Probability and Its Applications* **1**, pp. 157–214.

Talagrand, M. (1995). Concentration of measure and isoperimetric inequalities in product spaces, *Inst. Hautes Études Sci. Publications Mathmatiques* **81**, pp. 73–205.

Chapter 2

Runtime Analysis of Evolutionary Algorithms for Discrete Optimization

Peter S. Oliveto and Xin Yao

School of Computer Science,
The University of Birmingham,
Birmingham, UK
P.S.Oliveto@cs.bham.ac.uk
X.Yao@cs.bham.ac.uk

Theoretical studies of evolutionary algorithms (EAs) have existed since the seventies when EAs started to become popular. These early studies contributed towards the understanding of EAs but did not explain their performance in terms of their dynamical or limit behaviour. Only in the nineties the first convergence and time complexity results appeared for simple algorithms and toy problems. Nowadays, it is possible to analyse the time complexity of more complicated algorithms for combinatorial optimization problems with practical applications. This chapter overviews the most popular mathematical techniques used in the runtime analysis of EAs and gives some simple examples of their application.

Contents

2.1. Introduction

Theoretical studies of Evolutionary Algorithms (EAs), albeit few, have always existed since the seventies when EAs started to become popular. Early studies were concerned with explaining the behaviour of EAs rather than analysing their performance. The *Schema theory* (Goldberg, 1989) is probably the most popular tool used in these early attempts. In particular, it was first proposed to analyse the behaviour of the *simple Genetic Algorithm* (sGA) (Holland, 1992). Although the schema theory was considered fundamental for understanding GAs up to the early nineties, it cannot explain the dynamical or limit behaviour of EAs.

In the nineties, with the advent of Markov Chain theory in the analysis of EAs, the first convergence results (related to their time-limit behaviour) appeared for optimization problems. Ideally, the EA should be able to find the solution to the problem it is tackling after a finite number of steps, regardless of its initialisation. In such a case the algorithm is said to *visit the global optimum in finite time.* If the algorithm holds the solution in the population forever after, then it is said to *converge* to the optimum. By using Markov Chains, Rudolph proved that canonical GAs using mutation, crossover and proportional selection do not converge to the global optimum, while *elitist*[a] variants do (Rudolph, 1994) and defined general conditions that, if satisfied by an EA, guarantee its convergence (Rudolph, 1998).

However, if an algorithm does converge, the analysis of the time limit behaviour does not give any hints about the expected time for the solution to be found, far less any precise statement. Aytug and Kohler performed an analysis of the number of generations which are sufficient to guarantee convergence with a fixed level of confidence (i.e., the algorithm converges in x generations with probability at least δ) independent of the optimization function (Aytug and Koehler, 1996; Greenhalgh and Marshall, 2000). Nevertheless, the best upper bound the analysis can guarantee is the same as that of a random algorithm choosing random individuals independently in each generation (i.e., Random Search (RS)). The failure of these approaches to obtain useful bounds confirmed the idea that when analysing the time complexity of problem-independent search heuristics the function to be optimised needs to be considered. Further confirmation came when Droste, Jansen and Wegener proved the existence of functions for which the (1+1)

[a]An EA has an *elitist* selection strategy if at least one of the current best solutions is never lost during the selection process.

EA finds the global optimum in expected time $\Theta(n^n)$ (Droste *et al.*, 2002) (versus $O(2^n)$ time guaranteed by Random Search on *pseudo-Boolean functions* i.e., real-valued functions defined over bitstrings). Hence, the need of measuring the performance of EAs on specific problems became evident in the late nineties. In an overview of the field of Evolutionary Computation in 1995, Yao states that to his best knowledge a 1991 paper by Hart and Belew is the only paper on the topic (Yao, 1996). In fact, in that paper Hart and Belew also proved that "*when analyzing the computational complexity of a Genetic Algorithm (GA) classes of functions should be analyzed relative to the algorithmic parameters chosen for the GA*" (Hart and Belew, 1991).

In particular, it appeared necessary to use "standard" measures such as a relationship between the size of the problem being tackled and the expected time needed for the solution to be found. Beyer et al. confirmed this stating that "*there were almost no results, before the mid-nineties, estimating the worst-case expected optimization time of an evolutionary algorithm working on some problem or estimating the probability of obtaining within t(n) steps a solution fulfilling a minimum demand of quality*" (Beyer *et al.*, 2002). The only work available before the late nineties with such flavour was Mühlenbein's analysis of the (1+1) EA for OneMax (Mühlenbein, 2002). As a consequence, an attempt was finally made towards a systematical time complexity analysis that turns the theory of EAs into *"a legal part of the theory of efficient algorithms"* (Beyer *et al.*, 2002).

This chapter overviews the major techniques used in the runtime analysis of EAs for discrete optimisation. It is structured as follows. Section 2.2 introduces EAs. In Section 2.3 definitions of computational time complexity of EAs are given while common test functions used to study the performance of EAs are introduced in Section 2.4. Section 2.5 is an overview of the most popular mathematical methods used in the runtime analysis of EAs. In the same section example applications to the classical test functions are given. Finally, in the conclusion a brief discussion of the state of the art is given.

2.2. Evolutionary Algorithms

A general EA derived from Evolution Strategies (ESs) is the **($\mu + \lambda$) EA**. The algorithm keeps μ candidate solutions (*individuals*) collectively called the *parent population*. A fitness function that assigns values to individuals is used to describe the quality of the candidate solutions. At each step (a *generation*), individuals of the parent population are mutated to create λ new individuals called the *offspring population.* At the end of each generation the algorithm chooses the best μ individuals out of both the offspring and the parent population to survive for the next generation. This process iteratively goes on from generation to generation until a termination condition is satisfied. The standard algorithm, with candidate solutions represented as bitstrings, works as described in Algorithm 1.

Algorithm 1 $(\mu+\lambda)$-EA_p

Let $t := 0$.
Initialize P_0 with μ individuals chosen uniformly at random.
repeat
  for $i = 1$ to λ **do**
    {Create λ offspring}
    choose x_i from μ uniformly at random;
    flip each bit in x_i with probability p;
  end for
  Create the new population by choosing the best μ individuals out of the $\mu + \lambda$ parent and offspring populations;
  Let $t = t + 1$;
until a stopping condition is fulfilled.

For the selection step, in case of more than μ individuals with strictly better fitness than the rest, generally a random choice among individuals with same fitness occurs to determine which are the "best" μ. Most EAs can be derived from Algorithm 1.

(1+1) EA. Obtained by setting $\mu = \lambda = 1$, the (1+1) EA is the most simple and theoretically analysed EA. Here, however, selection accepts the offspring if and only if its fitness is at least as good as that of its parent. The mutation probability of $p = 1/n$ is generally considered the best choice (Bäck, 1993). Droste, Jansen and Wegener, have proved that a mutation probability of $p = \Theta(1/n)$ is optimal for OneMax and for separable functions in general (Droste *et al.*, 1998b).

Random Local Search (RLS). RLS is similar to the (1+1) EA except that exactly one bit flips in each iteration.

(μ,λ)-EA. A variant of Algorithm 1 is the (μ,λ)-EA where the new parent population of μ individuals is chosen only from the offspring population. For the so called *comma strategy* algorithms, λ needs to be chosen greater than μ.

Genetic Algorithms (GAs) Genetic Algorithms (GAs), probably the most famous subclass of EAs, use different selection schemes, based on probability measures such as *fitness proportional selection* or *tournament selection* and a crossover operator such as *one point crossover* and *uniform crossover* (Goldberg, 1989).

2.3. Computational Complexity of EAs

Analysing the computational complexity of an algorithm generally means predicting the resources the algorithm requires. The considered resources may be of various kinds, for example the amount of memory, but usually the computational time is measured (Cormen *et al.*, 1990). The time required by an algorithm usually grows with the size of the input, so traditionally its runtime is described as a function of the size of its input. Just like for the computational complexity of deterministic algorithms, most of the results in the runtime analysis of EAs are expressed using asymptotic notation. In classical computational complexity the runtime is the number of primitive operations, generally called *steps*, that are executed before the output is given. In practical applications of EAs the time required to perform a fitness function evaluation is much higher than the time required by the rest of the operations performed in one generation. Hence, the computational complexity of an EA implies counting the number of fitness function evaluations performed to find the optimal solution.

In contrast to deterministic algorithms, randomised algorithms make random choices during their execution, so they do not perform the same operations in every run, even if the input is the same. Also, they do not necessarily output the same result on a given input if they are run more than once. Hence, the runtime of the algorithm is a random variable. The *expected runtime* (i.e., the expected number of fitness function evaluations until the optimum is found once) of the algorithm is used as a measure of their performance. In particular, if T_f is a random variable measuring the time required by the algorithm to find the solution for a certain function f, the runtime analysis consists of estimating $E(T_f)$, the expected value of T_f.

Sometimes $E(T_f)$ is not sufficient to give an idea of how likely it is that the algorithm will be efficient (i.e., will return the optimal solution in polynomial time with respect to the size of the problem). For this reason results about $p(T_f \leq t)$, i.e., the *success probability* given a certain number of steps t, are also desired. In general, if T_f is the random variable measuring the time for the solution to be found by an EA for a certain function f, then the runtime analysis of a randomised algorithm for optimising f consists of (Wegener, 2001a):

(1) The estimation of the expected runtime $E(T_f)$;
(2) The calculation of the success probability $p(T_f \leq t)$;

For the calculation of success probabilities usually tail inequalities are used (see Chapter 1 for an overview).

2.4. Test Problems

Computational complexity results of EAs first appeared in the nineties with analyses of simple EAs for pseudo-Boolean functions with significant structures for theoretical analyses called *toy problems*. These functions were artificially designed to highlight characteristics of the studied EAs when tackling optimisation problems and to build up a first basis of mathematical techniques for the runtime analysis of EAs. By building up on these first results, nowadays it is possible to analyse more complicated EAs on combinatorial optimisation problems having practical applications.

The most popular test function used to study EAs is OneMax. The function counts the number of ones in the bitstring.

$$\textsc{OneMax}(x) = \sum_{i=1}^{n} x_i$$

The goal of the optimization process is that of finding the bitstring with the maximal number of one-bits (i.e., maximization). Obviously the global optimum is the bitstring containing only one-bits, i.e., the 1^n bitstring. The (1+1) EA optimises OneMax in expected $\Theta(n \ln n)$ time (Droste *et al.*, 2002) (by symmetry, the same result holds for ZeroMax defined as $\textsc{ZeroMax}(x) := n - \textsc{OneMax}(x)$).

Both OneMax and ZeroMax belong to the class of functions called *unimodal* that only have one local optimum. It was believed that EAs were efficient on all unimodal functions. Long path problems were designed to disprove this conjecture (Rudolph, 1997).

Another popular *unimodal* test function is LEADINGONES which is defined as follows:

$$\text{LEADINGONES}(x) = \sum_{i=1}^{n} \prod_{j=1}^{i} x_j$$

It simply counts the number of consecutive ones at the beginning of the bitstring (i.e., the number of *leading ones*). Again the global optimum is the 1^n bitstring and the (1+1) EA optimises the function in $\Theta(n^2)$ expected steps (Droste *et al.*, 2002).

Deceptive functions were introduced as functions that disproved the *building block hypothesis* of Schema theory (Goldberg, 1989). A class of well-known bimodal deceptive functions is that of *trap functions.*

$$\text{TRAP}(x) = \begin{cases} n+1 & \text{if } x = 0^n \\ \text{ONEMAX}(x) & \text{otherwise.} \end{cases}$$

The function is deceptive for EAs because the algorithms climb up the the ONEMAX slope heading towards the 1^n bitstring getting further and further away from the 0^n bitstring, the global optimum. With overwhelming probability the (1+1) EA requires $n^{\Theta(n)}$ steps to optimise TRAP.

Another exponential time pseudo-Boolean toy problem for EAs is the needle-in-a-haystack function.

$$\text{NEEDLE}(x) = \begin{cases} 1 \text{ if } x = 1^n \\ 0 \text{ otherwise.} \end{cases}$$

The function consists of a huge plateau of constant fitness and only one optimal point called the *needle.*

Short path functions with plateaus of constant fitness were introduced by Jansen and Wegener to prove that if the plateau is not too large then the (1+1) EA and RLS can walk to the end of the plateau efficiently.

$$\text{SPC}(x) = \begin{cases} 2n & \text{if } x = 1^n \\ n & \text{if } x = 1^i 0^{n-i} \\ \text{ZEROMAX} & \text{otherwise} \end{cases}$$

It consists of a ZEROMAX slope followed by a plateau of n points of constant fitness $\{1^i 0^{n-i}\}$ for $0 \leq i < n$ leading to the 1^n bitstring, the global optimum. The (1+1) EA requires $O(n^3)$ steps to optimise SPC (Jansen and Wegener, 2001a).

2.5. Mathematical Techniques

When the first runtime analyses of EAs were performed, the study of the (1+1) EA was a natural choice. Although it is different compared to EAs often used in practice, its analysis is important for the following reasons (Wegener, 2001b):

(1) The (1+1) EA is very efficient for a large number of functions;
(2) The (1+1) EA can be interpreted as a randomised hill-climber that can not get stuck forever on a local optimum;
(3) The analysis of the (1+1) EA reveals tools that can be used in the analysis of more complex EAs.

First, a general upper bound on the runtime of the (1+1) EA that holds for any pseudo-Boolean function is given. The theorem was first presented by Droste *et al.* (2002).

Theorem 2.1. *The expected runtime of the (1+1) EA for an arbitrary function defined in* $\{0,1\}^n$ *is at most* n^n.

Proof. If the function contains multiple global optima, then we consider one of them, say x^*. Let i be the number of bit positions in which the current solution and the global optimum differ. Hence, in $n-i$ bit positions they have the same value. In order to reach the global optimum the algorithm has to mutate the i bits and leave the $n-i$ bits unchanged. Each bit flips with probability $1/n$, hence does not flip with probability $(1-1/n)$. Then the probability of reaching the global optimum, x^* from any current solution x is lower bounded by:

$$p(x^*|x) = \left(\frac{1}{n}\right)^i \left(1 - \frac{1}{n}\right)^{n-i} \geq \left(\frac{1}{n}\right)^n = n^{-n}$$

Since the above probability bound holds for all current solutions, it implies an upper bound on the expected runtime of n^n. □

In Subsection 2.5.4 it will be shown that there exist functions for which the bound is asymptotically tight. This emphasises the need of performing the runtime analysis on classes of functions. The subsequent subsections illustrate the most popular methods used in the analysis of EAs and example applications on classical test functions.

2.5.1. *An Analytic Markov Chain Framework*

A general framework for analysing the average hitting times of EAs by using the theory of Markov chains was presented by He and Yao (2003). Let $(X_t; t = 0, 1, ..)$ be an homogeneous absorbing Markov chain, defined on state-space S. Then the matrix of the probabilities of each state i to reach each state j in one step, (i.e., the *transition matrix*), can be written in the following canonical form (Iosifescu, 1980) :

$$\mathbf{P} = \begin{pmatrix} \mathbf{I} & \mathbf{0} \\ \mathbf{R} & \mathbf{T} \end{pmatrix}. \tag{2.1}$$

Let T denote the set of transient states and $S - T$ the set of absorbing states. Then, each entry $\mathbf{T_{i,j}}$ of the submatrix $\mathbf{T}$ denotes the probability of going in one step from a transient state i to a transient state j, while $\mathbf{R}$ is the sub-matrix of the probabilities of going in one step from each state i, with $i \in T$ to each state j with $j \in S - T$. $\mathbf{I}$ is a unit submatrix for the $S - T$ absorbing states.

The matrix $\mathbf{N} = (\mathbf{I} - \mathbf{T})^{-1}$ is called the *fundamental matrix* of the absorbing Markov chain, and $\mathbf{m} = \mathbf{N}\mathbf{1} = (\mathbf{I} - \mathbf{T})^{-1}\mathbf{1}$ is the vector of the mean absorption times where each entry m_i is the expected time to reach the recurrent states when the chain starts from state i (Isaacson and Madsen, 1976).

Unfortunately for most transition matrices $\mathbf{P}$, it is difficult if not impossible to invert the matrix $(\mathbf{I} - \mathbf{T})$ to obtain the fundamental matrix, $\mathbf{N}$. It is possible to obtain an explicit expression for the vector $\mathbf{m}$ when the matrix $\mathbf{T}$ is in a *simple form* (i.e., a bi-diagonal, tridiagonal or a lower triangular matrix). This is the case for simple EAs such as the (1+1) EA and RLS. In the following the canonical matrix for the (1+1) EA and RLS will be derived.

Let the search space $S = \{0, 1\}^n$ be partitioned into $N + 1$ subspaces $S_0, \ldots, S_N$ according to their fitness values such that for each subspace S_i, $0 \le i \le N$ $f(x_i) = f_i$ and $f_{max} = f_0 > f_1 > \cdots > f_N = f_{min}$[b]. In general, for the (1+1) EA and RLS the following matrix is obtained where

[b]This idea will be defined more precisely and generally in Section 2.5.2 when the *artificial fitness levels* method will be introduced.

the sub-matrix $\mathbf{T}$ of the canonical transition matrix $\mathbf{P}$ is lower triangular.

$$\mathbf{P} = \begin{pmatrix} 1 & 0 & 0 & 0 & \dots & 0 \\ p_{(1,0)} & p_{(1,1)} & 0 & 0 & \dots & 0 \\ p_{(2,0)} & p_{(2,1)} & p_{(2,2)} & 0 & \dots & 0 \\ \vdots & \vdots & \vdots & \vdots & \ddots & \vdots \\ p_{(N,0)} & p_{(N,1)} & p_{(N,2)} & p_{(N,3)} & \cdots & p_{(N,N)} \end{pmatrix}$$

So the first hitting time of the EA (i.e., the time to reach the absorbing states) is given by $\mathbf{m} = (\mathbf{I} - \mathbf{T})^{-1}\mathbf{1}$, and the following exact expressions for each m_i can be derived:

$$\begin{cases} m_0 = & 0, & i \in S_0 \\ m_1 = & 1/p_{(1,0)}, & i \in S_1 \\ m_i = & \frac{1+p_{(i,1)}m_1+p_{(i,2)}m_2+\dots+p_{(i,i-1)}m_{i-1}}{p_{(i,0)}+p_{(i,1)}+p_{(i,2)}+\dots+p_{(i,i-1)}}, & i = 2, \dots, N \end{cases}$$

However, deriving the exact probability values for each $p_{(i,j)}$ and then calculating the above expressions is complex and tedious even for simple functions such as OneMax. Since RLS only flips one bit per iteration the canonical matrix for OneMax is less complicated. In the following theorem the Markov chain framework will be used to calculate the expected runtime of RLS for OneMax.

Theorem 2.2. *The RLS algorithm requires expected runtime $n \cdot H_i$ steps to optimise* OneMax *when initialised with i zero-bits.*

In the statement of the theorem, H_i is the i_{th} harmonic number.

Proof. The search space of the OneMax function consists of $n+1$ different fitness values according to the number of remaining zero-bits to be changed into one-bits. Hence, the dimension of the matrix is $N \times N$ with $N = n+1$. Let the current search point consist of i zeroes and $n-i$ ones. Since RLS flips one random bit at each step, the probability of flipping a zero is i/n and that of flipping a one is $(n-i)/n$. If a one is flipped into a zero, fitness will decrease and the new point will not be accepted by selection. Hence the canonical matrix of RLS for OneMax is:

$$\begin{pmatrix} 1 & 0 & 0 & 0 & \dots & 0 & 0 \\ 1/n & (n-1)/n & 0 & 0 & \dots & 0 & 0 \\ 0 & 2/n & (n-2)/n & 0 & \dots & 0 & 0 \\ \vdots & \vdots & \vdots & \ddots & \vdots & \vdots & \vdots \\ 0 & 0 & 0 & 0 & (n-1)/n & 1/n & 0 \\ 0 & 0 & 0 & 0 & \dots & 1 & 0 \end{pmatrix}$$

the solutions to the system are the following:

$$\begin{cases} m_0 = & 0 \\ m_j = & \frac{1+p_{(j,j-1)}m_{j-1}}{p_{(j,j-1)}} = \frac{1+(j/n)\cdot m_{j-1}}{j/n} = n/j + m_{j-1};\ 0 < j \leq n \end{cases}$$

and the expected runtime starting from a bitstring with i zeroes is:

$$\sum_{k=1}^{i}(n/k) = n\sum_{k=1}^{i}(1/k) = n \cdot H_i$$

□

He and Yao also derived the transition matrix for population based EAs such as (N+N)-EAs both using elitist and non-elitist selection schemes. However, in general, the matrices are not in simple form so no explicit solutions may be derived. Again, only if the mutation operator flips one bit at a time will the matrices be in simple form (i.e., tridiagonal). He and Yao took advantage of this situation by analysing various selection schemes in a comparison between EAs using only one individual against population based EAs on different pseudo-Boolean functions (He and Yao, 2002). Nevertheless, the calculations are not simple. The general limitations, described above, in deriving expected runtime expressions from the transition matrix of the Markov chain highlight the necessity of introducing other randomised algorithm analysis tools that may be used as "tricks" to gather information about the Markov process without having to build the exact Markov chain model.

2.5.2. *Artificial Fitness Levels*

The most natural method spawned in the first runtime analyses is *artificial fitness levels*. The class of unitation functions contains all the functions where the fitness value depends on the number of ones in the bitstring representing an individual, but not on their position. When analysing the (1+1) EA on such functions, it was natural to partition the space in sets depending on the number of ones of a bitstring. Later, the method was extended to be used in more general settings rather than just with *unitation* functions. A description of the general method follows.

The search space is partitioned into subsets based on the fitness function. Rather than considering the whole search space S as a set of $|S| = 2^n$ different states, it is divided into $m < 2^n$ states $A_1, \ldots, A_m$, such that for all points $a \in A_i$ and $b \in A_j$ it happens that $f(a) < f(b)$ if $i < j$. We also require that A_m contains only optimal search points. In such a way, the Markov chain can be constructed considering only m different states rather than n. Apart from providing a smaller Markov chain transition matrix, artificial fitness levels lead to simplified calculations.

Let the search space be decomposed into $m < 2^n$ partitions $A_1, \ldots, A_m$ as described above. Furthermore, let $p(A_i)$ be the probability that a randomly chosen search point belongs to A_i and p_i be a lower bound on the probability that an individual belonging to A_i is mutated into an individual belonging to A_j for some $j > i$. Then the expected runtime $E(T_f)$ is bounded from above as follows (Wegener, 2001a):

$$E(T_f) \leq \sum_{1 \leq i \leq m-1} p(A_i) \cdot (p_i^{-1} + \cdots + p_{m-1}^{-1}) \leq (p_1^{-1} + \cdots + p_{m-1}^{-1}). \quad (2.2)$$

The proof of the first inequality is based on the law of total probability and the second is trivial. Now, as an example application, the method will be applied to the analysis of the (1+1) EA for OneMax.

Theorem 2.3. *The expected runtime of the (1+1) EA on* OneMax *is* $O(n \ln n)$.

Proof. We consider the current solution to be in fitness level A_i if it has i ones. We first calculate the probability p_i of reaching a fitness level A_j from level A_i with $j > i$. It is sufficient that one of the $n - i$ zeroes is flipped into a one-bit and that all the other bits remain unchanged. Since each bit flips with probability $1/n$ and is not flipped with probability $(1 - 1/n)$, the probability that one of the $n - i$ bits is flipped is $(n - i)/n$ and the probability that the remaining $n - 1$ bits are not changed is $(1 - 1/n)^{n-1}$. Hence, the probability is bounded as follows

$$p_i = (n-i)\frac{1}{n}\left(1 - \frac{1}{n}\right)^{n-1} \geq \frac{1}{e}\frac{n-i}{n}$$

The inequality follows because $(1 - 1/n)^{n-1} \geq 1/e$ (see Lemma 1.3 in Chapter 1). Then, by using the artificial fitness levels Equation (2.2), the expected time to reach the optimum is bounded by

$$E(T) \leq \sum_{i=0}^{n-1} p^{-i} \leq \sum_{i=0}^{n-1} \frac{e \cdot n}{n-i} = e \cdot n \sum_{i=1}^{n} \frac{1}{i} \leq e \cdot n \cdot (\ln n + 1) = O(n \ln n)$$

□

The theorem can also be proved with other techniques, such as drift analysis (see Section 2.5.6) and the *multiplicative weight decrease* method. The latter method was introduced for the analysis of EAs for the minimum spanning tree problem (Neumann and Wegener, 2007) while it was applied for the analysis of the (1+1) EA for OneMax in (Neumann, 2006). The proof of the matching lower bound for OneMax is less straightforward and inspired by the *coupon collector's problem* presented in the next section.

By considering a partition A_i to contain all candidate solutions with exactly i leading ones, the artificial fitness levels method can be used in a similar way to obtain an upper bound of $O(n^2)$ for LeadingOnes. A drift analysis proof will be presented in Section 2.5.6.

2.5.3. *Coupon Collector Problem*

The coupon collector problem is used in the analysis of EAs when it is necessary to derive a lower bound for a given number of bits to be chosen at least once. The problem is defined in the following way (Motwani and Raghavan, 1995):

There are n types of coupons and at each trial one coupon is chosen at random. Each coupon has the same probability of being extracted. The goal is to find the expected number of trials for the collector to obtain all the n types of coupons.

The problem together with relevant results has been introduced in Chapter 1. By Theorem 1.21 (1), the expected time to collect all coupons is $(1 + o(1))n \ln n$.

So by considering zero-bits as missing coupons and one bits as collected ones, another proof of Theorem 2.2 on the expected runtime of RLS for OneMax follows directly from the coupon collector's theorem. Furthermore, by applying Theorem 1.23 of Chapter 1 also a statement of the success probability within $cn \ln n$ can be easily obtained (i.e., $\Pr(T \geq cn \ln n) \leq n^{-c+1}$ for any c). Analogously by applying Theorem 1.24 of the same chapter, the probability that the runtime will be lower than the expectation can be immediately derived (i.e., $\Pr(T < (1 - \epsilon)(n - 1) \ln n) \leq \exp(-n^{\epsilon})$).

The (1+1) EA may flip more than one bit in each generation, meaning that in each trial more than one coupon may be obtained (or coupons swapped for others), hence a direct application of the above method will not suffice for the obtainment of a lower bound on the runtime of the (1+1) EA for OneMax. To this end, Droste, Jansen and Wegener first prove that at most $n/2$ one-bits are created during initialisation with constant probability 1/2 (by the symmetry of the binomial distribution), then they show that there is a constant probability that in $cn \ln n$ steps one of the $n/2$ remaining zero-bits have never been flipped (Droste *et al.*, 2002). We formalise this in the following theorem.

Theorem 2.4. *The expected runtime of the (1+1) EA on* ONEMAX *is* $\Omega(n \ln n)$.

Proof. With probability $(1-1/n)^t$ a given bit does not flip at all in a phase of t steps. Hence, with probability $1-(1-1/n)^t$ it flips at least once in the phase of t steps. Taking this probability to the power of $n/2$ we get the probability that $n/2$ given bits flip at least once, and finally the complementary probability is that of the event that at least one bit does not flip. Hence, the probability that at least one of the $n/2$ bits does not flip in $t=(n-1)\ln n$ steps is bounded as follows:

$$1-\left(1-(1-1/n)^{(n-1)\ln n}\right)^{n/2} \geq 1-\left(1-e^{-\ln n}\right)^{n/2} =$$
$$=1-(1-1/n)^{n/2} \geq 1-e^{-1/2}$$

and the expected runtime is at least

$$E(T) \geq (n-1)\ln n \cdot 1/2 \cdot (1-e^{-1/2}) = \Omega(n \ln n)$$ □

2.5.4. *Typical Run Investigations*

Typical runs were introduced in the analysis of EAs following the consideration that the "global behaviour of a process" is predictable with high probability contrary to the local behaviour which, instead, is quite unpredictable. For instance, the result of just one coin toss is not easy to state (i.e., if it will be a head rather than a tail), but it is possible to get very tight bounds on the number of heads that have appeared after a large number of coin tosses, allowing for an exponentially small error probability.

The idea behind *typical runs* is that of dividing the process into k phases which are long enough to assure that some event happens with probability p_k and does not happen with probability $1-p_k$, i.e., the *failure probability.* Following such a sequence of events, the last phase should lead the EA to the global optimum with a failure probability of $1-\sum_{i=1}^{k} p_i$ and a runtime depending on the sum of each phase time. Typical run investigations occur very often in runtime analyses of EAs, combined with Chernoff bounds used to obtain the failure probabilities. The following theorem is an example application for the TRAP function. See Chapter 1 for an introduction to Chernoff bounds and Markov's inequality.

Theorem 2.5. *The expected runtime of the (1+1) EA on* TRAP *is* $\Omega(n^n)$. *With probability at least* $1-e^{-\Omega(n)}$ *the (1+1) EA requires* $n^{\Omega(n)}$ *steps to optimise* TRAP.

Proof. We use the typical run investigations method and split the analysis into the following three phases:

(1) This phase ends when at least $n/3$ one-bits are in the current solution (i.e., $\sum_{i=1}^{n} x_i \geq n/3$).
(2) This phase ends when the 1^n bitstring is the current solution (i.e., $x = 1^n$).
(3) This phase ends when the global optimum is found (i.e., $x = 0^n$).

To use the typical run investigations method we need to show that the runtime required to complete the three phases is $n^{\Omega(n)}$ and that the failure probability of each phase is at most $e^{-\Omega(n)}$. Since the expected runtime conditional to the first two phases happening is n^n, we will also prove the first statement along the way.

By using Chernoff bounds we calculate the probability that the algorithm is initialised with less than $n/3$ one-bits. Let X be the random variable indicating the number of one-bits after initialisation. The probability of each bit being set to one is $q = 1/2$ leading to $n/2$ expected one-bits after initialisation (i.e., $E(X) = n \cdot q = n/2$). Setting $\delta = 1/3$ we get

$$\Pr(X \leq n/3) \leq e^{-E(X)\delta^2/2} = e^{-(n/2)\cdot(1/9)\cdot(1/2)} = e^{-n/36} = e^{-\Omega(n)}$$

Hence, Phase (1) ends after initialisation with an exponentially small failure probability.

After Phase (1), no points with less than $n/3$ one-bits except for the 0^n bitstring will be accepted because of their lower fitness values. Hence, from Phase (2) on, in order to reach the global optimum at least $n/3$ precise one-bits have to flip into zero-bits in each step. Since each of them flips with probability $1/n$, the probability of reaching the 0^n bitstring in one step is less than $n^{-n/3}$ [c].

The upper bound on the runtime of the (1+1) EA on OneMax also holds for reaching the 1^n bitstring of Trap as long as the 0^n bitstring is not found first. From Theorem 2.3 we know that the expected time for the algorithm to climb up OneMax is less than $c \cdot n \ln n$. By Markov's inequality, the probability that the algorithm will take more than $e \cdot c \cdot n \ln n$ steps is

$$P(T \geq e \cdot c \cdot n \ln n) \leq \frac{c \cdot n \ln n}{e \cdot c \cdot n \ln n} = 1/e$$

[c]This implies an expected time of at least $n^{n/3} = n^{\Omega(n)}$ for the optimum to be found. Together with a union bound proving that with exponentially small probability $n^{-n/12} = n^{-\Omega(n)}$ the optimum is reached in less than $n^{n/4}$ steps, we can already prove the second statement of the theorem.

and by iteratively applying Markov's inequality we get a probability of at most e^{-n} for the (1+1) EA to take more than $e \cdot c \cdot n^2 \ln n$ steps. The probability that in $e \cdot c \cdot n^2 \ln n$ steps $n/3$ precise bits flip is $e \cdot c \cdot n^2 \ln n \cdot n^{-n/3} = n^{-\Omega(n)}$ by the union bound. Hence, phase (2) ends in $O(n^2 \ln n)$ time with exponentially small failure probability.

Once the local optimum has been reached, the expected time to reach the global optimum is n^n because all the bits have to be flipped at the same time. This is conditional to the first two phases occurring. Hence, we get a lower bound on the expected runtime of $(1 - e^{-\Omega(n)}) \cdot n^n = \Omega(n^n)$ which proves the first statement.

Again, by the union bound the probability that that the optimum has been reached in less than $n^{n/2}$ steps is

$$\Pr(T \leq n^{n/2}) \leq n^{n/2} \cdot n^{-n} = n^{-n/2} = n^{-\Omega(n)}$$

Summing up the failure probabilities of each phase, we get that the algorithm requires at least exponential runtime $n^{n/2} = n^{\Omega(n)}$ with probability at least $1 - e^{-\Omega(n)}$. □

2.5.5. *Gambler's Ruin Problem*

When the function's landscape contains a plateau of constant fitness, and the EA needs to cross it to reach the global optimum, then it is forced to do a random walk on the plateau. This occurs because the algorithm cannot be driven towards the optimum by increasing fitness values. Big plateaus, such as those of the Needle-in-a-haystack problem, lead to exponential run times (Droste *et al.*, 1998a). Surprisingly, if the plateau is not too large the (1+1) EA has a good performance. This result has been obtained by applying the gambler's ruin theory (Feller, 1968) to the analysis of EAs.

The gambler's ruin problem, derived from classical probability theory, was introduced in the analysis of EAs by Jansen and Wegener when analysing the (1+1) EA on plateaus of constant fitness of short path problems (SPC) (Jansen and Wegener, 2001a) described in Section 2.4. The problem is formalised as follows (Feller, 1968).

Consider a gambler who wins or loses a dollar with probability p and q respectively. Let his initial capital be z and his adversary's capital be $a - z$. The game continues until the gambler or his adversary is ruined. The goal of the ruin problem is to find the probability of the gambler's ruin and the probability distribution of the game.

If q_z is the probability of the gambler's ruin, and $p_z = 1 - q_z$, the probability of his adversary's ruin is (Feller, 1968):

$$\begin{cases} q_z = \frac{(q/p)^a - (q/p)^z}{(q/p)^a - 1}, & p \neq q, \\ q_z = 1 - \frac{z}{a}, & p = q = 1/2. \end{cases} \tag{2.3}$$

Let T_z be the expected duration of the game. Then:

$$\begin{cases} T_z = \frac{z}{q-p} - \frac{a}{q-p}\frac{1-(q/p)^z}{1-(q/p)^a}, & p \neq q, \\ T_z = z(a-z), & p = q = 1/2. \end{cases} \tag{2.4}$$

A direct application of the method to RLS for SPC follows.

Theorem 2.6. *The expected runtime of RLS for the* SPC *function is* $O(n^3)$.

Proof. The expected runtime to reach the 0^n string is less than $c \cdot n \ln n$ (see Theorem 2.2 and consider ZeroMax rather than OneMax) unless a better point is found before. We pessimistically assume the 0^n string is found before a plateau point. The expected runtime to reach the point 10^{n-1} is n since the first bit has to flip. Hence our initial capital is $z = 1$ leading ones and we win the game when our capital consists of $a = n$ leading ones (i.e., the optimum). We now consider a step to be *relevant* if a point on the plateau is created and accepted. The probability of a relevant step is $2/n$ since there is only one relevant move heading towards the optimum (i.e., an extra leading one is added) and one heading away (i.e., a leading one is removed). This gives $n/2$ expected steps for a relevant move to occur. The conditional probability that a relevant step heads towards the optimum is $(1/n)/(2/n) = 1/2$. By applying the gambler's ruin Equation (2.3) for $p = q = 1/2$ with $z = 1$ and $a = n$, we obtain $q_z = 1 - \frac{z}{a} = 1 - 1/n$ for the probability of being ruined and $p_z = 1/n$ of winning. Hence, n gambler ruin games are necessary in expectation to reach the optimum. By Equation (2.4) the expected duration of each game is $T_z = z(a-z) = 1 \cdot (n-1) = n-1$. Since each relevant step occurs in $n/2$ expected steps, the expected number of steps to reach the end of the plateau, is $n/2 \cdot (n-1) = n^2/2 - n/2$. Then the expected runtime for n gambler ruin games is $n \cdot (n^2/2 - n/2) = n^3/2 - n^2/2$ and hence to reach the end of the plateau (i.e., the optimum). Finally we have to remember to consider the n steps required to obtain the first leading one at the beginning of each game. In total this is $n \cdot n = n^2$ in expectation. Summing up, the total expected runtime of RLS is $c \cdot n \ln n + n^2 + n^3/2 - n^2/2 = O(n^3)$. □

Attention has to be paid when applying the ruin ideas to the analysis of the (1+1) EA because the algorithm may flip more than one bit at a time, hence more than one dollar may be won or lost in each step. The proof for the (1+1) EA on SPC was first presented in (Jansen and Wegener, 2001a). The proof first shows that, with constant probability, cn^2 successful steps are sufficient to reach the end of the plateau with moves of at most length 3. Finally, an upper bound of $O(n^3)$ is achieved by considering that the expected time for each successful move is lower than $e \cdot n$. Recently a Global Gambler's Ruin Theorem, based on drift analysis, has been introduced by Happ et al. to deal with lower bounds for algorithms that flip more than one bit per iteration (Happ *et al.*, 2008).

2.5.6. *Drift Analysis*

Drift Analysis dates back to Hajek (1982) and was introduced in the study of the computational complexity of EAs by He and Yao (2001).

Given a Markov process $\{X_t\}_{t\geq 0}$ over a search space S and a distance *distance function* $g\colon S \to \mathbb{R}_0^+$ mapping each state to a non-negative real number, the *one-step drift* at time t of the Markov process is defined as

$$\Delta(t) = g(X_t) - g(X_{t+1}).$$

The drift $\Delta(t)$ represents the random decrease in distance to the optimum obtained by the algorithm in one step at time t. The idea behind drift analysis is quite simple. If the current process is at distance d from the optimum and at each step there is an improvement (i.e., a positive drift) towards the optimum of at least $\delta > 0$, then the optimal solution will be found in at most d/δ steps.

2.5.6.1. *Drift Analysis for Upper Bounds*

From the previous idea the following drift theorem for obtaining upper bounds on the runtime of EAs is derived.

Theorem 2.7 (Drift theorem for upper bounds). *Let $\{X_t\}_{t\geq 0}$ be a Markov process over a finite set of states S, and $g : S \to \mathbb{R}_0^+$ a function that assigns a non-negative real number to every state. Let the time to reach the optimum be $T := \min\{t \geq 0 : g(X_t) = 0\}$. If there exists $\delta > 0$ such that at any time step $t \geq 0$ and at any state $X_t > 0$ the following condition holds:*

$$E(\Delta(t)|g(X_t) > 0) = E(g(X_t) - g(X_{t+1}) \mid g(X_t) > 0) \geq \delta, \tag{2.5}$$

then

$$E(T \mid X_0) \le \frac{g(X_0)}{\delta} \tag{2.6}$$

and

$$E(T) \le \frac{E(g(X_0))}{\delta}. \tag{2.7}$$

The first statement was proved by He and Yao (2001). The second statement of the theorem follows from the first one by using the law of total expectation. So if the drift condition, (i.e., Condition (2.5)), can be proved for $\delta = 1/\text{poly}(n)$, then a polynomial upper bound on the expected runtime can be achieved by applying Theorem 2.7. As an example application, the (1+1) EA will be analysed for LEADINGONES.

Theorem 2.8. *The expected time for the (1+1) EA to optimise* LEADINGONES *is* $O(n^2)$.

Proof. Let the distance function $g(X_t) = i$ where i is the number of missing leading ones. The negative drift is 0 since if a leading one is removed from the current solution the new point will not be accepted. A positive drift (i.e., of length 1) is achieved as long as the first 0 is flipped and the leading ones remain unchanged.

$$E(\Delta^+(t)) = \sum_{k=1}^{n-i} k \cdot (p(\Delta^+(t)) = k) \ge 1 \cdot 1/n \cdot (1 - 1/n)^{n-i} \ge 1/(en)$$

Hence, $\Delta(t) = E(\Delta^+(t)) - E(\Delta^-(t)) \ge 1/(en) = \delta$ and the expected runtime is (cf. Equation (2.7)):

$$E(T) \le \frac{E(g(X_0))}{\delta} \le \frac{n}{1/(en)} = e \cdot n^2 = O(n^2)$$

□

Using a similar distance function as in that of the previous proof (i.e., counting the missing ones rather than the missing leading ones) for a drift analysis of the runtime of the (1+1) EA for ONEMAX leads to an upper bound of $O(n^2)$ which is not tight. The tight $O(n \ln n)$ bound can be achieved with similar calculations by using $g(X_t) = \ln(i + 1)$ as distance function where i is the number of missing ones in the bitstring at time t.

We consider the more general class of pseudo-Boolean *linear functions* which also contains ONEMAX. Pseudo-Boolean linear functions are defined as $f(x) = w_0 + \sum_{i=1}^{n} w_i x_i$, where we assume $w_0 = 0$ since constant additive terms do not have any influence on the behaviour of the (1+1) EA and the weights w_i are considered to be non-negative (otherwise the x_i may be replaced by $1 - x_i$). Hence, the 1^n bitstring is always the global optimum and is the unique optimum if all the weights are positive.

Theorem 2.9. *The expected runtime of the (1+1) EA for the class of linear functions is* $O(n \ln n)$.

The presented bound is tight since an $\Omega(n \ln n)$ bound has been presented for the BinVal (binary values) linear function (Droste *et al.*, 2002) or for OneMax in Theorem 2.4. The theorem was first proved already by using a "potential function" in (Droste *et al.*, 2002) before drift analysis was defined. However the proof was very long and involved. A considerably simpler proof is obtained through drift analysis (He and Yao, 2004).

Proof. W.l.o.g. we assume the weights are sorted, i.e., $w_1 \geq w_2 \geq \cdots \geq w_n$. We consider the distance function introduced in (He and Yao, 2004):

$$g(X_t) = \ln\left(1 + 2\sum_{i=1}^{n/2}(1 - x_i) + \sum_{i=n/2+1}^{n}(1 - x_i)\right)$$

Hence, the distance is at most $\ln(1 + (3/2)n) = \Theta(\ln n)$ obtained for the 0^n bitstring. Apart from an upper bound on the distance, we need a lower bound on the expected drift. We observe that the drift is minimal when only one 0-bit is in the current string (minimising the positive drift) and all the other bits are set to one. More precisely the drift is minimised for $x_t = 01111\ldots1$ (case 1) or $x_t = 1^{n/2}01^{n/2-1}$ (case 2). We consider the two cases separately and we will call a bit a *left bit*, b_l if it is one of the first $n/2$ bits of bitstring. Otherwise, it will be called a *right bit*, b_r.

Case 1. A positive drift is obtained if the 0-bit is flipped to one and no other bit is flipped or if the 0-bit is flipped to one and a *right bit* is flipped to zero while everything else remains unchanged. The positive drift is bounded as follows:

$$\begin{aligned} E(\Delta^+(t)) &= \frac{1}{n}\left(1 - \frac{1}{n}\right)^{n-1}(\ln(3) - \ln(1)) \\ &\quad + \binom{n/2}{1}\left(\frac{1}{n}\right)^2\left(1 - \frac{1}{n}\right)^{n-2}(\ln(3) - \ln(2)) \\ &\geq \frac{\ln(3)}{en} + \frac{\ln(3/2)}{2en} \end{aligned}$$

A negative drift will happen if the 0-bit flips together with at least three *right bits* or if the 0-bit flips together with at least a *left bit* and any number of *right bits* (except for $b_l = 1$ and $b_r = 0$ where the distance is zero). The negative drift is bounded as follows:

$$E(\Delta^-(t)) \leq \sum_{b_r=3}^{n/2} \binom{n/2}{b_r} \left(\frac{1}{n}\right)^{1+b_r} (\ln(1+b_r) - \ln(3))$$
$$+ \sum_{b_l=1}^{n/2-1} \sum_{b_r=0}^{n/2} \binom{n/2-1}{b_l} \binom{n/2}{b_r} \left(\frac{1}{n}\right)^{1+b_l+b_r} (\ln(1+2b_l+b_r) - \ln(3))$$

By using $\ln(x+1) \leq x$ for all x and $\binom{n}{i} \leq n^i/i!$ the first term is bounded from above by

$$\frac{1}{n} \sum_{b_r=3}^{n/2} \frac{(1/2)^{b_r}}{b_r!} \cdot \frac{b_r}{3} \leq \frac{1}{2 \cdot 3 \cdot n} \sum_{b_r=3}^{n/2} \frac{(1/2)^{b_r-1}}{(b_r-1)!} = \frac{1}{6 \cdot n} \sum_{b_r=2}^{(n/2)-1} \frac{(1/2)^{b_r}}{b_r!}$$

$$\leq \frac{1}{6 \cdot n} \left([-1 - 1/2] + \sum_{b_r=0}^{\infty} \frac{(1/2)^{b_r}}{b_r!} \right) \leq \frac{1}{6 \cdot n} [-3/2 + e^{1/2}] \leq \frac{1}{36 \cdot n}$$

because the sum $\sum_{i=0}^{\infty} n^i/i!$ converges to e^n.

Similarly the second negative drift term is bounded from above as follows:

$$\frac{1}{n} \sum_{b_l=1}^{(n/2)-1} \sum_{b_r=0}^{n/2} \frac{(1/2)^{b_l}}{(b_l)!} \cdot \frac{(1/2)^{b_r}}{(b_r)!} \ln\left(\frac{1+2b_l+b_r}{3}\right)$$

The terms such that the distance is less than $\ln(20)$ sum up to at most 0.28. By using $\ln(x+1) \leq x/6$ for $x > 20$ we get $\ln((1+2b_l+b_r)/3) \leq \ln((3+2b_l+b_r)/3) \leq (2b_l+b_r)/(6 \cdot 3)$.

For $b_r = 0$ we get

$$\frac{1}{n} \sum_{b_l=29}^{n/2-1} \frac{(1/2)^{b_l}}{b_l!} \cdot \frac{2b_l}{6 \cdot 3} \leq \frac{1}{3 \cdot 36 \cdot n}$$

where the sum starts from $b_l = 29$ because it should be $\ln((1+2 \cdot b_l)/3) \geq \ln(20)$ and is bounded with similar calculations as those for the first term.

We bound the remaining terms as

$$\frac{1}{n}\sum_{b_l=1}^{(n/2)-1}\sum_{b_r=1}^{n/2}\frac{(1/2)^{b_l}}{(b_l)!}\cdot\frac{(1/2)^{b_r}}{(b_r)!}\left(\frac{2b_l+b_r}{6\cdot 3}\right)$$

$$\leq \frac{1}{6\cdot 3\cdot n}\sum_{b_l=1}^{(n/2)-1}\sum_{b_r=1}^{n/2}\frac{3(1/2)^{b_l}}{(b_l-1)!}\cdot\frac{(1/2)^{b_r}}{(b_r-1)!}$$

$$\leq \frac{1}{6\cdot 2\cdot n}\sum_{b_l=1}^{(n/2)-1}\frac{(1/2)^{b_l}}{(b_l-1)!}\cdot\sum_{b_r=1}^{n/2}\frac{(1/2)^{b_r-1}}{(b_r-1)!}$$

$$\leq \frac{e^{1/2}}{6\cdot 4\cdot n}\sum_{b_l=1}^{(n/2)}\frac{(1/2)^{b_l-1}}{(b_l-1)!}\leq\frac{e}{6\cdot 4\cdot n}$$

Summing up, the total expected drift for **Case 1** is

$$E(\Delta(t)) = E(\Delta^+(t)) - E(\Delta^-(t))$$
$$\geq \frac{1}{n}\left(\frac{\ln(3)}{e}+\frac{\ln(3/2)}{2e}-\frac{1}{36}-0.28-\frac{1}{108}-\frac{e}{24}\right)=\Omega\left(\frac{1}{n}\right)$$

Case 2. A positive drift is obtained if the 0-bit is flipped to one and no other bit is flipped. A negative drift will happen if the 0-bit is flipped to one and a *left bit* is flipped to zero while everything else remains unchanged or if the 0-bit is flipped together with any number of *right bits*. Hence, the drift is bounded by

$$E(\Delta(t)) \geq \frac{1}{en}[\ln(2)-\ln(1)] - \binom{n/2}{1}\left(\frac{1}{n}\right)^2\left(1-\frac{1}{n}\right)^{n-2}[ln(3)-\ln(2)]$$

$$-\sum_{b_r=2}^{n/2-1}\binom{n/2-1}{b_r}\left(\frac{1}{n}\right)^{1+b_r}[ln(1+b_r)-\ln(2)]=\Omega\left(\frac{1}{n}\right)$$

with similar calculations to those of Case 1.

Finally, by Theorem 2.7 we get

$$E(T)\leq\frac{E(g(X_0))}{\delta}\leq\frac{\Theta(\ln n)}{\Omega(1/n)}=O(n\ln n)$$

□

Recently two new proofs have appeared. Firstly, an explicit bound of $3.8n\log_2 n + 7.6\log_2 n$ has been proved in (Jägerskupper, 2008). Secondly, a new drift theorem (i.e., *Multiplicative Drift Theorem*) inspired by the multiplicative weight decrease method has been introduced and used to slightly improve the constants (Doerr *et al.*, 2010a,b). In this chapter we preferred to present a proof using the classical drift theorem (Theorem 2.7) for the standard (1+1) EA, which to our knowledge was not available in literature[d]. However, since the Multiplicative Drift Theorem is tailored towards applications where there is a logarithmic factor in the runtime bound (a slowdown that seems to appear naturally as EAs approach the optimum) we believe it is promising for future applications of EAs. Doerr and Goldberg have extended the Multiplicative drift Theorem to also include a probability tail bound in addition to an upper bound on the expected runtime. This latest version is stated as follows (Doerr and Goldberg, 2010).

Theorem 2.10 (Multiplicative drift theorem). *Let* $\{X_t\}_{t\geq 0}$ *be a sequence of random variables taking values in some set* S. *Let* $g : S \to \{0\} \cup \mathbb{R}_{\geq 1}$ *and assume that* $g_{\max} := \max\{g(x) | x \in S\}$ *exists. Let* $T := \min\{t \geq 0 : g(X_t) = 0\}$. *If there exists* $\delta > 0$ *such that*

$$E(g(X_{t+1}) \mid g(X_t)) \leq (1-\delta) g(X_t)$$

then

$$E(T) \leq \frac{1}{\delta}(1 + \ln g_{\max})$$

and for every $c > 0$

$$\Pr\left(T > \frac{1}{\delta}(\ln g_{\max} + c)\right) \leq e^{-c}.$$

2.5.6.2. *Drift Analysis for Lower Bounds*

By changing the direction of Inequalities (2.5, 2.6, 2.7) of the classical Drift Theorem for upper bounds (i.e., Theorem 2.7), a drift theorem for polynomial lower bounds is obtained (He and Yao, 2001). We will apply it to prove a matching lower bound on the runtime of the (1+1) EA for LeadingOnes.

Theorem 2.11. *The expected time for the (1+1) EA to optimise* LeadingOnes *is* $\Omega(n^2)$.

Proof. Let the current solution have $n - i$ leading ones (i.e., $1^{n-i}0*$). We define the distance function as the number of missing leading ones, i.e.,

[d]The proof presented in (He and Yao, 2004) considers a (1+1) EA that only accepts new individuals that have strictly better fitness than their predecessor.

$g(X) = i$. The $(n - i + 1)$th bit is a zero and let Y be a random variable representing the number of consecutive one-bits after the first zero (i.e., the *free riders*). The $i - 1$ bits after the zero-bit are uniformly distributed at initialisation and still are uniformly distributed because the mutation operator mutates each bit uniformly and the bits have never had any influence on the selection operator. Hence, the expected number of free riders is

$$E(Y) = \sum_{k=1}^{i-1} k \cdot \Pr(Y = k) = \sum_{k=1}^{i-1} \Pr(Y \geq k) = \sum_{k=1}^{i-1} (1/2)^k \leq 1$$

The negative drift is 0 since a point with less than i leading ones will not be accepted by selection. Let A be the event that the first zero-bit flips into a one-bit. Hence, the positive drift (i.e., the decrease in distance) is bounded as follows:

$$E(\Delta^+(t)) \leq p(A) \cdot E(\Delta^+(t)|A) = 1/n \cdot (1 + E(Y)) \leq 2/n$$

Since, also at initialisation the expected number of free riders is less than 2, it follows that $E(g(X_0)) \geq n - 2$, and by applying inequality (2.7) with inverted sign we get

$$E(T) \geq \frac{E(g(X_0))}{\delta} = \frac{n-2}{2/n} = \Omega(n^2)$$

□

If the expected drift is negative, then the algorithm is moving in expectation away from the optimum rather than towards it. However, this is not enough to prove exponential lower bounds on the runtime. To this end, also the probability that the process may perform large jumps towards the optimum should be proven to be low. Various drift theorems for the obtainment of exponential lower bounds have been devised (i.e., (He and Yao, 2001), (Giel and Wegener, 2003)). Compared to previous versions, the following theorem with simplified drift conditions has recently been presented (Oliveto and Witt, 2008):

Theorem 2.12 (Simplified/negative drift theorem). *Let X_t, $t \geq 0$, be the random variables describing a Markov process over the state space $S := \{0, 1, \ldots, N\}$ and denote $\Delta_t(i) := (X_{t+1} - X_t \mid X_t = i)$ for $i \in S$ and $t \geq 0$. Suppose there exist an interval $[a, b]$ of the state space and three constants $\delta, \epsilon, r > 0$ such that for all $t \geq 0$*

(1) $E(\Delta_t(i)) \geq \epsilon$ for $a < i < b$
(2) $\Pr(\Delta_t(i) = -j) \leq 1/(1+\delta)^{j-r}$ for $i > a$ and $j \geq 1$

then there is a constant $c^* > 0$ *such that for* $T^* := \min\{t \geq 0\colon X_t \leq a \mid X_0 \geq b\}$ *it holds* $\Pr(T^* \leq 2^{c^*(b-a)}) = 2^{-\Omega(b-a)}$.

Here a and b identify an interval where the drift $\Delta_t(i)$ typically is negative, i.e., $X_{t+1} > X_t$. Then Condition 1 guarantees that when the process is inside the interval it drifts away from the target. Since the probability of a step of length j decreases exponentially with j, Condition 2 assures that there are very low chances that the algorithm performs long jumps to the other side of the interval. If the two conditions hold then an upper bound on the success probability in B steps is obtained by applying Theorem 2.12. If the interval $[a, b]$ depends on n and B is exponential in n then the probability that the runtime is not exponential can be proved to be overwhelmingly small. The (1+1) EA for NEEDLE will be used as an example.

Theorem 2.13. *Let* $0 < \eta \leq 1/2$ *be constant. Then there is a constant* $c > 0$ *such that with probability* $1-2^{-\Omega(n)}$ *the (1+1) EA on* NEEDLE *creates only search points with at most* $n/2 + \eta n$ *ones in* 2^{cn} *steps.*

Proof. To use the simplified drift theorem, we need to define the interval $[a, b]$ and show that conditions 1 and 2 of the theorem hold for the interval. Then the theorem will imply that with overwhelming probability the algorithm requires exponential time to cross the interval.

Let $\gamma := \eta/2$. By Chernoff bounds the probability that the initial bit string has less than $n/2 - \gamma n$ zeroes is $e^{-\Omega(n)}$. Hence, we set $b := n/2 - \gamma n$ and $a := n/2 - 2\gamma n$. Let X_t denote the number of zeroes in the bit string at time step t. Now we have to check that the two conditions of the simplified drift theorem hold.

Given a bitstring in state $i < n/2 - \gamma n$, i.e., with i zeroes, let $\Delta(i)$ denote the random increase of the number of zeroes (i.e., the drift). Condition 1 holds if $E(\Delta(i)) \geq \epsilon$ for some constant $\epsilon > 0$. The (1+1) EA flips 0-bits and 1-bits independently with probability $1/n$. Since there are i zero-bits and $n - i$ one-bits in the current string, the expected number of zero-bits flipped into one-bits is i/n and the viceversa is $(n - i)/n$. Hence,

$$E(\Delta(i)) = \frac{n-i}{n} - \frac{i}{n} = \frac{n-2i}{n} \geq 2\gamma$$

and Condition 1 holds with $\epsilon := 2\gamma$.

To prove Condition 2 we have to show that the probability of performing flips of j bits decreases exponentially with j. In particular, Condition 2 is: $\Pr(\Delta(i) = -j) \leq 1/(1+\delta)^{j-r}$ for $j \in \mathbb{N}_0$. Since to reach state $i - j$ or less from state i, at least j bits have to flip, the probability of a drift of length j can be bounded as follows:

$$\Pr(\Delta(i) = -j) \leq \binom{n}{j}\left(\frac{1}{n}\right)^j \leq \frac{1}{j!} \leq \left(\frac{1}{2}\right)^{j-1}$$

This proves Condition 2 by setting $\delta = r = 1$. Since every point on NEEDLE except for the optimum has the same fitness value, all the moves will be accepted. So we do not need to further analyse the effects of selection. Finally, the Simplified Drift Theorem implies that for a constant $c^* > 0$ the global optimum is found in $2^{c^*(b-a)} = 2^{cn}$ steps, where $c := c^*(b-a)/n > 0$ is a different constant, with probability at most $2^{-\Omega(b-a)} = 2^{-\Omega(n)}$. □

Theorem 2.12 is usually sufficient to prove results for the (1+1) EA. Recently, a more general version of the Simplified Drift Theorem tailored towards the analysis of population-based EAs has been introduced and applied to the analysis of an EA with fitness proportional selection (Neumann *et al.*, 2009).

2.6. Conclusion

An introduction to the most common methods used in the runtime analysis of EAs has been given. Although these tools were initially developed to analyse the (1+1) EA for toy problems, thanks to their generality, nowadays they are applied to the analysis of the same algorithm on more sophisticated problems. In fact, both positive and negative results concerning the (1+1) EA are now available for a wide range of classical combinatorial optimisation problems with practical applications (see Oliveto *et al.*, 2007) for a detailed overview of results.

However, the (1+1) EA is very different from the EAs used by practitioners in real-world applications. The main challenge in the field is to be able to analyse *realistic* EAs using populations and more sophisticated operators such as crossover and stochastic selection mechanisms. Indeed it is still not possible to analyse the sGA even for ONEMAX. This challenge might require the extension of the available mathematical tools or new ones might have to be devised.

Only preliminary work in this direction has been accomplished. Work on toy problems has been carried out to prove the existence of instance classes for which crossover is beneficial (Jansen and Wegener, 2002, 2005) and of those where population based EAs without crossover outperform the

(1+1) EA (Jansen and Wegener, 2001b; Witt, 2008; Storch, 2004; He and Yao, 2002).

Analyses of population based EAs such as the (1+λ)-EA (Jansen *et al.*, 2005), the (1,λ)-EA (Jägerskupper and Storch, 2007) and the (μ+1)-EA (Witt, 2006) reveal no advantages of using a population for the "classical" toy problems compared to using a (1+1) EA. Promising techniques have spawned from these works such as *family trees* (Witt, 2006) that allow to capture the status of the entire population at a time step. However, further work needs to be done to understand why populations are beneficial like reported by practitioners.

Preliminary results concerning diversity maintenance are also available (Friedrich *et al.*, 2008) proving it is a very important issue when using populations.

A further approach towards realistic GAs has recently been accomplished with the analyses of population based EAs using stochastic selection operators such as *rank selection* (Lehre and Yao, 2009b) and *fitness proportional selection* (Neumann *et al.*, 2009), however crossover is not considered. In these works extensions of both the *family trees* and the *simplified drift theorem* methods were required.

Concerning populations and classical combinatorial problems only preliminary work is available for minimum spanning trees (Neumann and Wegener, 2007), cliques for sparse graphs (Storch, 2007), subset sum problems (He and Yao, 2001) and vertex cover (Oliveto *et al.*, 2008). A proof is still not available that for a classic combinatorial problem it is beneficial to use a realistic population-based EA rather than the (1+1) EA.

Recently, the methods have been extended to the analyses of EAs for practical software engineering problems (Lehre and Yao, 2007, 2009a) and other nature inspired meta-heuristics such as ant colony optimisation (Doerr *et al.*, 2007), particle swarm optimisation (Sudholt and Witt, 2008), artificial immune system inspired algorithms (Zarges, 2008, 2009; Horoba *et al.*, 2009) and estimation of distribution algorithms (Chen *et al.*, 2007, 2009). Such analyses are still in their infancy and results for *realistic* versions of these other nature inspired meta-heuristics are still lacking.

Acknowledgments

Small parts of this chapter have previously appeared in Oliveto *et al.* (2007). This work was supported by the following EPSRC grants: EP/P502322/1 and EP/D052785/1. The authors thank Daniel Johannsen and Carola

Winzen for reviewing the chapter and are grateful to Dr. Per Kristian Lehre, Dr. Jun He, Dr. Thomas Jansen and Ass. Prof. Carsten Witt for insightful discussions.

References

Aytug, H. and Koehler, G. J. (1996). Stopping criteria for finite length genetic algorithms. *ORSA J. Comp.* **8**, pp. 183–191.

Bäck, T. (1993). Optimal mutation rates in genetic search, in *In Proceedings of the Fifth International Conference on Genetic Algorithms (ICGA)*, pp. 2–8.

Beyer, H. G., Schwefel, H. and Wegener, I. (2002). How to analyse evolutionary algorithms, *Theoretical Computer Science* **287**, pp. 101–130.

Chen, T., Lehre, P. K., Tang, K. and Yao, X. (2009). When is an estimation of distribution algorithm better than an evolutionary algorithm? in *Proceedings of the 2009 IEEE Congress on Evolutionary Computation (CEC'2009)* (18-21 May, Trondheim, Norway,), pp. 1470–1477.

Chen, T., Tang, K., Chen, G. L. and Yao, X. (2007). On the analysis of average time complexity of estimation of distribution algorithms, in *Proceedings of the 2007 IEEE Congress on Evolutionary Computation (CEC'2007)* (25-28 September, Singapore), pp. 453–460.

Cormen, T. H., Rivest, R. L. and Leiserson, C. E. (1990). *Introduction to Algorithms* (McGraw-Hill Inc., New York, NY, USA).

Doerr, B. and Goldberg, L. A. (2010). Drift analysis with tail bounds, in *Proceedings of the Parallel Problem Solving from Nature 2010 (PPSN 2010)*, pp. 174–183.

Doerr, B., Johannsen, D. and Winzen, C. (2010a). Drift analysis and linear functions revisited, in *Proceedings of the 2010 IEEE World Congress on Computational Intelligence (WCCI 2010)*, pp. 1967–1974.

Doerr, B., Johannsen, D. and Winzen, C. (2010b). Multiplicative drift analysis, in *Proceedings of the 2010 Genetic and Evolutionary Computation Conference (GECCO 2010)*, pp. 1449–1456.

Doerr, B., Neumann, F., Sudholt, D. and Witt, C. (2007). On the runtime analysis of the 1-ant ACO algorithm, in *Proceedings of the Genetic and Evolutionary Computation Conference (GECCO 2007)* (ACM Press), pp. 33–40.

Droste, S., Jansen, T. and Wegener, I. (1998a). On the optimization of unimodal functions with the (1 + 1) evolutionary algorithm, in *PPSN V: Proceedings of the 5th International Conference on Parallel Problem Solving from Nature* (Springer-Verlag, London, UK), pp. 13–22.

Droste, S., Jansen, T. and Wegener, I. (1998b). A rigorous complexity analysis of the (1 + 1) evolutionary algorithm for separable functions with Boolean inputs. *Evolutionary Computation* **6**, 2, pp. 185–196.

Droste, S., Jansen, T. and Wegener, I. (2002). On the analysis of the (1+1) evolutionary algorithm, *Theoretical Computer Science* **276**, 1-2, pp. 51–81.

Feller, W. (1968). *An introduction to probability theory and its applications. - Vol. 1* (Wiley).

Friedrich, T., Oliveto, P. S., Sudholt, D. and Witt, C. (2008). Theoretical analysis of diversity mechanisms for global exploration, in *Proc. of Genetic and Evolutionary Computation Conference - GECCO 2008* (ACM Press), pp. 945–952.

Giel, O. and Wegener, I. (2003). Evolutionary algorithms and the maximum matching problem, in *STACS '03: Proceedings of the 20th Annual Symposium on Theoretical Aspects of Computer Science* (Springer-Verlag, London, UK), pp. 415–426.

Goldberg, D. E. (1989). *Genetic Algorithms for Search, Optimization, and Machine Learning* (Addison-Wesley).

Greenhalgh, D. and Marshall, S. (2000). Convergence criteria for genetic algorithms. *SIAM Journal of Computing* **30**, 1, pp. 269–282.

Hajek, B. (1982). Hitting-time and occupation-time bounds implied by drift analysis with applications, *Advances in Applied Probability* **13**, 3, pp. 502–525.

Happ, E., Johannsen, D., Klein, C. and Neumann, F. (2008). Rigorous analyses of fitness-proportional selection for optimizing linear functions, in *Proceedings of the annual conference on Genetic and Evolutionary Computation (GECCO '08)* (ACM Press), pp. 953–960.

Hart, W. E. and Belew, R. K. (1991). Optimizing an arbitrary function is hard for the genetic algorithm, in *In proceedings of the Fourth International Conference on Genetic Algorithms (ICGA)*, pp. 190–195.

He, J. and Yao, X. (2001). Drift analysis and average time complexity of evolutionary algorithms, *Artificial Intelligence* **127**, 1, pp. 57–85.

He, J. and Yao, X. (2002). From an individual to a population: an analysis of the first hitting time of population-based evolutionary algorithms. *IEEE Transactions on Evolutionary Computation* **6**, 5, pp. 495–511.

He, J. and Yao, X. (2003). Towards an analytic framework for analysing the computation time of evolutionary algorithms, *Artificial Intelligence* **145**, 1-2, pp. 59–97.

He, J. and Yao, X. (2004). A study of drift analysis for estimating computation time of evolutionary algorithms, *Natural Computing: an international journal* **3**, 1, pp. 21–35.

Holland, J. H. (1992). *Adaptation in Natural and Artificial Systems: An Introductory Analysis with Applications to Biology, Control, and Artificial Intelligence* (The MIT Press).

Horoba, C., Jansen, T. and Zarges, C. (2009). Maximal age in randomized search heuristics with aging, in *Proc. of Genetic and Evolutionary Computation Conference (GECCO 2009)* (ACM Press), pp. 803–810.

Iosifescu, M. (1980). *Finite Markov Processes and Their Applications* (John Wiley & Sons).

Isaacson, D. L. and Madsen, R. W. (1976). *Markov Chains: Theory and Applications* (John Wiley & Sons).

Jägerskupper, J. (2008). A blend of Markov-chain and drift analysis, in *Procceedings of the 10th International Conference on Parallel Problem Solving From Nature (PPSN X)*, pp. 41–51.

Jägerskupper, J. and Storch, T. (2007). When the plus strategy outperforms the comma strategy - and when not, in *Proceedings of The First IEEE Symposium on Foundations of Computational Intelligence (FOCI 2007)*, pp. 25–32.

Jansen, T., Jong, K. A. D. and Wegener, I. A. (2005). On the choice of the offspring population size in evolutionary algorithms, *Evolutionary Computation* **13**, 4, pp. 413–440.

Jansen, T. and Wegener, I. (2001a). Evolutionary algorithms: How to cope with plateaus of constant fitness and when to reject strings of the same fitness, *IEEE Transactions on Evolutionary Computation* **5**, 6, pp. 589–599.

Jansen, T. and Wegener, I. (2001b). On the utility of populations in evolutionary algorithms, in *GECCO '01: Proceedings of the 3rd annual conference on Genetic and evolutionary computation*, pp. 1034–1041.

Jansen, T. and Wegener, I. (2002). On the analysis of evolutionary algorithms - a proof that crossover really can help, *Algorithmica*, pp. 47–66.

Jansen, T. and Wegener, I. (2005). Real royal road functions: where crossover provably is essential, *Discrete Applied Mathematics* **149**, 1-3, pp. 111–125.

Lehre, P. K. and Yao, X. (2007). Runtime analysis of (1+1) ea on computing unique input output sequences, in *Proceedings of the 2007 IEEE Congress on Evolutionary Computation (CEC'2007)* (25-28 September, Singapore), pp. 1882–1889.

Lehre, P. K. and Yao, X. (2009a). Crossover can be constructive when computing unique input output sequences, in *In Proceedings of the 7th International Conference on Simulated Evolution and Learning* (Dec 7-10 Melbourne, Australia), pp. 595–604.

Lehre, P. K. and Yao, X. (2009b). On the impact of the mutation-selection balance on the runtime of evolutionary algorithms, in *Proceedings of Foundations of Genetic Algorithms (FOGA'2009)* (Morgan Kaufmann Publishers Inc.), pp. 47–58.

Motwani, R. and Raghavan, P. (1995). *Randomized Algorithms* (Cambridge University Press).

Mühlenbein, H. (2002). How genetic algorithms really work: mutation and hillclimbing, in *Proc. of the Second Conference on Parallel Problem Solving from Nature (PPSN II)*, pp. 15–25.

Neumann, F. (2006). *Combinatorial Optimization and the Analysis of Randomized Search Heuristics*, Ph.D. thesis, Christian-Albrechts University of Kiel.

Neumann, F., Oliveto, P. S. and Witt, C. (2009). Theoretical analysis of fitness-proportional selection: Landscapes and efficiency, in *Proceedings of the 2009 Genetic and Evolutionary Computation Conference (GECCO 2009)*, pp. 835–842.

Neumann, F. and Wegener, I. (2007). Randomized local search, evolutionary algorithms, and the minimum spanning tree problem, *Theoretical Computer Science* **378**, 1, pp. 32–40.

Oliveto, P. S., He, J. and Yao, X. (2007). Time complexity of evolutionary algorithms for combinatorial optimization: A decade of results, *International Journal of Automation and Computing* **4**, 3, pp. 281–293.

Oliveto, P. S., He, J. and Yao, X. (2008). Analysis of population-based evolutionary algorithms for the vertex cover problem, in *Proceedings of the 2008 World Congress on Computational Intelligence (WCCI2008)* (Hong Kong), pp. 1563–1570.

Oliveto, P. S. and Witt, C. (2008). Simplified drift analysis for proving lower bounds in evolutionary computation, in *Proccedings of the 10th International Conference on Parallel Problem Solving From Nature (PPSN X)*, pp. 82–91.

Rudolph, G. (1994). Convergence analysis of canonical genetic algorithms, *IEEE Transactions on Neural Networks* **5**, 1, pp. 96–101.

Rudolph, G. (1997). How mutations and selection solve long path problems in polynomial expected time, *Evolutionary Computation* **4**, pp. 195–205.

Rudolph, G. (1998). Finite Markov chain results in evolutionary computation: A tour d'horizon, *Fundamenta Informaticae* **35**, 1–4, pp. 67–89.

Storch, T. (2004). Real royal road functions for constant population size, *Theoretical Computer Science* **320**, pp. 123–134.

Storch, T. (2007). Finding large cliques in sparse semi-random graphs by simple randomised search heuristics, *Theoretical Computer Science* **386**, 1-2, pp. 114–131.

Sudholt, D. and Witt, C. (2008). Runtime analysis of binary PSO, in *Proceedings of the Genetic and Evolutionary Computation Conference (GECCO 2008)* (ACM Press), pp. 135–142.

Wegener, I. (2001a). Methods for the analysis of evolutionary algorithms on pseudo-Boolean functions, in R. Sarker, M. Mohammadian and X. Yao (eds.), *Evolutionary Optimization* (Kluwer Academic Publishers, Dordrecht, The Netherlands), pp. 349–369.

Wegener, I. (2001b). Theoretical aspects of evolutionary algorithms, in *Proceedings of the 28th International Colloquium on Automata, Languages and Programming* (Springer-Verlag, London, UK), pp. 64–78.

Witt, C. (2006). Runtime analysis of the (μ+1) EA on simple pseudo-Boolean functions evolutionary computation, in *GECCO '06: Proceedings of the 8th annual conference on Genetic and evolutionary computation* (ACM Press, New York, NY, USA), pp. 651–658.

Witt, C. (2008). Population size versus runtime of a simple evolutionary algorithm, *Theoretical Computer Science* **403**, 1, pp. 104–120.

Yao, X. (1996). An overview of evolutionary computation, *Chinese Journal of Advanced Software Research* **3**, 1, pp. 12–29.

Zarges, C. (2008). Rigorous runtime analysis of inversely fitness proportional mutation rates, in *Proceedings of the 10th international conference on Parallel Problem Solving from Nature, LNCS 5199* (Springer-Verlag, Berlin, Heidelberg), pp. 112–122.

Zarges, C. (2009). On the utility of the population size for inversely fitness proportional mutation rates, in *Proc. of Genetic and Evolutionary Computation Conference (GECCO 2009)* (ACM Press), pp. 39–46.

Chapter 3

Evolutionary Computation in Combinatorial Optimization

Daniel Johannsen

Max-Planck-Institut für Informatik,
Campus E1 4,
66123 Saarbrücken,
Germany
johannse@mpi-inf.mpg.de

In this chapter, we study the runtime of the (1+1) Evolutionary Algorithm on classical problems in combinatorial optimization. In particular, we investigate the problems of finding minimum spanning trees, maximum weight bases, maximum matchings, single-source and all-pair shortest paths, and Euler tours. On these problems, we demonstrate prominent techniques in runtime analysis like drift analysis, the study of typical runs, and the use of large deviation bounds.

Contents

3.1. Introduction

Classical combinatorial optimization problems play a central role in the theoretical study of evolutionary algorithms. On the one hand, these problems are general enough to make meaningful comparisons among different evolutionary algorithms. On the other hand, combinatorial optimization problems have enough structural properties to make the theoretical analysis of such algorithms possible.

This chapter gives a concise overview over the runtime analysis of the (1+1) Evolutionary Algorithm (EA) on polynomially solvable problems in combinatorial optimization. We conduct these analyses for the minimum spanning tree problem in the context of maximum weight bases of matroids; for the single-source and all-pair shortest path problem; for the maximum matching problem; and for the problem of finding an Euler tour.

An extensive discussion of these problems and the corresponding problem-specific polynomial time algorithms can be found in Mehlhorn and Sanders (2009) and in Cormen, Leiserson, Rivest and Stein (2001), for example.

We have selected these problems for two main reasons. First, they are among the most studied combinatorial problems in the theory of evolutionary algorithms. Second, these problems are particularly suited to demonstrate typical techniques used in the runtime analysis of evolutionary algorithms. For example, we will present the concepts of drift analysis; typical runs; dominance of stochastic processes; and large deviation bounds.

3.2. The Basic Combinatorial (1+1) Evolutionary Algorithm

A type of problem that occurs in classical combinatorics is to find a specific subgraph in a given graph (we consider all graphs to be finite, simple, and undirected unless explicitly stated otherwise). In particular, such problems assign an objective value to each subgraph and ask to maximize or minimize this value over all subgraphs of a given class. Prominent examples which we discuss in this chapter are the problems of finding a minimum spanning tree (MST), a single-source shortest path tree (SSSP), or a maximum matching (MM) in an edge-weighted graph.

Given a graph $G = (V, E)$, we identify a subgraph F of G with its edge set, that is, with slight abuse of notation we let F be an element of the power set 2^E of E. A subgraph optimization problem on G then consists of the space of *feasible* subgraphs $\mathcal{F} \subseteq 2^E$ and a partial order $\succeq$ on $\mathcal{F}$. A subgraph $F_{\text{opt}} \in \mathcal{F}$ is *optimal* if $F_{\text{opt}} \succeq F$ for all $F \in \mathcal{F}$. For all problems we consider in this chapter, there exists such an optimum.

Often, we define $\succeq$ implicitly by an objective function $f\colon 2^E \to \mathbb{R}$. Then, $F \succeq H$ if and only if $f(F) \geq f(H)$ in case f is to be maximized and $F \succeq H$ if and only if $f(F) \leq f(H)$ in case f is to be minimized. In

Algorithm 2 The Basic Combinatorial (1+1) EA

Let E be a finite set. Furthermore, let $\mathcal{F} \subseteq 2^E$ be a set of feasible search points, let $\succeq$ be a partial order relation on $\mathcal{F}$, and let $x^{(0)} \in \mathcal{F}$ be an initial search point. The basic combinatorial (1+1) EA corresponding to the tuple $(E, \mathcal{F}, \succeq, x^{(0)})$ iteratively generates a sequence of search points $\{x^{(t)}\}_{t \in \mathbb{N}}$ in $\mathcal{F}$ by the following procedure.

For all $t \in \mathbb{N}$ with $t \geq 1$, the basic combinatorial (1+1) EA generates a random candidate search point $y^{(t)} \subseteq E$ such that $y^{(t)}$ differs from $x^{(t-1)}$ in each edge $e \in E$ with probability $1/|E|$. In other words, $\Pr[e \in x^{(t-1)} \triangle y^{(t)}] = 1/|E|$ independently for all $e \in E$ (where $\triangle$ is the symmetric difference between two sets). Afterwards,

$$x^{(t)} := \begin{cases} y^{(t)} & \text{if } y^{(t)} \in \mathcal{F} \text{ and } y^{(t)} \succeq x^{(t-1)}, \\ x^{(t-1)} & \text{otherwise.} \end{cases}$$

both cases, all optimal solutions have the same objective value f_{opt}. Such an objective function may be given by edge-weights $w \colon E \to \mathbb{R}^+$. In this case, the objective value of a subgraph $F \subseteq E$ is the cumulative weight of its edges $w(F) := \sum_{e \in F} w(e)$. As first example of a subgraph optimization problem, we consider the MST problem.

Problem 3.1 (The minimum spanning tree problem).
Let $G = (V, E)$ be a connected graph and $w \colon E \to \mathbb{R}^+$ a weight function on E. The minimum spanning tree (MST) problem *asks for a spanning tree $F \subseteq E$ that minimizes $w(F) := \sum_{e \in F} w(e)$.*

For the MST problem, the space of feasible subgraphs $\mathcal{F}$ contains all spanning trees of G. For two spanning trees F and H, we let $F \succeq H$ (that is, F is "better" than H) if and only if $w(F) \leq w(H)$.

We are interested in how the basic combinatorial (1+1) EA solves subgraph optimization problems like the MST problem.

In particular, we are interested in the *optimization time* of the basic combinatorial (1+1) EA. This is the random variable T that describes the first point in time $t \in \mathbb{N}$ for which $x^{(t)}$ is optimal.

Clearly, the basic combinatorial (1+1) EA is not only applicable to subgraph optimization problems but to all combinatorial problems that can be formulated in terms of a space $\mathcal{F}$ of feasible subsets over a ground set E and a partial order relation $\succeq$ that represents an optimization objective.

For example, we can formulate the (1+1) EA on bit strings (Algorithm 1) as a basic combinatorial (1+1) EA: Let $\{0,1\}^E$ be the set of $|E|$-dimensional Boolean vectors with index set E. We associate every Boolean vector $x \in \{0,1\}^E$ with a set $F \in 2^E$ by identifying the events "$x_e = 1$" and "$e \in F$" for all $e \in E$. If we now set E to $\{1,\ldots,n\}$, $\mathcal{F}$ to 2^E and choose $x^{(0)}$ uniformly at random from $\mathcal{F}$ then the basic combinatorial (1+1) EA is equivalent to the bit string (1+1) EA.

Traditionally, the basic combinatorial (1+1) EA for subgraph optimization problems is often formulated in the terminology of the bit string (1+1) EA (confer the references in the respective sections). However, this comes at the loss of the expressiveness introduced by the set of feasible solutions. In the case of the bit string (1+1) EA, this loss has to be compensated by introducing an additional optimization criterion.

We briefly discuss this compensation on the example of the MST problem. If we define the MST problem over the whole space 2^E, then the empty set is the only optimal search point. To counteract this, we introduce a penalty criterion $c(F)$ that equals the number of connected components in a subgraph F of G. We set $F \succeq H$ for two subgraphs F and H if $(c(F), w(F))$ is lexicographically smaller than $(c(H), w(H))$. Then, the basic combinatorial (1+1) EA also conducts a proper search on all infeasible search points (in contrast to starting with $x^{(0)} \notin \mathcal{F}$ and then rejecting all $y^{(t)}$'s until $y^{(t)} \in \mathcal{F}$).

However, often the analysis of this additional phase of finding the first feasible search point does not result in much additional insight. For the MST problem, for example, a simple coupon collector argument shows that the expected time until a search point in $\mathcal{F}$ is found is $O(|E| \ln |E|)$. This is of lower order compared to the expected overall optimization time (see Theorem 3.2).

For the other problems discussed in this chapters, similar arguments apply. The expected time to find the first feasible search point is of lower order compared to the time to find an optimum afterwards (confer the references in the respective sections). Therefore, we consider as feasible search points only those that are essential in understanding the runtimes of the basic combinatorial (1+1) EA on these problems (that is, spanning trees, in our example).

A further difference between the bit string (1+1) EA and the basic combinatorial (1+1) EA is that in the former the initial search point is chosen uniformly at random and in the latter given as part of the input. This is sensible in the light of the previous discussion since we no longer want to dictate how the initial search point is generated. Moreover, this strengthens existing results on the optimization times of the discussed problems since we may assume worst case initial search points.

From now on, we only consider the basic combinatorial (1+1) EA and refer to it simply as (1+1) EA.

We conclude this section with a theorem on the optimization time of the (1+1) EA for the MST problem.

Algorithm 3 (1+1) EA_{MST}

The basic combinatorial (1+1) EA for the minimum spanning tree problem, the (1+1) EA_{MST}, is an instance of Algorithm 2.
Let $G = (V, E)$ be a graph and $w \colon E \to \mathbb{R}^+$ a weight function on E. Then, the space of feasible search points $\mathcal{F}$ consists of all spanning trees of G and $F \succeq H$ holds for $F, H \in \mathcal{F}$ if $w(F) \leq w(H)$ holds.

The following result follows from the analysis in Neumann and Wegener (2007) which is based on Raidl and Julstrom (2003). We prove it in the more general setting of maximal weight bases in the following sections (the upper bound in Theorem 3.7 and the lower bound in Theorem 3.11). In these results we also specify leading constants of the given bounds and drop the restriction to integral edge weights.

Theorem 3.2 (Neumann and Wegener (2007)). *Let $G = (V, E)$ be a connected graph and let $w \colon E \to \mathbb{N}^+$ be a weight function on E with maximum weight $w_{\max}$.*

The expected optimization time of the (1+1) EA_{MST} on (G, w) starting with an arbitrary initial search point $x^{(0)}$ is $O\left(|E|^2 (\ln |V| + \ln w_{\max})\right)$.

Furthermore, for every $n \in \mathbb{N}$ there exists a connected graph $G = (V, E)$ on n vertices and a weight function $w \colon E \to \mathbb{N}^+$ such that expected optimization time of the (1+1) EA_{MST} (Algorithm 3) on (G, w) with the initial search point chosen uniformly at random from all spanning trees on G is $\Theta\left(|E|^2 \ln |V|\right)$.

Neumann and Wegener (2005) also analyzed multi-objective evolutionary algorithms for the MST problem. They showed for two such algorithms a superior expected optimization time of $O\left(|V|\,|E|\,(\ln |V| + \ln w_{\max})\right)$.

3.3. Matroids — The Realm of the Greedy Algorithm

As a consequence of Theorem 3.2 from the previous section, the (1+1) EA_{MST} solves the MST problem in expected time $O(|E|^2 \ln |V|)$ for polynomially bounded integer weights. In comparison, the two non-evolutionary greedy algorithms by Kruskal (1956) and by Jarník and Prim (Jarník (1930); Prim (1957)) solve this problem in times $O(|E| \ln |V|)$ and $O(|E| + |V| \ln |V|)$, respectively.

As hill-climber strategies, greedy algorithms are one of the most basic tools in black-box optimization. One of the most general problems known to be tractable for greedy algorithms is the problem of finding a maximum weight basis of a matroid.

Definition 3.3 (Matroid). *Let E be a finite set and $\mathcal{F} \subseteq 2^E$ be a set of subsets of E which are called* independent sets *of M. Then the pair $M = (E, \mathcal{F})$ is called an* independence system *if the two conditions*

(i) $\emptyset \in \mathcal{F}$ *and*
(ii) $\forall F \in \mathcal{F}, H \subseteq F \colon H \in \mathcal{F}$

hold and is called a matroid *if in addition the condition*

(iii) $\forall F, H \in \mathcal{F}, |H| < |F| \colon \exists e \in F \setminus H$ *such that* $H \cup \{e\} \in \mathcal{F}$

holds. An inclusion maximal set $B \in \mathcal{F}$ of an independence system is called a basis *of M.*

Condition (iii) in the previous definition implies that all bases of a matroid M have the same size $r(M)$, the *rank* of M. Another consequence of condition (iii) is the following *exchange property* (confer Reichel and Skutella (2007)).

Lemma 3.4. *Let $M = (E, \mathcal{F})$ be a matroid, F an independent set of M, and B a basis of M. Furthermore, let $H_F := F \setminus B$ and $H_B := B \setminus F$. Then there exist an injective map $\varphi \colon H_F \to H_B$ such that*

(i) $F \cup \{\varphi(f)\} \setminus \{f\}$ *is independent for all* $f \in H_F$ *and*
(ii) $F \cup \{b\}$ *is independent for all* $b \in H_B \setminus \varphi(H_F)$.

Proof. If F is not a basis then by condition (iii) of Definition 3.3 we can successively add elements from H_B to F until F is a basis. These elements then satisfy condition (ii) of the statement. Thus, we may suppose F is a basis. In this case $|F| = |B|$ and we only have to verify condition (i) of the statement.

Consider the bipartite graph on the two partitions H_F and H_B such that for $f \in H_F$ and $b \in H_B$ the pair (f, b) is an edge if $F \cup \{b\} \setminus \{f\}$ is independent. Suppose for all $H \subseteq H_F$ the inequality $|N(H)| \geq |H|$ holds where $N(H)$ is the neighborhood of H in B. Then by the Theorem of Hall (see, e.g., Diestel (2005)) there exists a bipartite matching that covers H_F which defines the function φ satisfying condition (i) of the statement.

For proof by contradiction, let us assume that there exists a subset H of H_B such that $|N(H)| < |H|$. Then, by condition (iii) of Definition 3.3, there exists a $b \in H$ such that $(F \setminus H_F) \cup N(H) \cup \{b\}$ is independent. Again by (iii), we find $f_1, \ldots, f_{|F_H|-|N(H)|-1}$ in $F_H \setminus N(H)$ such that the set $F' = (F \setminus F_H) \cup N(H) \cup \{b, f_1, \ldots, f_{|F_H|-|N(H)|-1}\}$ remains independent. Since $|F'| = |F|$, there exists an element $f \in F_H \setminus N(H)$ that is not in F'. But then (f, h) is an edge — a contradiction to $f \notin N(H)$. □

For a graph $G = (V, E)$, let $\mathcal{F}$ be the set of edge sets $F \subseteq E$ such that the subgraph (V, F) is acyclic. Then $M_G = (E, \mathcal{F})$ is a matroid, called the *graph matroid* of G. In M_G, the spanning trees of G correspond to the bases of M_G which have all size $n - 1$. Thus, the *maximum** spanning tree problem is equivalent to the problem of finding a maximum weight basis of the corresponding graph matroid.

Problem 3.5 (The maximum weight basis problem).
Let $M = (E, \mathcal{F})$ be an independence system and let $w \colon E \to \mathbb{R}^+$ be a weight function on E. The maximum weight basis problem *asks for a set $F \in \mathcal{F}$ that maximizes $w(F) := \sum_{e \in F} w(e)$. Note that such a set is necessarily a basis.*

Given an independence system $M = (E, \mathcal{F})$, the greedy algorithm for the maximum weight basis problem starts with the empty set. It then iteratively adds the element of largest weight that does not violate the independence of the current set. This greedy algorithm finds a maximum weight basis of M for every weight function $w \colon E \to \mathbb{R}^+$ if and only if M is a matroid (see, e.g., Cormen, Leiserson, Rivest and Stein (2001)).

For a matroid $M = (E, \mathcal{F})$ with known weights, the greedy algorithm can be implemented by first sorting the elements in E and then checking independence in decreasing order. The computation time of this procedure

*The minimum spanning tree and the maximum spanning tree problems can be easily transformed into each other by replacing $w(e)$ by $w_{\max} + 1 - w(e)$ for all edges $e \in E$.

is $O(|E| \ln |E| + f(|E|))$ where $f(k)$ is the time needed to check k sets for independence.

Unlike the greedy algorithm, the basic combinatorial (1+1) EA solves the maximum weight basis problem also for non-matroidal independence systems. Since it is more generic, we do not expect it to outperform the greedy algorithm if the problem instance is a matroid. Still, in the next section we show an upper bound of $O(|E|^3 \ln |E|)$ on the expected optimization time of the basic combinatorial (1+1) EA which reduces to $O(|E|^2 \ln |E|)$ for polynomially bounded integer weights.

3.4. Multiplicative Drift Analysis

We next show an upper bound on the optimization time of the (1+1) EA on the maximum weight basis problem.

Algorithm 4 (1+1) EA_{MWB}

The basic combinatorial (1+1) EA for the maximum weight basis problem, the (1+1) EA_{MWB}, is an instance of Algorithm 2.
Let $M = (E, \mathcal{F})$ be a matroid and $w\colon E \to \mathbb{R}^+$ a weight function on E. Then, $\mathcal{F}$ is the space of feasible search points and $F \succeq H$ holds for $F, H \in \mathcal{F}$ if $w(F) \geq w(H)$ holds.

We study the average weight increase in each iteration of this algorithm, confer Reichel and Skutella (2007).

Proposition 3.6. *Let $M = (E, \mathcal{F})$ be a matroid and let $w\colon E \to \mathbb{R}^+$ be a weight function on E. Let w_{opt} be the weight of a maximum weight basis of M. Furthermore, let $\{x^{(t)}\}_{t\in\mathbb{N}}$ be the sequence of search points generated by the (1+1) EA_{MWB} on (M, w). Then,*

$$\mathrm{E}\big[w(x^{(t+1)}) - w(x^{(t)}) \,\big|\, w(x^{(t)})\big] \geq \frac{w_{\text{opt}} - w(x^{(t)})}{\mathrm{e}\,|E|^2},$$

for all $t \in \mathbb{N}$ where $\mathrm{e} = 2.718\ldots$ is the Euler constant.

Proof. Let $t \in \mathbb{N}$ and $m := |E|$. Let B be a maximum weight basis ($w(B) = w_{\text{opt}}$) and let $F := x^{(t)}$ with $t \in \mathbb{N}$. Furthermore, let $\varphi\colon H_F \to H_B$ with $H_F := B \backslash F$ and $H_B := F \backslash B$ be the injective function that is provided by Lemma 3.4.

For all $f \in H_F$ we define the indicator variable I_f by

$$I_f := \begin{cases} 1 & \text{if } x^{(t+1)} = x^{(t)} \cup \{\varphi(f)\} \setminus \{f\}, \\ 0 & \text{otherwise,} \end{cases}$$

and for all $b \in H_B \setminus \varphi(H_F)$ the indicator variable I_b by

$$I_b := \begin{cases} 1 & \text{if } x^{(t+1)} = x^{(t)} \cup \{b\}, \\ 0 & \text{otherwise.} \end{cases}$$

Since the variables I_f with $f \in H_F$ and I_b with $b \in H_B \setminus \varphi(H_F)$ indicate disjoint events,

$$\sum_{f \in H_F} I_f + \sum_{b \in H_B \setminus \varphi(H_F)} I_b \leq 1$$

holds. Moreover, according to the definition of the (1+1) EA, it holds that $w(x^{(t+1)})$ is at least as large as $w(x^{(t)})$. Thus,

$$w(x^{(t+1)}) - w(x^{(t)}) \geq \sum_{f \in H_F} (w(b) - w(f))\, I_f + \sum_{b \in H_B \setminus \varphi(H_F)} w(b)\, I_b\,.$$

With this inequality at hand, we bound the expected weight increase. For all $f \in H_F$ and all $b \in H_B \setminus \varphi(H_F)$ let $p_f := \Pr[I_f = 1 \mid w(x^{(t)})]$ and $p_b := \Pr[I_b = 1 \mid w(x^{(t)})]$. Then,

$$\mathrm{E}\big[w(x^{(t+1)}) - w(x^{(t)}) \,\big|\, w(x^{(t)})\big] \geq \sum_{f \in H_F} (w(b) - w(f))\, p_f + \sum_{b \in H_B \setminus \varphi(H_F)} w(b)\, p_b\,.$$

To conclude the proof of this proposition, it suffices to show that p_f and p_b are at least $p := \mathrm{e}^{-1}\, m^{-2}$ for all $f \in H_F$ with $w(f) \leq w(\varphi(f))$ and for all $b \in H_B \setminus \varphi(H_F)$.

Let f be an element of H_F such that $w(f) \leq w(\varphi(f))$. Then p_f is the probability of $x^{(t+1)} = x^{(t)} \cup \{\varphi(f)\} \setminus \{f\}$. That is, p_f is the probability of $x^{(t)} \Delta y^{(t+1)} = \{f, \varphi(f)\}$ where $y^{(t+1)}$ is the respective candidate search point of the (1+1) EA. Thus,

$$p_f = \left(1 - \frac{1}{m}\right)^{m-2} \frac{1}{m^2} \geq \frac{1}{\mathrm{e}\, m^2}\,.$$

Similarly, if $b \in H_B \setminus \varphi(H_F)$ then p_b is the probability of $x^{(t)} \Delta y^{(t+1)} = \{b\}$. Thus

$$p_b = \left(1 - \frac{1}{m}\right)^{m-1} \frac{1}{m} \geq \frac{1}{\mathrm{e}\, m}\,.$$

□

The previous proposition shows that on average the distance in weights towards an optimal search point decreases at least proportional to its current value in each iteration. In such a situation, the expected optimization time can be bounded from above using the theorem on multiplicative drift (Theorem 2.10) from the previous chapter.

Let $M = (E, \mathcal{F})$ be a matroid and let $w\colon E \to \mathbb{R}^+$ be a weight function on E. Then, the previous theorem allows us to bound the expected optimization time of the (1+1) EA$_{\text{MWB}}$ on a weighted matroid (M, w). Let w_{opt} be the weight of a maximum weight basis and let $g(F) := w_{\text{opt}} - w(F)$ for all $F \in \mathcal{F}$ be a potential function on $\mathcal{F}$. Then the following statement directly follows from Proposition 3.6.

Theorem 3.7 (confer Reichel and Skutella (2007)).
Let $M = (E, \mathcal{F})$ be a matroid and let $w\colon E \to \mathbb{R}^+$ be a weight function on E. Let w_{opt} be the weight of a maximum weight basis and $w_{\text{2nd-opt}}$ be the maximum weight over all bases that do not have weight w_{opt}. Furthermore, let T be the optimization time of the (1+1) EA$_{\text{MWB}}$ (Algorithm 4) on (M, w). Then,

$$\mathrm{E}[T \mid x^{(0)}] \leq \mathrm{e}\,|E|^2 \left(1 + \ln \frac{w_{\text{opt}} - w(x^{(0)})}{w_{\text{opt}} - w_{\text{2nd-opt}}}\right)$$

and the probability to exceed this bound on the expectation by an additive term of order $|E|^2$ is decreasing exponentially.

As a consequence of this theorem, the expected optimization time of (1+1) EA$_{\text{MWB}}$ on (M, w) is $O(|E|^2 \ln |E|)$ for integer weights that polynomially bounded in $|E|$. For non-integer weights a weight-independent upper bound can be inferred from the following theorem.

Theorem 3.8 (Reichel and Skutella (2009)). *Let E be a finite index set, $\mathcal{F} \subseteq 2^E$ a search space, and $w\colon E \to \mathbb{N}^+$ a weight function on E. Then there exists a bounded weight function $\widetilde{w}\colon E \to \{0, \ldots, |E|^{|E|/2}\}$ on E such that the two corresponding sequences of search points $\{x^{(t)}\}_{t \in \mathbb{N}}$ and $\{\widetilde{x}^{(t)}\}_{t \in \mathbb{N}}$ generated by the basic combinatorial (1+1) EA (Algorithm 2) are the same for all choices of $x^{(0)}$.*

This theorem also holds for arbitrary weight functions $w\colon E \to \mathbb{R}^+$. Since the number of weights is finite, we can scale all weights by a very large constant until rounding to the next integer does not change the behavior of the (1+1) EA anymore. Then we apply the previous theorem. This implies

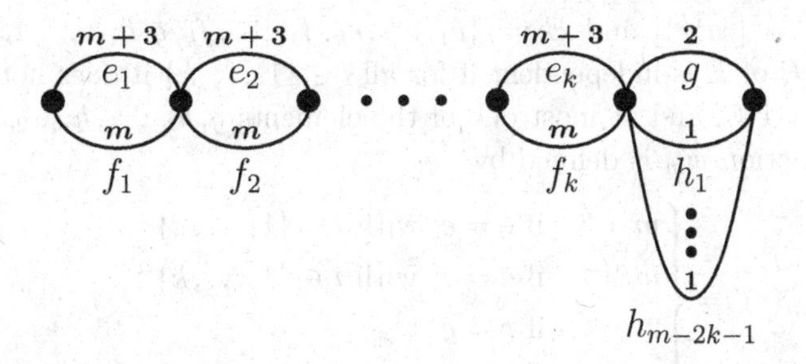

Fig. 3.1. The above (multi-)graph is a chain of k pairs of parallel edges terminated by a multi-edge consisting of $m - 2k$ parallel edges. Each pair has an edge of weight $m + 3$ and one of weight m. Of the $m - 2k$ parallel edges, one edge has weight 2 and the others have weight 1. The independent sets of the corresponding graph matroid are the cycle-free subgraphs of this graph, that is, an independent set contains at most one edge of each pair and at most one of the $m - 2k$ parallel edges. For $k := \lfloor \sqrt[3]{m} \rfloor$, the expected time for the (1+1) EA to find the optimal search point for the minimum weight basis problem (which constitutes of the row of upper edges) starting with a random feasible set of edges is $\Omega(m^2 \ln m)$.

that the expected optimization time of the (1+1) EA_{MWB} on (M, w) with an arbitrary weight function $w\colon E \to \mathbb{R}^+$ is $O(|E|^3 \ln |E|)$.

Corollary 3.9 (confer Reichel and Skutella (2009)).
Let $M = (E, \mathcal{F})$ be a matroid and let $w\colon E \to \mathbb{R}^+$ be a weight function on E. Furthermore, let T be the optimization time of the (1+1) EA_{MWB} (Algorithm 4) on (M, w). Then for $|E| \geq 2$, independent of $x^{(0)} \in \mathcal{F}$,

$$\mathrm{E}[T] \leq \tfrac{e}{2}\, |E|^3 (1 + \ln |E|)$$

and the probability to exceed this bound on the expectation by an additive term of order $|E|^3$ is decreasing exponentially.

3.5. Lower Bounds and Typical Runs

In the previous section we have proven upper bounds on the optimization time of the maximum weight basis problem. The following example will serve as instance for a lower bound. It can be perceived as a graph matroid corresponding to the graph with multiple edges depicted in Figure 3.1.

Definition 3.10. For all $m \in \mathbb{N}$ with $m \geq 3$, let the matroid $M_m = (E, \mathcal{F})$ and the weight function $w_m\colon E \to \{1, 2, m, m + 3\}$ be defined as follows.

Let $k := \lfloor m^{1/3} \rfloor$ and $E := \{e_1, \ldots, e_k, f_1, \ldots, f_k, g, h_1, \ldots, h_{m-2k-1}\}$. A subset F of E is independent if for all $i \in \{1, \ldots, k\}$ it does not contain both, e_i and f_i, and at most one of the elements $g, h_1, \ldots, h_{m-2k-1}$. The weight function w_m is defined by

$$w_m(e) := \begin{cases} m+3 & \text{if } e = e_i \text{ with } i \in \{1, \ldots, k\}, \\ m & \text{if } e = f_i \text{ with } i \in \{1, \ldots, k\}, \\ 2 & \text{if } e = g, \\ 1 & \text{if } e = h_i \text{ with } i \in \{1, \ldots, m-2k-1\}. \end{cases}$$

For M_m with w_m, we consider a *typical run*. This means, we show that the sequence $\{x^{(t)}\}_{t \in \mathbb{N}}$ generated by the (1+1) EA has certain properties with a probability that is bounded from below by a positive constant. These properties then imply an optimum is not found within $\Omega(|E|^2 \ln |E|)$ iterations.

Theorem 3.11 (confer Neumann and Wegener (2007)). *Let $m \in \mathbb{N}$ with $m \geq 3$. Furthermore, let M_m be the matroid and $w_m \colon E \to \mathbb{R}^+$ be the weight function from Definition 3.10. Then the expected optimization time of the (1+1) EA$_{\text{MWB}}$ (Algorithm 4) on (M_m, w_m) with $x^{(0)}$ chosen uniformly at random from $\mathcal{F}$ is $\Omega(|E|^2 \ln |E|)$.*

Proof. Since we prove an asymptotic result, we may suppose that in the following m is sufficiently large. Let $x^{(0)}$ be chosen uniformly at random from $\mathcal{F}$. Let $\{x^{(t)}\}_{t \in \mathbb{N}}$ be the sequence of search points generated by the (1+1) EA$_{\text{MWB}}$ on (M_m, w_m). Furthermore, let T be the random variable that denotes the earliest point in time $t \in \mathbb{N}$ for which $x^{(t)}$ is the maximum weight basis $\{e_1, \ldots, e_k, g\}$ of M_m.

Let $I := \{1, \ldots, k\}$. For $i \in I$ and $t \in \mathbb{N}$ we say that at time t position i is *free* if neither e_i nor f_i is in $x^{(t)}$ and *occupied* otherwise. We call it *critical* if $f_i \in x^{(s)}$ for all $s \leq t$. Clearly, a critical position is never free and $T > t$ if there exist any critical positions at time t.

For the remainder of the proof, we neglect whether or not any one of the elements $g, h_1, \ldots, h_{m-2k-1}$ is in $x^{(t)}$. We can do this, since these events have no influence on the number of occupied and critical positions.

Since $x^{(0)}$ is chosen uniformly at random, in expectation $k/3$ positions are critical at time $t_0 := 0$. Consider the event that at time t_0 there are at least $k/4$ positions that are critical. Markov's inequality (Theorem 1.7) guarantee this event with a probability bounded from below by a positive constant.

First, we show that there are no free positions at time $t_1 := \lfloor 2\,\mathrm{e}\,m \ln m \rfloor$ with a probability that tends to one as m tends to infinity. .

Let $i \in I$ be a position that is free at time t. Let p be the probability that i is occupied at time $t+1$. Then p is at least the probability that $x^{(t)} \triangle y^{(t+1)} = \{e_i\}$. Thus,

$$p \geq \left(1 - \frac{1}{m}\right)^{m-1} \frac{1}{m} \geq \frac{1}{\mathrm{e}\,m}.$$

Now, let the potential function $g\colon \mathcal{F} \to \mathbb{N}$ measure the number of free positions in an independent set. Since there are $g(x^{(t)})$ free positions at time t, the expected decrease in g between t and $t+1$ is at least $g(x^{(t)})/em$. By Theorem 2.10, the expected time until no free positions remain is at most $em \ln m$. Thus, by the Chernoff bounds (Corollary 1.10), the probability that there are no free positions after t_1 iterations tends to one as m tends to infinity.

Next, we show that at time t_1 all positions that were critical at time t_0 are still critical with a probability that tends to one as m tends to infinity.

Let $i \in I$ be critical at time t. The only way for i to become non-critical at time $t+1$ is if $x^{(t)} \triangle y^{(t+1)}$ contains f_i and at least one e_j with $j \in I$ or if it contains f_j' with $j \in I \backslash \{i\}$. This happens with probability at most $2\,k\,m^{-2}$. Thus, by the union bound, the probability that any position becomes non-critical at any point in time prior to t_1 is at most $4\,\mathrm{e}\,k^2\,m^{-1} \ln m$ which tends to zero as m tends to infinity.

Finally, we show that with a probability bounded from below by a positive constant, at time $t_2 := \lfloor \frac{1}{3} m^2 \ln m \rfloor$ there still exists a critical position.

We have already seen that with probabilities that tend to one as m tends to infinity, the events (i) that there are at least $k/4$ critical positions at time t_1 and (ii) that there are no free positions at time t_1 happen. Thus, by the union bound, (i) and (ii) occur simultaneously with a probability that is bounded away from zero by a positive constant if m is sufficiently large.

Suppose at time t_1 there exists at least $k/4$ critical and no free positions. Then, between the times t_1 and t_2 a critical position $i \in I$ can only become non-critical if the event "$x^{(t)} \triangle y^{(t+1)} \supseteq \{f_i, e_i\}$" occurs. This happens with probability at most m^{-2}. Therefore, with probability at least

$$\left(1 - \frac{1}{m^2}\right)^{\frac{1}{3} m^2 \ln m} \sim \frac{1}{m^{1/3}}$$

position i remains critical until time t_2.

Fig. 3.2. Maximizing a linear pseudo-Boolean function $f(x) = \sum_{i=1}^{m} w(i)\,x_i$ with strictly positive weight $w(1), \ldots, w(m)$ is equivalent to finding a maximum spanning tree on the path of length m with edges $1, \ldots, m$ and weights $w(1), \ldots, w(m)$. For this problem, the (1+1) EA finds an optimal search point in expected time $\Theta(m \ln m)$, independent of the weights $w(1), \ldots, w(m)$, see Droste, Jansen and Wegener (2002).

Since the events "$x^{(t)} \Delta y^{(t+1)} \supseteq \{f_i, e_i\}$" and "$x^{(s)} \Delta y^{(s+1)} \supseteq \{f_j, e_j\}$" are mutually independent for $i \neq j$ and all combinations of s and t, the probability that there exist no critical positions at time t_2 is at most $(1 - m^{-1/3})^k \sim 1/\mathrm{e}$. Thus, with a probability bounded from below by a positive constant, the optimization time is at least t_2 which concludes the proof. □

With the previous theorem, we have seen that for matroids with an integer weight function polynomially bounded in $|E|$ the $O(|E|^2 \ln |E|)$ bound on the expected optimization time of the (1+1) $\mathrm{EA}_{\mathrm{MWB}}$ from the previous section is tight. However, for unbounded and real weights we only know the weight dependent bound from Theorem 3.7 and the general bound of $O(|E|^3 \ln |E|)$ from Corollary 3.9.

Open questions. It is a central open problem to close the gap between these two bounds. For the special case of linear pseudo-Boolean functions this question can be answered positively (see Figure 3.2). This leads us to conjecture that there exists a weight-independent upper bound of $O(|E|^2 \ln |E|)$ on the expected optimization time of the (1+1) $\mathrm{EA}_{\mathrm{MWB}}$.

3.6. A Hard Problem for the (1+1) Evolutionary Algorithm

In the previous sections we have seen that the basic combinatorial (1+1) EA solves the MST and the SSSP problems in expected polynomial time if the weights are integral and polynomially bounded. A third classical problem which is known to be polynomially solvable is the *maximum matching problem.* Micali and Vazirani (1980) showed that this problem can be solved in time $O(|V|^{1/2}\,|E|)$.

Problem 3.12 (The maximum matching problem).
Let $G = (V, E)$ be a (multi-)graph. The maximum matching *problem asks for a maximum set of vertex disjoint edges.*

The basic combinatorial (1+1) EA for this problem is defined as follows.

Algorithm 5 (1+1) $\mathrm{EA_{MM}}$

The basic combinatorial (1+1) EA for the maximum matching problem, the (1+1) $\mathrm{EA_{MM}}$, is an instance of Algorithm 2.
Let $G = (V, E)$ be a graph. Then, the space of feasible search points $\mathcal{F}$ consists of all matchings on G and $F \succeq H$ holds for $F, H \in \mathcal{F}$ if $|F| \geq |H|$ holds.

This algorithm was studied extensively by Giel and Wegener (2003, 2006) and later by Oliveto, He and Yao (2008) in the context of processes with negative drift. For example, Giel and Wegener (2003) showed that the expected optimization time (1+1) $\mathrm{EA_{MM}}$ on paths is polynomially bounded in $|E|$.

Theorem 3.13 (Giel and Wegener (2003)). *Let $m \in \mathbb{N}$. The expected optimization time of the (1+1) $\mathrm{EA_{MM}}$ (Algorithm 5) on a path of m edges is $O(m^4)$ independent of the initial search point $x^{(0)}$.*

Moreover, Giel and Wegener (2003) have shown that the (1+1) $\mathrm{EA_{MM}}$ is a PRAS for the maximum matching problem. However, they also studied an instance of a bipartite graph due to Sasaki and Hajek (1988) for which the (1+1) $\mathrm{EA_{MM}}$ has exponential expected optimization time.

In this section we show such an exponential lower bound on the expected optimization time for the multi-graph[†] in Figure 3.3.

Definition 3.14. Let $k \in \mathbb{N}_{\geq 3}$ and let P_k be the multi-edged path on the set of vertices $\{u_1, w_1, \ldots, u_k, w_k\}$ such that there is a single edge d_i between u_i and w_i for all $1 \leq i \leq k$ and nine parallel edges $e_i^1, \ldots, e_i^9$ between w_i and u_{i+1} for all $1 \leq i \leq k-1$. Then, P_k has $m = 10k - 9$ edges in total. For simplicity, we call an edge d_i *odd* and an edge e_i^j *even.*

The choice of this graph over the original graph in Giel and Wegener (2003) strongly simplifies the analysis while preserving the main proof ideas. Furthermore, it exposes a difficulty that exists for the (1+1) $\mathrm{EA_{MM}}$ but not for problem-specific algorithms: We can adapt any problem-specific algorithm for simple graphs to an equally efficient algorithm for multi-graphs. The algorithm simply performs an initialization step where all multi-edges are replaced by single edges. Then, a maximum matching on

[†] A multi-graph is a graph with parallel edges.

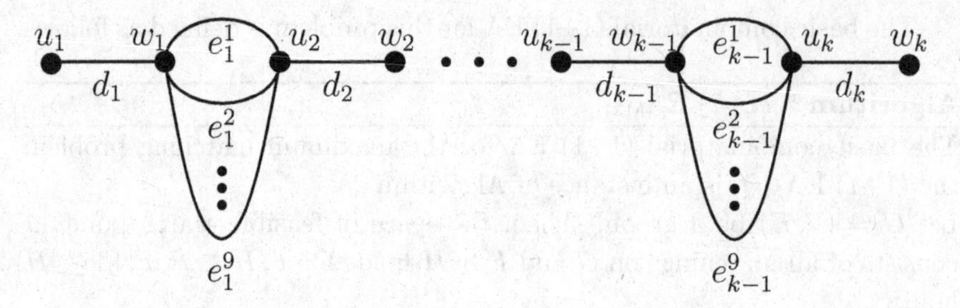

Fig. 3.3. The multi-edged path P_k on $2k$ vertices and $10k - 9$ edges for $k \geq 3$. Every *odd* edge d_i is a single edge and every *even* edge e_i^j is one of nine parallel edges. In an iteration of the (1+1) EA for the maximum matching problem it is roughly nine times more likely to replace a specific odd edge by an arbitrary neighboring even edge than vice versa. Because of this, augmenting paths on even edges have a strong tendency to grow.

this modified graph is also a maximum matching on the original multigraph. In comparison, the (1+1) EA_{MM} solves the maximum matching problem on paths in expected polynomial time. But, if we replace every second edge of a path of odd length by a multi-edge with nine parallel edges, then the expected optimization time of the (1+1) EA_{MM} becomes exponential.

Theorem 3.15 (confer Giel and Wegener (2003)). *Let $k \in \mathbb{N}_{\geq 3}$ and let P_k be the multi-edged path on $m = 10k - 9$ edges defined in Definition 3.14. Then, the expected optimization time of the (1+1) EA_{MM} (Algorithm 5) on P_k starting with a non-optimal initial search point is $2^{\Omega(m)}$.*

Before proving this theorem, we make a couple of preliminary observations. In particular, we first show three propositions that then lead to the proof of the previous theorem.

Let $P := P_k$ with $k \geq 3$ and $m = 10k - 9$. The set of odd edges of P forms the only perfect matching M^* on P. Thus, M^* is the unique maximum matching on P. It is of size k.

Let $\{x^{(t)}\}_{t \in \mathbb{N}}$ be the sequence of search points generated by the (1+1) EA_{MM} for the maximum matching problem on P initialized with an arbitrary matching $x^{(0)}$ other than M^*. Furthermore, let T^* be the random variable that denotes the first point in time $t \in \mathbb{N}$ for which $x^{(t)} = M^*$, that is the optimization time of the (1+1) EA_{MM}.

We call a matching *almost perfect* if it contains $k - 1$ edges. For example, let M° be the almost perfect matching $M^* \triangle \{d_1, e_1^1, d_2\}$. We first show that we only loose a factor of $\Theta(1/m^3)$ on any lower bound for the expected optimization time of the (1+1) EA_{MM} if we condition on the event that $x^{(0)} = M^\circ$.

Proposition 3.16.

$$\mathrm{E}[T^* \mid x^{(0)} \neq M^*] \geq \frac{1}{2m^3}\,\mathrm{E}[T^* \mid x^{(0)} = M^\circ]\,.$$

Proof. Let T be the random variable that denotes the first point in time $t \in \mathbb{N}$ for which $x^{(t)}$ is either M° or M^*. We show that

$$\Pr[x^{(T)} = M^\circ] \geq \frac{1}{2m^3}\,. \tag{3.1}$$

This is sufficient as the statement then follows from the law of total expectation and from

$$\mathrm{E}[T^* \mid x^{(T)} = M^\circ] = \mathrm{E}[T \mid x^{(T)} = M^\circ] + \mathrm{E}[T^* \mid x^{(0)} = M^\circ]\,.$$

Let $t \in \mathbb{N}$ with $x^{(t)} \neq M^*$. Then, $|x^{(t)}| \leq k-1$ and $y^{(t+1)}$ is accepted by the (1+1) EA_{MM} if $y^{(t+1)} = M^\circ$. Since $x^{(t)} \triangle M^\circ$ and $x^{(t)} \triangle M^*$ differ exactly by the three edges d_1, e_1^1, d_2, it holds that

$$\Pr[x^{(t+1)} = M^\circ] \geq \frac{1}{m^3}\,\Pr[x^{(t+1)} = M^*]$$

and inequality (3.1) follows from $\Pr[x^{(T)} = M^\circ] + \Pr[x^{(T)} = M^*] = 1$. □

Justified by this result, from now on we condition on the event that $x^{(0)} = M^\circ$. Since M° is an almost perfect matching, for $t < T^*$ all matchings $x^{(t)}$ are also almost perfect.

Every almost perfect matching M defines a unique *augmenting subpath* S of P. More precisely, there exists two indices $a, b \in \{1, \ldots, k\}$ with $a \leq b$ such that the path $S := (u_a, d_a, w_a, e_a, \ldots, e_{b-1}, u_b, d_b, w_b)$ consist of all odd edges $d_a, \ldots, d_b$ that are unmatched and all even edges $e_a, \ldots, e_{b-1}$ that are matched. Let $\ell(S) := b - a$ be the number of even edges in S. Thus, S is of odd length $2\,\ell(s)+1$, since S always starts and ends with an odd edge (where $e_a = e_b$ is possible). For example, the alternating path in M° is $S^+ := (u_1, d_1, w_1, e_1^1, u_2, d_2, w_2)$ with $\ell(S^+) = 1$.

With slight abuse of notation, we associate with the matching M^* the *empty augmenting path* $S^* := (v_0)$ such that $\ell(S^*) = 0$. Note that in general an augmenting path S with $\ell(S) = 0$ can also correspond to an almost perfect matching and contain a single unmatched odd edge. We are not interested in this distinction. Instead, we let T_0 be the random variable that denotes the first point in time t such that $\ell(S^{(t)}) = 0$ and use it as a suitable lower bound of T^*.

Proposition 3.17.

$$\mathrm{E}[T^* \mid x^{(0)} = M^\circ] \geq \mathrm{E}[T_0 \mid \ell(S^{(0)}) = 1] .$$

The remainder of this section is devoted to proving a lower bound on $\mathrm{E}[T_0]$. On the one hand, we start with $\ell(S^{(0)}) = 1$ and it is quite likely that $\ell(S^{(t)})$ reaches zero for early points in time. On the other hand, we will see that the drift of $\ell(S^{(t)})$ is positive and thus $\ell(S^{(t)}$ has the tendency to increase up to $k-1$. The following proposition shows that the latter event happens with a probability that is bounded away from zero by a constant.

Proposition 3.18. *There exists a constant* $\delta > 0$ *such that*

$$\mathrm{E}[T_0 \mid \ell(S^{(0)}) = 1] \geq \delta\, \mathrm{E}[T_0 \mid \ell(S^{(0)}) = k-1] .$$

Proof. Let $\ell(S^{(0)}) = 1$ and let T be the random variable that denotes the first point in time $t \in \mathbb{N}$ for which $\ell(S^{(t)})$ is either 0 or $k-1$. Similar to Proposition 3.16, we show that there exists a constant $\delta > 0$ such that

$$\Pr[\ell(S^{(T)}) = k] \geq \delta .$$

Let $t < T$ and let $S^{(t)} = (v_{a(t)}, \ldots, w_{b(t)})$ be the augmenting path corresponding to the almost perfect matching $x^{(t)}$.

Since $t < T$, we have $a(t) > 1$ or $b(t) < k$, say $b(t) < k$ without loss of generality. Then, adding any of the edges $e^1_{b(t)}, \ldots, e^9_{b(t)}$ to $x^{(t)}$ and removing $d_{b(t)+1}$ increases the length of $S^{(t)}$ by two (and thus $\ell(S^{(t)})$ by one). Independent of $S^{(t)}, \ldots, S^{(0)}$, this increase happens with probability at least $9\,(1-1/m)^{m-2}\, m^{-2}$. Thus,

$$\Pr\left[\ell(S^{(t+1)}) - \ell(S^{(t)}) = 1 \,\middle|\, S^{(t)}, \ldots, S^{(0)}\right] \geq \frac{3}{m^2} . \tag{3.2}$$

On the other hand, shortening $S^{(t)}$ by two is not equally likely. We either have to add $d_{a(t)}$ to $x^{(t)}$ and remove $e_{a(t)}$ in return or have to add $d_{b(t)}$ to $x^{(t)}$ and remove $e_{b(t)-1}$. Thus, again independent of $S^{(t)}, \ldots, S^{(0)}$,

$$\Pr\left[\ell(S^{(t+1)}) - \ell(S^{(t)}) = -1 \,\middle|\, S^{(t)}, \ldots, S^{(0)}\right] \leq \frac{2}{m^2} . \tag{3.3}$$

More generally, to shorten $S^{(t)}$ by at least $2j$ edges, we have to change $x^{(t)}$ by a total of j edge pairs split up between the two ends of the augmenting path given by $x^{(t)}$. We may change $x^{(t)}$ further but the above change is necessary to reduce $\ell(S^{(t)})$ by $2j$ edges. There are $j+1$ ways to split up the j edge pairs between the two ends of the augmenting path given by $x^{(t)}$. Hence, independent of $S^{(t)}, \dots, S^{(0)}$,

$$\Pr\left[\ell(S^{(t+1)}) - \ell(S^{(t)}) \le -j \mid S^{(t)}, \dots, S^{(0)}\right] \le \frac{j+1}{m^{2j}}. \tag{3.4}$$

Consider the sequence $\{X^{(s)}\}_{s\in\mathbb{N}}$ with $X^{(s)} \in \mathbb{Z} \cup \{-\infty\}$ which is a pessimistic view on the sequence $\{\ell(S^{(t)})\}_{t\in\mathbb{N}}$ in the sense that $\{X^{(s)}\}_{s\in\mathbb{N}}$ is at most as likely as $\{\ell(S^{(t)})\}_{t\in\mathbb{N}}$ to reach k before 0.

Informally speaking, we construct $\{X^{(s)}\}_{s\in\mathbb{N}}$ from $\{\ell(S^{(t)})\}_{t\in\mathbb{N}}$ by (i) ignoring time steps where $\ell(S^{(t)})$ does not change at all; (ii) increasing $X^{(s)}$ by one whenever $\ell(S^{(t)})$ increases by at least one; (iii) decreasing $X^{(s)}$ by one whenever $\ell(S^{(t)})$ decreases by exactly one; and (iv) setting $X^{(s)}$ to $-\infty$ whenever $\ell(S^{(t)})$ decreases by at least two.

Formally, we define $X^{(s)}$ recursively. Let $t(0) := 0$ and $x^{(0)} := 1$. For $s \ge 1$, let $t(s)$ be the s-th point in time such that $\ell(S^{t(s)}) \ne \ell(S^{t(s)-1})$ and s_T the point in time such that $t(s_T) = T$. Then, for $s \in \mathbb{N}$ with $s < s_T$ let

$$X^{(s+1)} := \begin{cases} X^{(s)} + 1 & \text{if } \ell(S^{t(s)}) \ge \ell(S^{t(s)-1}) + 1\,, \\ X^{(s)} - 1 & \text{if } \ell(S^{t(s)}) = \ell(S^{t(s)-1}) - 1\,, \\ -\infty & \text{if } \ell(S^{t(s)}) \le \ell(S^{t(s)-1}) - 2\,. \end{cases}$$

We can infer from (3.2), (3.3), and (3.4) by increasing the bound in (3.4) to $5/m^4$ that for all $s < s_T$,

$$X^{(s+1)} := \begin{cases} X^{(s)} + 1 & \text{with probability at least } \frac{3}{5}(1 - \frac{1}{m^2+1})\,, \\ X^{(s)} - 1 & \text{with probability at most } \frac{2}{5}(1 - \frac{1}{m^2+1})\,, \\ -\infty & \text{with probability at most } \frac{1}{m^2+1}\,. \end{cases}$$

For $s \ge s_T$, we let the random variables $\{X^{(s)}\}_{s \ge s_T+1}$ be mutually independent such that

$$X^{(s+1)} := \begin{cases} X^{(s)} + 1 & \text{with probability } \frac{3}{5}(1 - \frac{1}{m^2+1})\,, \\ X^{(s)} - 1 & \text{with probability } \frac{2}{5}(1 - \frac{1}{m^2+1})\,, \\ -\infty & \text{with probability } \frac{1}{m^2+1}\,. \end{cases}$$

Let R be the random variable that denotes the first point in time $s \in \mathbb{N}$ such that $X^{(s)} \le 0$ or $X^{(s)} = k$. By definition of $X^{(s)}$, it holds that $\ell(S^{t(s)}) \geq X^{(s)}$ for all $s \leq s_T$. Thus,

$$\Pr\left[\ell(S^{(T)}) = k\right] \geq \Pr\left[X^{(R)} = k\right].$$

In the remainder of this proof we show that $\Pr\left[X^{(R)} = k\right] \geq \delta$ for some constant $\delta > 0$.

By Markov's inequality (Theorem 1.7), it holds that $X^{(s)} > -\infty$ for all $s \leq \frac{m^2+1}{2}$ with probability at least 1/2. Suppose that indeed $X^{(s)} > -\infty$ for all $s \leq \frac{m^2+1}{2}$. Then, until time $\frac{m^2+1}{2}$, the $X^{(s)}$'s perform a random walk on $\mathbb{Z}$ such that $X^{(s)}$ increases with probability at least 3/5 and decreases with probability at most 2/5.

For $i \in \{1, \ldots, k-1\}$, let $b_i := \binom{i}{2}$ and let A_i be the event that $X^{(b_i)} \geq i$ and $X^{(s)} > 0$ for all $s \leq b_i$.

Let $i \in \{1, \ldots, k-1\}$. We bound the probability of A_{i+1} conditioned on A_i. Since $b_{i+1} - b_i = \binom{i+1}{2} - \binom{i}{2} = i$, the event $X^{(b_{i+1})} \geq i+1$ conditioned on A_i implies that $X^{(b_i+1)}, \ldots, X^{(b_{i+1})}$ are all positive.

Let $\mu := \mathrm{E}[X^{(b_{i+1})} \mid A_i]$ be the expected value of $X^{(b_{i+1})}$ conditioned on $X^{(b_i)} \geq i$ and $X^{(s)} > 0$ for all $s \leq b_i$. By the linearity of expectation,

$$\mu \geq i - \tfrac{2}{5}\, i + \tfrac{3}{5}\, i = \tfrac{6}{5}\, i\,.$$

By Lemma 1.18, we can apply the Chernoff bounds (Corollary 1.10) to probability that $X^{(b_{i+1})} \leq i$. Therefore,

$$\Pr\left[X^{(b_{i+1})} \leq i \,\middle|\, A_i\right] = \Pr\left[X^{(b_{i+1})} \leq \left(1 - \tfrac{1}{6}\right)\mu \,\middle|\, A_i\right] \leq \mathrm{e}^{-i/60}\,.$$

Thus, $\Pr[A_{i+1} \mid A_i] \geq 1 - \mathrm{e}^{-i/60}$ and

$$\Pr[A_k] = \Pr[A_0] \cdot \prod_{i=1}^{k-1} \Pr[A_{i+1} \mid A_i] \geq \prod_{i=1}^{k-1} \left(1 - \mathrm{e}^{-i/60}\right).$$

Let $L \in \mathbb{N}$ be minimal such that $\mathrm{e}^{-L/60} \leq 1/2$. Then, since $1 - a \geq e^{-2a}$ holds for all $a < 1/2$,

$$\begin{aligned}
\Pr[A_k] &\geq \left(1 - \mathrm{e}^{-1/60}\right)^L \cdot \prod_{i \in \mathbb{N}} \mathrm{e}^{-2\mathrm{e}^{-i/60}} \\
&= \left(1 - \mathrm{e}^{-1/60}\right)^L \cdot \mathrm{e}^{-2\sum_{i \in \mathbb{N}} \mathrm{e}^{-i/60}} \\
&= \left(1 - \mathrm{e}^{-1/60}\right)^L \cdot \mathrm{e}^{-\frac{2}{1-\mathrm{e}^{-1/60}}}\,.
\end{aligned}$$

Thus, $\Pr[A_k] > 2\delta$ where $\delta > 0$ is a constant. Since $b_k = \binom{k}{2} \leq \frac{m^2+1}{2}$, this implies that $\Pr[X^{(R)} = k] > \delta$ which concludes the proof. □

Using the previous two propositions, we finally show Theorem 3.15.

Proof of Theorem 3.15. We want to give a lower bound on $\mathrm{E}[T^*]$, the expected optimization time of the (1+1) $\mathrm{EA}_{\mathrm{MM}}$. By Proposition 3.16, Proposition 3.17, and Proposition 3.18, we know that there exists a constant $\delta > 0$ such that

$$\begin{aligned}\mathrm{E}[T^* \mid x^{(0)} \neq M^*] &\geq \frac{1}{2m^3}\,\mathrm{E}[T^* \mid x^{(0)} = M^\circ] \\ &\geq \frac{1}{2m^3}\,\mathrm{E}[T_0 \mid \ell(S^{(0)}) = 1] \\ &\geq \frac{\delta}{2m^3}\,\mathrm{E}[T_0 \mid \ell(S^{(0)}) = k-1]\,.\end{aligned}$$

We have seen in the proof of Proposition 3.18, that for all $1 \leq i \leq k-2$,

$$\mathrm{E}\big[\ell(S^{(t+1)}) - \ell(S^{(t)}) \mid \ell(S^{(t)}) = i\big] \geq \frac{3}{m^2} - \frac{2}{m^2} - \frac{k-1}{m^4} \geq \frac{1}{2m^2}\,,$$

Furthermore, by (3.4) we have for all $1 \leq i \leq k-1$ and $j \geq 1$,

$$\Pr\big[\ell(S^{(t+1)}) - \ell(S^{(t)}) = -j \mid \ell(S^{(t)}) = i\big] \leq \frac{j+1}{m^{2j}} \leq \frac{1}{3^j}\,.$$

We apply Theorem 2.12 with $n = k-1$, $\varepsilon := 1/(2m^2)$, $\delta := 2$ and $r := 0$. Then, there exists two constants $\alpha, \beta > 0$ such that

$$\Pr[T_0 \leq \mathrm{e}^{\alpha(k-1)} \mid \ell(S^{(0)}) = k-1] \leq \mathrm{e}^{-\beta(k-1)}$$

and Theorem 3.15 follows by the law of total expectation. □

Open questions. Among the classical combinatorial problems that are solvable in polynomial time, the maximum matching problem stands out because of its exponential optimization time for the (1+1) EA. This leads to a number of questions. What are the combinatorial properties that make this problem so difficult for the (1+1) EA? How do we recognize these properties in other problems, preferably only by evaluating the objective function? Are there generic randomized search heuristics other than the (1+1) EA that solve this problem efficiently?

3.7. Shortest Path Problems

Another type of problems for which evolutionary algorithms have been extensively studied are *shortest path problems.* In particular, much attention has been paid to the *single-source shortest path (SSSP) problem* (Scharnow, Tinnefeld and Wegener (2004); Doerr, Happ and Klein (2007a); Baswana, Biswas, Doerr, Friedrich, Kurur and Neumann (2009)) and the *all-pair shortest path (APSP) problem,* (Doerr, Happ and Klein (2008); Doerr and Theile (2009); Horoba and Sudholt (2009)).

Problem 3.19 (The single-source shortest path problem). *Let $G = (V, E)$ be a strongly connected directed graph, let s be a distinguished* source vertex *in V, and let $w\colon E \to \mathbb{R}^+$ be a weight function. For $v \in V$, an* optimal s-v-path *in G is a directed path $P \subseteq E$ from s to v that minimizes $w(P) := \sum_{e \in P} w(e)$. The* single-source shortest path problem *asks for an optimal s-v-path in G for every $v \in V$.*

A *directed spanning tree* $F \subseteq E$ with root s is a tree in G spanning all vertices of G such that the edges are directed away from s. It is well-known that there exist a set of optimal s-v-paths that forms a directed spanning tree with root s. Such a tree is called a *shortest path tree* of s (see, e.g., Mehlhorn and Sanders (2009); Cormen, Leiserson, Rivest and Stein (2001)).

From now on, we interpret the SSSP problem as the problem of finding a shortest path tree. That is, the search space $\mathcal{F}$ of the basic combinatorial (1+1) EA for the SSSP problem is the space of all directed spanning trees $F \subseteq E$ of G with root s.

We study two ways to define the order relation on $\mathcal{F}$ required by the basic combinatorial (1+1) EA. In this section, we analyze the single-criterion objective function that sums the weights of all paths. In Section 3.8, we analyze the multi-criteria optimization problem that optimizes the different path weights independently.

Let $F \in \mathcal{F}$. For all vertices $v \in V$, let P_v be the directed s-v-path in F and let $w(v, F) := w(P_v)$ be its weight (with $P_s := (s)$ and $w(s, F) := 0$). Then, the single-criterion objective function $f\colon 2^E \to \mathbb{R}_0^+$ for the SSSP problem on G is defined by $f(F) := \sum_{v \in V} w(v, F)$ for all directed spanning trees $F \subseteq E$ with root s.

Algorithm 6 (1+1) $\text{EA}_{\text{SC-SSSP}}$

The basic combinatorial (1+1) EA for the single-criterion single-source shortest path problem, the (1+1) $\text{EA}_{\text{SC-SSSP}}$, is an instance of Algorithm 2. Let $G = (V, E)$ be a strongly connected directed graph, s a source vertex of G, and $w\colon E \to \mathbb{R}^+$ a weight function on E. Then, the space of feasible search points $\mathcal{F}$ consists of all directed spanning trees of G with root s and $F \succeq H$ holds for $F, H \in \mathcal{F}$ if $f(F) \leq f(H)$ holds.

Note that $f(F)$ is not the same weight function as for the MST problem. In particular, $f(F)$ strongly depends on the choice of s. Thus, the analysis of the (1+1) $\text{EA}_{\text{SC-SSSP}}$ is not covered by that of the (1+1) EA_{MWB} in Section 3.4. Again, we determine the drift in the objective function in each iteration of the algorithm (confer Baswana *et al.* (2009)).

Proposition 3.20. *Let $G = (V, E)$ be a strongly connected directed graph, s be a distinguished source vertex in V, and $w \colon E \to \mathbb{R}^+$ a weight function on E. Furthermore, let f_{opt} be the single-criterion objective value of a shortest path tree and let $\{x^{(t)}\}_{t \in \mathbb{N}}$ be the sequence of search points generated by the (1+1) $\text{EA}_{\text{SC-SSSP}}$ on (G, s, w). Then, regardless of $x^{(0)}$,*

$$\mathrm{E}[f(x^{(t)}) - f(x^{(t+1)}) \mid f(x^{(t)})] \geq \frac{f(x^{(t)}) - f_{\text{opt}}}{\mathrm{e}\,|E|^2\,|V|}\,.$$

Proof. Let $p := (1 - 1/|E|)^{|E|-2}\,|E|^{-2}$ and let S be a shortest path tree, that is, $f(S) = f_{\text{opt}}$. Furthermore, let $F := x^{(t)}$ with $t \in \mathbb{N}$. If $f(F) = f_{\text{opt}}$ the statement follows trivially. Thus, suppose $f(F) > f(S)$. Then,

$$f(F) - f(S) = \sum_{v \in V} \big(w(v, F) - w(v, S)\big)\,.$$

Thus, there exists a $v \in V$ such that

$$w(v, F) - w(v, S) \geq \frac{f(F) - f(S)}{|V|}\,.$$

Let $P := (u_0, e_1, u_1, \ldots, u_{k-1}, e_k, u_k)$ with $k \in \mathbb{N}$ be the s-v-path in S with $s = u_0$ and $v = u_k$. To each edge e_j of P, we assign a contribution $c(e_j)$ to $w(v, F) - w(v, S)$ by defining

$$c(e_j) := \big(w(u_j, F) - w(u_j, S)\big) - \big(w(u_{j-1}, F) - w(u_{j-1}, S)\big)$$

for all $j \in \{1, \ldots, k\}$. This implies

$$\sum_{j=1}^{k} c(e_j) = w(v, F) - w(v, S)\,.$$

Note that since $e_j \in P$, it holds for all $j \in \{1, \ldots, k\}$ that

$$c(e_j) = w(u_j, F) - w(u_{j-1}, F) - w(e_j)\,. \tag{3.5}$$

Among the edges of P, we are particularly interested in those with strictly positive contribution. We call their index set J, that is,

$$J := \{j \in \{1, \ldots, k\} \mid c(e_j) > 0\}\,.$$

Let $j \in J$. Then it follows from Equation 3.5 that $e_j \notin F$ and u_{j-1} is not in the directed subtree of F which has u_j as its root. Let $\tilde{e}_j := (\tilde{u}_{j-1}, u_j)$ be the last edge of the s-u_j-path in F. Let $F_j = F \cup \{e_j\} \setminus \{\tilde{e}_j\}$ be the directed spanning tree with root s that is obtained by detaching the subtree of F rooted at u_j from $\tilde{u}_{j-1}$ and re-attaching it at u_{j-1}.

Then, $w(u_j, F_j) = w(u_{j-1}, F) + w(e_j)$ and thus again by Equation 3.5,

$$c(e_j) = w(u_j, F) - w(u_j, F_j)\,.$$

Since $w(u, F)$ is at least $w(u, F_j)$ for all $u \in V$, we obtain

$$f(F) - f(F_j) \geq c(e)\,.$$

Now the remainder of the proof is analogue to that of Proposition 3.6. For $j \in J$ let

$$I_j := \begin{cases} 1 & \text{if } x^{(t+1)} = F_j\,, \\ 0 & \text{otherwise.} \end{cases}$$

Then $\Pr[I_j = 1] = p$,

$$f(x^{(t+1)}) - f(x^{(t)}) \geq \sum_{j \in J} c(e_j)\, I_j\,,$$

and

$$\mathrm{E}[f(x^{(t+1)}) - f(x^{(t)}) \mid f(x^{(t)})] \geq \sum_{j \in J} c(e_j)\, p \geq \frac{f(F) - f(S)}{\mathrm{e}\, |E|^2\, |V|}\,.$$

□

This result indicates that $g\colon \mathcal{F} \to \mathbb{R}_0^+$ defined by $g(F) = f(F) - f_{\text{opt}}$ for all $f \in \mathcal{F}$ is a suitable potential function. Using Theorem 2.10, we bound the expected optimization time of the (1+1) $\text{EA}_{\text{SC-SSSP}}$.

Theorem 3.21 (confer Baswana *et al.* (2009)). *Let $G = (V, E)$ be a strongly connected directed graph, s be a distinguished source vertex in V, and let $w\colon E \to \mathbb{R}^+$ be a weight function on E. Let f_{opt} be the single-criterion objective value of a shortest path tree and $f_{\text{2nd-opt}}$ be the minimal single-criterion objective value over all directed spanning trees with root s that do not have objective value f_{opt}.*

Furthermore, let T be the optimization time of the (1+1) $\text{EA}_{\text{SC-SSSP}}$ on (G, s, w). Then,

$$\mathrm{E}[T \mid x^{(0)}] \leq \mathrm{e}\, |E|^2\, |V| \left(1 + \ln \frac{f(x^{(0)}) - f_{\text{opt}}}{f_{\text{2nd-opt}} - f_{\text{opt}}}\right)$$

and the probability to exceed this bound on the expectation by an additive term of order $|E|^2 |V|$ is decreasing exponentially.

For integer weights polynomially bounded in $|E|$, the previous theorem shows that the expected optimization time of the (1+1) $\text{EA}_{\text{SC-SSSP}}$ is $O(|E|^2\,|V|\,\ln|V|)$. Note that for general weights we cannot apply Theorem 3.8 to give a weight-independent upper bound.

We will see in Section 3.10 that by using a more sophisticated problem representation this bound can be decreased to $O(|E|\,|V|\,\ln|V|)$. In comparison, the runtime of the problem-specific algorithm by Dijkstra (1959) is $O(|E| + |V| \ln|V|)$.

Open questions. As for the minimum spanning tree and maximum weight basis problems, it is an open question whether the asymptotic optimization time depends on the choice of the weights and of the graph for non-trivial instances. In this context, it would be interesting to see whether it is possible to adapt Theorem 3.8 to the case of the SSSP problem.

3.8. Multi-Criteria Optimization

In the previous section, we analyzed how the basic combinatorial (1+1) EA optimizes the SSSP problem using a single-criterion objective function. Let us briefly recall the setting in Section 3.7. Given a graph $G = (V, E)$ and a source $s \in V$, the search space $\mathcal{F}$ consists of all directed spanning trees of G with root s. For a tree $F \in \mathcal{F}$ and a vertex $v \in V$, we denote by $w(v, F)$ the sum of the weights of the edges of the unique s-v-path in F.

In Section 3.8, we considered the single-criterion objective function $f\colon \mathcal{F} \to \mathbb{R}_0^+$ with $f(F) := \sum_{v \in V} w(v, F)$. Now, we study the multi-criterion approach that optimizes each $w(v, \cdot)$ independently.

Algorithm 7 (1+1) $\text{EA}_{\text{MC-SSSP}}$

The (1+1) $\text{EA}_{\text{MC-SSSP}}$ is basic combinatorial (1+1) EA (Algorithm 2) for the multi-criteria single-source shortest path problem.
Let $G = (V, E)$ be a strongly connected directed graph, s a source vertex of G, and $w\colon E \to \mathbb{R}^+$ a weight function on E. Then, the space of feasible search points $\mathcal{F}$ consists of all directed spanning trees of G with root s and $F \succeq H$ holds for $F, H \in \mathcal{F}$ if $w(v, F) \le w(v, H)$ for all $v \in V$ holds.

The multi-criteria setting strongly simplifies the analysis. Once the (1+1) EA finds an optimal s-v-path, the objective functions ensures that it is never replaced by a sub-optimal path (which in general is not true in the single-criterion setting).

In the following theorem, the optimization time depends on the *unweighted edge radius* $\ell_G(s)$. This is the minimal value $\ell \in \mathbb{N}$ such that for every vertex $v \in V$ there exists an optimal s-v-path on at most ℓ edges. That is,

$$\ell_G(s) := \max_{v \in V} \left(\min_{P\colon P \text{ is optimal } s\text{-}v\text{-path}} \big(|E(P)|\big) \right).$$

Theorem 3.22 (confer Doerr *et al.* (2007a)). *Let $G = (V, E)$ be a strongly connected directed graph, s be a distinguished source vertex in V, and let $w\colon E \to \mathbb{R}^+$ be a weight function on E.*

Furthermore, let T be the optimization time of the (1+1) EA$_{\text{MC-SSSP}}$ (Algorithm 7) on (G, s, w). Then for all $\varepsilon > 0$,

$$\Pr\left[T \leq (1+\varepsilon)(2+\sqrt{3})\,\mathrm{e}\,|E|^2 \max\{\ell_G(s)\,,\, \ln|V|\} \,\middle|\, x^{(0)}\right] \geq 1 - |V|^{-\varepsilon}.$$

Proof. Let $\varepsilon > 0$ and $x^{(0)}$ be fixed. Let $p := (1 - 1/|E|)^{|E|-2}\,|E|^{-2}$, let $n := |V|$, and let $\ell^* := \max\{\ell_G(s)\,,\, \ln n\}$. For all $v \in V$, let $w_{\text{opt}}(v)$ be the weight of an optimal s-v-path and let T_v be the random variable that denotes the first point in time $t \in \mathbb{N}$ for which the s-v-path in $x^{(t)}$ is optimal. We first show that

$$\Pr[T_v > (1+\varepsilon)\,(2+\sqrt{3})\,p^{-1}\,\ell^*] \leq n^{-(1+\varepsilon)}$$

holds for all $v \in V$. Let $v \in V$ and let $P := (u_0, e_1, u_1, \ldots, u_{k-1}, e_k, u_k)$ be an optimal s-v-path on $k \leq \ell_G(s)$ edges with $s = u_0$ and $v = u_k$.

Consider a point in time $t \leq T_v$ and let $j(t) \in \{1, \ldots, k\}$ be minimal such that $w(u_{j(t)}, x^{(t-1)}) \neq w_{\text{opt}}(u_{j(t)})$. Like in the single-criterion case, we know that the edge $e_{j(t)} := (u_{j(t)-1}, u_{j(t)})$ is not in $x^{(t-1)}$. Moreover, there has to exist another edge $\tilde{e}_{j(t)} = (\tilde{u}_{j(t)-1}, u_{j(t)})$ with $x^{(t-1)} \cup \{e_{j(t)}\} \setminus \{\tilde{e}_{j(t)}\} \in \mathcal{F}$ (see proof of Proposition 3.20). We say $x^{(t)}$ is a *pessimistic improvement* if $x^{(t)} = x^{(t-1)} \cup \{e_{j(t)}\} \setminus \{\tilde{e}_{j(t)}\}$.

If $x^{(t)}$ is a pessimistic improvement, then $w(u_i, x^{(t)}) = w_{\text{opt}}(u_i)$ for all $i < j(t)$. Let $j \in \{1, \ldots, k\}$. Then there exists at most one point in time $t \in \mathbb{N}$ for which $j(t) = j$ and $x^{(t+1)}$ is a pessimistic improvement.

Thus, if k pessimistic improvements occur then $w(v, x^{(t)}) = w_{\text{opt}}(v)$ holds with certainty. Note that in general the converse is not true. At no time $t \in \mathbb{N}$, the s-u_i-sub-path of P can be guaranteed to be a subgraph of $x^{(t)}$ for $i > 1$.

Now we make use of the fact that for $t \leq T_v$, the events that the $x^{(t)}$'s are pessimistic improvements are mutually independent (although the positions $j(t)$ of these improvements may be highly dependent). More precisely, the probability that $x^{(t)}$ and $x^{(t-1)}$ differ only in the edges $e_{j(t)}$ and $\tilde{e}_{j(t)}$ is exactly p and this holds for all outcomes of $x^{(0)}, \ldots, x^{(t-1)}, x^{(t+1)}, \ldots, x^{(T_v)}$.

For $t \leq T_v$, let $I^{(t)} \in \{0,1\}$ be the indicator variable that is one if $x^{(t)}$ is a pessimistic improvement. Then $\Pr[I^{(t)} = 1] = p$ and the $I^{(t)}$ are mutually independent for $t \leq T_v$. Now, for $t > T_v$, we let $I^{(t)}$ be auxiliary random variables in $\{0,1\}$ which are mutually independent and also assume the value one with probability p.

Recall the observation that if there have been k pessimistic improvements at time $r := (1+\varepsilon)(2+\sqrt{3})p^{-1}\ell^*$ then $w(v, x^{(r)}) = w_{\text{opt}}(v)$. Let $Z := \sum_{t=1}^{r} I^{(t)}$. Then, the event "$Z \geq k$" implies the event "$T_v \leq r$". In terms of probabilities,

$$\Pr[Z \geq k] \leq \Pr[T_v \leq r] .$$

Thus, since $k \leq \ell_G(s) \leq \ell^*$,

$$\Pr[T_v > r] \leq \Pr[Z < k] \leq \Pr[Z < \ell^*] .$$

Now, $\mathrm{E}[Z] = r\,p = (1+\varepsilon)(2+\sqrt{3})\,\ell^*$. Hence,

$$\Pr[T_v > r] \leq \Pr\left[Z < (1+\varepsilon)^{-1}(2+\sqrt{3})^{-1}\mathrm{E}[Z]\right] .$$

We can now apply the Chernoff bounds (Corollary 1.10) to the right hand-side of the previous inequality. Since $\ell^* \geq \ln n$, we obtain

$$\Pr[T_v > r] \leq \mathrm{e}^{-(1+\varepsilon)\,\ell^*} \leq n^{-(1+\varepsilon)} .$$

Finally, since the previous inequality holds for all $v \in V$, the statement follows by the union bound argument

$$\Pr[T > r] = \Pr\left[\bigcup_{v \in V} (T_v > r)\right] \leq n^{-\varepsilon} .$$

□

Open questions. The previous theorem shows that we can give an upper bound on the CDF of the multiple-criteria SSSP problem that is independent of the weight function. It is an open problem to give such bounds for the single-criterion SSSP or the MST problem.

3.9. Permutation Based Search Spaces

All search spaces we investigated so far were composed of subsets of a given base set E, mostly the edge set of a graph. Because of this, we were able to apply the basic combinatorial (1+1) EA. However, sometimes other representations of the search space are more natural and, as we will see, can prove to be more efficient.

In this section, we introduce a generic (1+1) EA for general search spaces. On the example of the classical *Euler tour problem*, we discuss different ways of how this algorithm can be applied to the search space of permutations.

Problem 3.23 (The Euler tour problem).
Let $G = (V, E)$ be an Eulerian graph (G is connected and every vertex of G has even degree). The Euler tour *problem asks for a closed walk in G that uses each edge exactly once.*

We represent walks in G by permutations of the edge set E (confer Scharnow, Tinnefeld and Wegener (2004)). Let $E = \{e_1, \ldots, e_m\}$ with $m = |E|$ and let $\mathcal{S}_m$ be the space of all permutations of the numbers $1, \ldots, m$. Furthermore, let $W := (u_0, f_1, \ldots, f_k, u_k)$ be a walk in G of length $k \in \mathbb{N}$.

We identify W with all permutations $\sigma \in \mathcal{S}_m$ such that $u_k \notin e_{\sigma(k+1)}$ and such that $f_j = e_{\sigma(j)}$ holds for all $j \in \{1, \ldots, k\}$ (permutations with $u_k \in e_{\sigma(k+1)}$ correspond to longer walks). For a permutation σ, let $k = k(\sigma)$ be the maximal integer such that $W_\sigma := (u_0, e_{\sigma(1)}, \ldots, e_{\sigma(k)}, u_k)$ is a proper walk. Then W_σ is the walk that corresponds to σ.

Euler's Theorem (Euler (1741)) guarantees that in an Eulerian graph a walk of length m exists and moreover that all such walks are closed. Thus, we may rephrase the Euler tour problem as the problem to optimize the objective function $k \colon \mathcal{S}_m \to \mathbb{N}$ as defined above over the search space $\mathcal{S}_m$ (where k depends on the structure of G).

We now formulate the *generic* (1+1) EA which works on arbitrary search spaces. For this, we define how to randomly generate the candidate search point $y^{(t+1)}$ from the current search point $x^{(t)}$. We call the process *variation.*

Definition 3.24 (Variation operator ϕ). *Let $\mathcal{S}$ be a finite search space. A* variation operator *is a sampling procedure that generates a random search point in $\mathcal{S}$ according to a distribution that is based on a given search point in $\mathcal{S}$.*

The generic (1+1) EA mimics the two defining features of the basic combinatorial (1+1) EA. These are that (i) the candidate search point $y^{(t+1)}$ is likely to be chosen in a local neighborhood of $x^{(t)}$ and (ii) each point in the search space can be chosen as $y^{(t+1)}$ with positive probability. The first properties ensures that most of the time the algorithm explores the search space locally. The second property guarantees that the algorithm eventually finds a global optimum.

Let us recall how variation is performed by the basic combinatorial (1+1) EA. Each element $e \in E$ indicates a location for a potential variation by adding or removing e from $x^{(t)}$. The candidate solution $y^{(t+1)}$ is generated by performing a variation of $x^{(t)}$ at location e with probability $1/|E|$ independently for all edges $e \in E$.

For the search space of permutations $\mathcal{S}_m$, several notions of locality are possible. The canonical local variation operator is the *exchange* variation operator ϕ_{exchange}. Let $\sigma \in \mathcal{S}_m$ be a permutation. Then $\phi_{\text{exchange}}(\sigma)$ is generated by choosing two positions a, b uniformly at random from $\{1, \ldots, m\}$ and transposing them in σ.

A second local variation operator is the *jump* variation operator ϕ_{jump}. Again, let $\sigma \in \mathcal{S}_m$ be a permutation. This time, ϕ_{jump} is generated by choosing two positions a, b uniformly at random from $\{1, \ldots, m\}$, removing the element at position a from σ, and reinserting it so that it becomes position b.

In both cases, the respective variation operator is local in the informal sense of property (i). However, unlike the case of the basic combinatorial (1+1) EA, two different exchanges might change the same position of the permutation. Thus, we cannot perform several local variations simultaneously.

To obtain property (ii), we first choose a random number k according to Pois(1), the Poisson distribution with parameter 1. Then, we perform k sequential random local variations. By this, we transform a local variation operator into a variation operators that also satisfies property (ii). We specifically choose the Poisson distribution to simulate the basic combinatorial (1+1) EA. There, the number of local variations (changed elements) is governed by the Binomial distribution which is known to converge to the Poisson distribution[‡].

The *optimization time* T of the generic (1+1) EA for $(\mathcal{S}, \phi, \succeq, x^{(0)})$ is the random variable that describes the first point in time $t \in \mathbb{N}$ for which $x^{(t)}$ is optimal.

[‡]In the results referenced in this chapter, the number of successive variation operators applied in each single variation step is often distributed according to $1 + \text{Pois}(1)$ instead of Pois(1). From the analytic point of view, this choice is artificial as it deviates from the original purpose of simulating the basic combinatorial (1+1) EA. Furthermore, this difference does not have an effect on the order of magnitude of the expected runtimes considered in this chapter.

Algorithm 8 The Generic (1+1) EA

Let $\mathcal{S}$ be a finite search space. Furthermore, let ϕ be a local variation operator on $\mathcal{S}$, let $\succeq$ be a partial order relation on $\mathcal{S}$, and let $x^{(0)} \in \mathcal{S}$ be an initial search point. Given $(\mathcal{S}, \phi, \succeq, x^{(0)})$, the generic (1+1) EA iteratively generates a sequence of search points $\{x^{(t)}\}_{t \in \mathbb{N}}$ in $\mathcal{S}$ by the following procedure.

For all $t \in \mathbb{N}$ with $t \geq 1$, the generic (1+1) EA generates a random number k according to Pois(1). Next, the generic (1+1) EA generates the candidate search point $y^{(t)} := \phi^k(x^{(t-1)})$ by k successive applications of ϕ to $x^{(t-1)}$. Afterwards,

$$x^{(t)} := \begin{cases} y^{(t)} & \text{if } y^{(t)} \succeq x^{(t-1)}, \\ x^{(t-1)} & \text{otherwise.} \end{cases}$$

Neumann (2008) showed that the generic (1+1) EA for the Euler tour problem using the exchange variation operator ϕ_{exchange} has at least exponential expected optimization time.

Theorem 3.25 (Neumann (2008)). *Let $n \geq 3$. Then, there exists an Eulerian graph $G = (V, E)$ of size n such that the expected optimization time of the generic (1+1) EA for the Euler tour problem on G using the local exchange variation operator $\phi_{exchange}$ with $x^{(0)}$ chosen uniformly at random is $2^{\Omega(|V|)}$.*

However, the *jump* variation operator ϕ_{jump} results in a polynomially bounded expected optimization time of the generic (1+1) EA for the Euler tour problem.

Theorem 3.26 (Neumann (2008)). *Let $G = (V, E)$ be an Eulerian graph. Then the expected optimization time of the generic (1+1) EA for the Euler tour problem on G using the local exchange variation operator ϕ_{jump} is $O(|E|^5)$ independent of $x^{(0)}$. For all $n \geq 3$, there exists an Eulerian graph $G = (V, E)$ on n vertices such that the expected optimization time of the generic (1+1) EA for the Euler tour problem on G using the local exchange variation operator ϕ_{jump}. with $x^{(0)}$ chosen uniformly at random is $\Omega(|E|^5)$.*

In Doerr, Hebbinghaus and Neumann (2007b) it is shown that if $a = 1$ is fixed in ϕ_{jump}, then the expected optimization time of the generic (1+1) EA for the Euler tour problem drops to $\Theta(|E|^3)$.

From these three examples, we conclude that for non-Boolean search spaces (like the space of permutations) the choice of locality can severely influence the optimization time.

However, an expected optimization time of $\Theta(|E|^3)$ for the Euler tour problem is still far off from the linear runtime of the problem-specific algorithm by Hierholzer (1873). In the following Section, we will see how it is possible to further reduce the expected optimization time of the generic (1+1) EA to $O(|E| \ln |E|)$.

3.10. Asymmetric and Adjacency-Based Variation Operators

In Section 3.9, we have seen that the choice of the local variation operator can have significant influence on the optimization time of the generic (1+1) EA. In this section we revisit the problems of finding a maximum weight basis, a single-source shortest path tree, an Euler tour, or an all-pair shortest path set. We study how alternative representations and variation operators for these problems influence the optimization time.

Consider the basic (1+1) EA for the maximum weight basis problem and recall the proof of Proposition 3.6. There, the bound on the drift is dominated by the probability $p_f = (1 - m^{-1})^{m-2} m^{-2}$ (where $m = |E|$) to exchange a particular element in a basis of E by a particular element that is not in E. However, we know that all bases are of the same size, namely the rank $r = r(M)$ of the matroid (which is $|V| - 1$ in the case of spanning trees).

We modify the distribution of the variation operator in the basic combinatorial (1+1) EA (confer Jansen and Sudholt (2010); Reichel and Skutella (2007)). The *asymmetric variation operator* is defined as follows. Instead of choosing each element e in $x^{(t-1)} \Delta y^{(t)}$ independently with probability $1/m$, we independently choose e with probability $1/r$ if $e \in x^{(t-1)}$ with probability $1/(m - r)$ if $e \notin x^{(t-1)}$. This way, in expectation $x^{(t-1)} \Delta y^{(t)}$ contains exactly two elements, one in $x^{(t)}$ and in $E \setminus x^{(t)}$. Consequently, for a particular pair of such elements

$$p_f = \left(1 - \frac{1}{r}\right)^{r-1} \left(1 - \frac{1}{m-r}\right)^{m-r-1} \frac{1}{r(m-r)} \geq \frac{1}{\mathrm{e}^2 \, r \, m} .$$

Thus, following the proof of Proposition 3.6 the bound of expected optimization time in Theorem 3.7 drops by a factor of m/r.

Theorem 3.27 (confer Reichel and Skutella (2007)).
Let $M = (E, \mathcal{F})$ be a matroid and let $w \colon E \to \mathbb{R}^+$ be a weight function on E. Let w_{opt} be the weight of a maximum weight basis and $w_{\textit{2nd-}\text{opt}}$ be the maximum weight over all bases that do not have weight w_{opt}. Furthermore, let T be the optimization time of the (1+1) EA for the maximum weight basis problem on (M, w) using the asymmetric variation operator defined above. Then,

$$\mathrm{E}[T \mid x^{(0)}] \le \mathrm{e}^2\, r(M)\, |E| \left(1 + \ln \frac{w_{\text{opt}} - w(x^{(0)})}{w_{\text{opt}} - w_{\textit{2nd-}\text{opt}}}\right)$$

and the probability to exceed this bound on the expectation by an additive term of order $r(M)|E|$ is decreasing exponentially.

Let us now turn to the SSSP problem. Recall the problem of finding a single-source shortest path tree according to a source vertex s in a strongly connected directed graph $G = (V, E)$ (Section 3.7). As search space, we considered all directed spanning trees with root s. We represented them as subsets of E, rejecting all subsets that do not from such a tree.

Let us take a closer look at the proof of Theorem 3.21. Consider the factor $\mathrm{e}\,|E|^2$ in the upper bound on the expected optimization time of the basic combinatorial (1+1) EA. We observe that it results from the probability p to change the position of a subtree in the current search point (see proof of Proposition 3.21).

In the representation by edge sets, this is done by removing the edge that attaches the subtree and inserting a new edge that re-attaches the subtree while leaving the other edges unchanged. Thus,

$$p = \left(1 - \frac{1}{|E|}\right)^{|E|-2} \frac{1}{|E|^2} \ge \frac{1}{\mathrm{e}\,|E|^2}\,.$$

Hence, the expected time for a particular subtree relocation is at most $\mathrm{e}\,|E|^2$.

To reduce the upper bound on the expected time to relocate a subtree, we choose a more suited representation of the search points. For this, we regard a data-structure commonly used to represent graphs — adjacency lists. The *adjacency list* L of a directed graph $G = (V, E)$ stores for every vertex $v \in V$ the sub-list $L(v) \subseteq E$ of all *incoming* directed edges (w, v).

A natural way to represent directed rooted spanning trees is to assign to each vertex the edge by which it is attached to its predecessor in the tree. Thus, in such a representation, we distinguish for each vertex $v \in V$ one edge (w, v) in its sub-list $L(v)$.

Let $\mathcal{L}$ be the space that consists of all sets of edges F such that there exists exactly one edge from $L(v)$ in F for every vertex $v \in V \setminus \{s\}$. Note that not every set in $\mathcal{L}$ corresponds to a directed spanning tree. In particular, a set in $\mathcal{L}$ might define a graph which contains directed cycles.

The search space $\mathcal{S}_{\text{SSSP}}$ consists of all sets in $\mathcal{L}$ that correspond to directed spanning trees rooted at s.

The local variation operator ϕ_{SSSP} on a set $F \in \mathcal{S}_{\text{SSSP}}$ generates the random set $F' \in \mathcal{S}_{\text{SSSP}}$ as follows. First, an edge e is chosen uniformly at random from $E \setminus L(s)$. Then, there exists a unique vertex $v \in V \setminus \{s\}$ such that $e \in L(v)$ and a unique edge $e' \in F \cap L(v)$. If $F \cup \{e\} \setminus \{e'\}$ represents a directed spanning tree rooted at s, then $F' = F \cup \{e\} \setminus \{e'\}$. Otherwise, $F' = F$.

We can now analyze the generic (1+1) EA for the SSSP problem on the search space $\mathcal{S}_{\text{SSSP}}$ and the local variation operator ϕ_{SSSP}. Let $\succeq_s$ and $\succeq_m$ be the partial orders on $\mathcal{S}_{\text{SSSP}}$ defined by the single-criterion objective function f in Section 3.7 and the multi-criteria partial order $\succeq$ in Section 3.8, respectively. Then the (1+1) EA using adjacency lists to optimize the single-criterion or multi-criteria SSSP problem is the generic (1+1) EA on the search space $\mathcal{S}_{\text{SSSP}}$ with variation operator ϕ_{SSSP} and an objective defined by the order relation $\succeq_s$ or $\succeq_m, x^{(0)}$, respectively.

It turns out that the proofs in Section 3.7 and Section 3.8 are still correct if we change the probability p to relocate a fixed subtree to a specific position. Using the local variation operator ϕ_{SSSP}, the lower bound on this probability increases by a factor of $|E|$ to

$$p' = \frac{1}{\mathrm{e}\,|E|}\,.$$

Thus, we get the following results which are the equivalents of Theorem 3.21 and Theorem 3.22 for the generic (1+1) EA using the local mutation operator ϕ_{SSSP}.

Theorem 3.28 (Doerr and Johannsen (2010)). *Let $G = (V, E)$ be a strongly connected directed graph, s be a distinguished source vertex in V, and let $w\colon E \to \mathbb{R}^+$ be a weight function on E. Let f_{opt} be the single-criterion objective value of a shortest path tree and $f_{\text{2nd-opt}}$ be the minimal* *single-criterion objective value over all directed spanning trees with root s that do not have objective value* f_{opt}.

Let $x^{(0)} \in \mathcal{S}_{SSSP}$ *and let* T_s *and* T_m *be the optimization times of the (1+1) EA on the single-criterion (multi-criteria, respectively) SSSP problem using adjacency lists. Then, for all* $\varepsilon > 0$,

$$\mathrm{E}[T_s \mid x^{(0)}] \leq \mathrm{e}\,|E|\,|V| \left(1 + \ln \frac{f(x^{(0)}) - f_{\text{opt}}}{f_{\text{2nd-opt}} - f_{\text{opt}}}\right)$$

and

$$\Pr\left[T_m \leq (1+\varepsilon)(2+\sqrt{3})\,\mathrm{e}\,|E|\,\max\{\ell_G(s)\,,\ln|V|\} \mid x^{(0)}\right] \geq 1 - |V|^{-\varepsilon}\,.$$

Next, we consider the Euler tour problem. Also for this problem there exist a superior representation based on adjacency lists (Doerr, Klein and Storch (2007c); Doerr and Johannsen (2007)). As described in Section 3.9, this problem is defined on an undirected Eulerian graph $G = (V, E)$. The corresponding adjacency list L stores for every vertex $v \in V$ the sub-list $L(v)$ of edges incident with v. This way, the edge $\{v, w\}$ occurs twice, in $L(v)$ and in $L(w)$. Note, since G is Eulerian, all sub-lists are of even size.

Consider a walk $W = (u_0, f_1, \ldots, f_k, u_k)$ in G. Then W can be represented by all pairs of successive edges $\{f_1, f_2\}, \{f_2, f_3\}, \ldots, \{f_{k-1}, f_k\}$. For each pair $\{f_i, f_{i+1}\}$, there exists exactly one sub-list of L which contains the two edges f_i and f_{i+1}, namely $L(u_i)$. Thus, we identify the walk W with pairings of edges in the sub-lists of L.

The search space $\mathcal{S}_{\text{Euler}}$ consist of all complete pairings of edges in L. A *complete pairing* is a set of edge pairs such that (i) the edges of a pair are in the same sub-list of L and (ii) each edge in a sub-list of L belongs to exactly one pair. Such a pairing always exists since all sub-lists are of even lengths. Moreover, a random pairing is easy to generate: for each sub-list, we successively pair two random vertices and then remove them from the list.

Since each edge belongs to exactly two pairs, the pairs partition the edge sets into disjoint and cyclically ordered sets. We call these cyclically ordered sets *tours*. Note that if we distinguish some vertex of a tour as start-/end-vertex, then the tour becomes a closed walk.

We have just seen that a complete pairing corresponds to a disjoint partition of the edge set into tours. Let $k\colon \mathcal{S}_{\text{Euler}} \to \mathbb{N}$ be the function that counts the number of tours $k(x)$ in the partition corresponding to a search point $x \in \mathcal{S}_{\text{Euler}}$. Whenever $k(x) = 1$, the partition corresponding to x contains only one tour which has to be an Euler tour. Consequently, we reformulate the Euler tour problem over $\mathcal{S}_{\text{Euler}}$ as the problem of minimizing k over $\mathcal{S}_{\text{Euler}}$.

The local variation operator ϕ_{Euler} that generates a random search point $y \in \mathcal{S}_{\text{Euler}}$ based on the search point $x \in \mathcal{S}_{\text{Euler}}$ is defined as follows. We choose the two edges e and e' uniformly at random from all pairs of edges that are in the same sub-list of L. In x, each of these two edges is paired with a second edge, say $\{e, f\}$ and $\{e', f'\}$ are paired. Then y is generated by removing the pairs $\{e, f\}$ and $\{e', f'\}$ from x and adding the pairs $\{e, e'\}$ and $\{f, f'\}$ in return.

The local variation operator ϕ_{Euler} that generates a random search point $y \in \mathcal{S}_{\text{Euler}}$ based on the search point $x \in \mathcal{S}_{\text{Euler}}$ is defined as follows. We choose an edge e uniformly at random from E. Then, we choose with probability 1/2 one of its end-vertices v and w (say v). Finally, we choose a second edge e' uniformly at random from $L(v)$. In x, each of these two edges is paired with a second edge, say $\{e, f\}$ and $\{e', f'\}$. Then y is generated by removing the pairs $\{e, f\}$ and $\{e', f'\}$ from x and adding the pairs $\{e, e'\}$ and $\{f, f'\}$ in return.

The (1+1) EA using adjacency lists to solve the Euler tour problem is defined to be the generic (1+1) EA on the search space $\mathcal{S}_{\text{Euler}}$ with the variation operator ϕ_{Euler} and an objective defined by the partial order relation $\succeq_{\text{Euler}}$ given by the function $k\colon \mathcal{S}_{\text{Euler}} \to \mathbb{N}$ above.

Theorem 3.29 (Doerr and Johannsen (2007)). *Let $G = (V, E)$ be an Eulerian graph. Furthermore, let $x^{(0)} \in \mathcal{S}_{\text{Euler}}$ and T be the optimization time of the (1+1) EA using adjacency lists to solve the Euler tour problem. Then,*

$$\mathrm{E}[T \mid x^{(0)}] \leq \mathrm{e}\,|E| \ln |E|$$

and the probability to exceed this bound on the expectation by an additive term of order $|E|$ is decreasing exponentially.

Proof. Let $t < T$ and $x^{(t)}$ be the t-th search point of the (1+1) EA using adjacency lists to solve the Euler tour problem. Then $x^{(t)}$ corresponds to a partition of E into $g(x^{(t)})$ tours and $g(x^{(t)})$ is at least two.

Let τ be one of these tours. Since $t < T$ and G is connected, τ shares at least one vertex v with another tour in the partition. Suppose that ℓ is the size of the sub-list and k of edges in $L(v)$ belong to τ. Since there exist at least one pairing in $L(v)$ for each tour visiting v, we have that $\ell \geq 4$ and $2 \leq k \leq \ell - 2$.

The probability, that the (1+1) EA operator performs exactly one local variation and merges τ at v with a second tour by is

$$p = \frac{1}{\mathrm{e}}\frac{k}{|E|}\cdot\frac{\ell-k}{\ell} \geq \frac{1}{\mathrm{e}\,|E|}\,.$$

For each of the $g(x^{(t)})$ tours there exists at least one vertex v at which such a merging may occur. Thus, the probability to merge two tours is at least $\frac{g(x^{(t)})}{\mathrm{e}|E|}$. Note that we do not over-count since we choose ordered pairs of edges (e, e') for variation. Hence, in expectation the number of tours decreases by at least $\frac{g(x^{(t)})}{\mathrm{e}|E|}$. The statement follows by Theorem 2.10. □

3.11. Population and Recombination

In this section we study the *all-pair shortest path problem* (APSP). In the context of evolutionary algorithms, this problem has been studied by Doerr, Happ and Klein (2008) and Doerr and Theile (2009). The main insight of these works is that these algorithms perform better if they also apply recombination of two search points instead of variation of single search points only. Further theoretical work on the influence of recombination on the runtimes of evolutionary algorithms can be found in Fischer and Wegener (2004) and Watson and Jansen (2007).

Problem 3.30 (The All-Pair Shortest Path (APSP) problem). *Let $G = (V, E)$ be a strongly connected directed graph and let $w\colon E \to \mathbb{R}^+$ be a weight function on E. For all $v, w \in V$, an* optimal v-w-path *in G is a directed path $P \subseteq E$ from v to w minimizing $w(P) := \sum_{e\in P} w(e)$. The* all-pair shortest path *problem asks for an optimal v-w-path in G for every ordered pair (v, w) of distinct vertices $v, w \in V$.*

A solution to the APSP problem is a set of optimal paths. Consequently, we apply an evolutionary algorithm that maintains in each iteration a set or multi-set of search points rather than a single search point. Moreover, for the APSP problem we observe that the concatenation of an optimal v-u-path with an optimal u-w-path often results in an optimal v-w-path.

In the context of evolutionary algorithms, we call the multi-set of current search points the *population* and the generation of a (random) search point from two others *recombination*. In our case, we perform recombination by the concatenation of two paths.

Algorithm 9 EA_{APSP}

Let the recombination probability p_{R} be given. Let W be the set of all pairs of distinct vertices in V^2. The search space $\mathcal{S}$ of the EA_{APSP} consists of all directed v-w-paths in G with $(v, w) \in W$. The multi-set of initial search points $x^{(0)}$ contains for each edge $(v, w) \in E$ the path $(v, (v, w), w)$ on this single edge. For two paths P and Q in $\mathcal{S}$, let $P \succeq Q$ if and only if P and Q have the same start-vertex and end-vertex and if $w(P) \leq w(Q)$.
The EA_{APSP} iteratively generates a sequence of multi-sets $\{x^{(t)}\}_{t \in \mathbb{N}}$ of search points in $\mathcal{S}$ by the following procedure.
For all $t \in \mathbb{N}$ with $t \geq 1$, the population-based evolutionary algorithm generates a random candidate search point $y^{(t)}$. With probability p_{R}, the point $y^{(t)}$ is generated by recombination and with probability $1 - p_{\mathrm{R}}$ by variation. Afterwards, we set

$$x^{(t)} := \begin{cases} \{x \in x^{(t-1)} : y^{(t)} \not\succeq x\} \cup \{y^{(t)}\} & \text{if } \forall x \in x^{(t-1)} : x \not\succeq y^{(t)}, \\ x^{(t-1)} & \text{otherwise.} \end{cases}$$

Recombination

(1) Choose two paths P and Q uniformly at random from $x^{(t-1)}$.
(2) If $P \circ Q \in \mathcal{S}$ then return $P \circ Q$.
(3) Otherwise, return either P or Q, each with equal probability.

Variation

(1) Choose an path $P = (v, \ldots, w)$ uniformly at random from $x^{(t-1)}$.
(2) Repeat the following steps k times, where k is chosen according to Pois(1):
    (a) Choose $u \in \{v, w\}$ uniformly at random.
    (b) Choose an edge $e \in E$ incident to u uniformly at random.
    (c) If e connects u to P then return P without u and e.
    (d) If $u = v$, and $e = (x, v)$ is an incoming edge of v, and $(x, (x, v), v) \circ P \in \mathcal{S}$, then return $(x, (x, v), v) \circ P$.
    (e) If $u = w$, and $e = (w, y)$ is an outgoing edge of w, and $P \circ (w, (w, y), y) \in \mathcal{S}$, then return $P \circ (w, (w, y), y)$.
    (f) Otherwise, return P

The concatenation $P \circ Q$ of two paths $P = (v, \ldots, w)$ and $Q = (w, \ldots, u)$ is the walk $(v, \ldots, w, \ldots, u)$. Note, that $P \circ Q$ may visit vertices more than once and thus need not be a path.

The following evolutionary algorithm generates a single candidate search point in each iteration. This is done by using either variation or recombination. In this, the chosen method depends on the recombination probability $p_{\mathrm{R}} \in [0, 1]$. More precisely, the algorithm samples from the Bernoulli distribution with parameter p_{R} and performs recombination if the generated bit evaluates to one.

Let $G = (V, E)$ be a strongly connected directed graph let $w\colon E \to \mathbb{R}^+$ be a weight function on E. The $\mathrm{EA}_{\mathrm{APSP}}$ is the population-based evolutionary algorithm for the APSP problem on (G, w) with recombination probability p_{R}.

The *optimization time* T of the $\mathrm{EA}_{\mathrm{APSP}}$ is the first point in time $t \in \mathbb{N}$ for which $x^{(t)}$ contains an optimal path for every $(v, w) \in W$. With probability tending to one, the optimization time of the $\mathrm{EA}_{\mathrm{APSP}}$ is $O(|V|^{3+1/4} \ln^{1/4} |V|)$.

Theorem 3.31 (Doerr and Theile (2009)). *Let $p_{\mathrm{R}} \in (0, 1)$ be constant. Let the graph $G = (V, E)$ be directed and strongly connected and let $w\colon E \to \mathbb{R}^+$ be a weight function on E. Furthermore, let T be the optimization time of the $\mathrm{EA}_{\mathrm{APSP}}$ with recombination probability $p_{\mathrm{R}} \in (0, 1)$. Then for all $\lambda > 0$ there exist a constant C_λ such that*

$$\Pr\left[T \le C_\lambda |V|^{3+1/4} \ln^{1/4} |V|\right] \ge 1 - |V|^{-\lambda}.$$

To prove this theorem, we give some preliminary definitions and two propositions. In the following, let $n := |V|$.

Let $(v, w) \in W$. Then $T_{(v,w)}$ is the random variable that describes the first point in time $t \in \mathbb{N}$ for which there is an optimal v-w-path in $x^{(t)}$.

For $g \in \mathbb{R}_0^+$, we call the pair $(x, y) \in W$ a *g-approximation* of (u, w) if there exist an optimal v-w-path containing an x-y-subpath that is at most g edges shorter that the v-w-path. Furthermore, let $T_{g,(v,w)}$ be the random variable that denotes the first point in time $t \in \mathbb{N}$ for which $x^{(t)}$ contains an optimal x-y-path such that $(x, y) \in W$ is a g-approximation of (v, w). Thus, $T_{(v,w)} = T_{0,(v,w)}$

Proposition 3.32. *Let $(v, w) \in W$, and $g \in \mathbb{R}_0^+$ with $g \ge 4 \ln n$. Then, for all $\lambda \ge 2$ and n sufficiently large,*

$$\Pr\left[T_{(v,w)} \le T_{g,(v,w)} + \tfrac{6\lambda g n^3}{1-p_{\mathrm{R}}}\right] \ge 1 - n^{1-\lambda}.$$

Proof. We may condition on the event that $T_{(v,w)} > T_{g,(v,w)}$ holds since otherwise the event that $T_{(v,w)} \leq T_{g,(v,w)} + \frac{6\lambda g n^3}{1-p_{\mathrm{R}}}$ holds with certainty. Thus, without loss of generality, suppose that $T_{(v,w)} > T_{g,(v,w)}$.

For $t \geq T_{g,(v,w)}$, let $h(t)$ be the minimal integer such that $x^{(t)}$ contains an optimal x-y-path for which (x, y) is a $h(t)$-approximation of (v, w). Then, $h(T_{v,w}) = 0$ and also $1 \leq h(T_{g,(v,w)}) \leq g$.

Next, we define the indicator variables $I_t \in \{0, 1\}$ for all $t \in \mathbb{N}$. For all $t \leq T_{(v,w)}$, we let $I_t = 1$ if and only if $h(t) < h(t-1)$. For all $t > T_{(v,w)}$, we let I_t be drawn independently according to the Bernoulli distribution with parameter $\frac{1}{2\,e\,n^3}$, that is, $\Pr[I_t = 1] = \frac{1-p_{\mathrm{R}}}{2\,e\,n^3}$.

Let $T_{g,(v,w)} \leq t < T_{(v,w)}$. We show that

$$\Pr\left[I_{t+1} \,\middle|\, I_{T_{g,(v,w)}+1}, \ldots, I_t\right] \geq \frac{1-p_{\mathrm{R}}}{2\,e\,n^3}. \tag{3.6}$$

By the definition of $h(t)$, there is an optimal v-w-path $(v = u_0, \ldots, u_k = w)$ with $k \in \mathbb{N}$ and two indices $a(t), b(t) \in \{0, \ldots, k\}$ with $a(t) < b(t)$ such that $a(t) + (k - b(t)) = h(t)$ and $x^{(t)}$ contains an optimal $u_{a(t)}$-$u_{b(t)}$-path.

Since $t < T_{(v,w)}$, it holds that $a(t) \geq 1$ or $k - b(t) \geq 1$. Without loss of generality, suppose $a(t) \geq 1$. Then the variation that chooses the optimal $u_{a(t)}$-$u_{b(t)}$-path from $x^{(t)}$ and adds the edge $(u_{a(t)-1}, u_{a(t)})$ decreases $h(t)$.

The probability that this variation happens depends on $|x^{(t)}|$ and the degree of $u_{a(t)}$. However, independently of $x^{(t)}$ (and all $X^{(s)}$ with $s < t$), this probability is at least $\frac{1-p_{\mathrm{R}}}{2\,e\,n^3}$. Thus, (3.6) holds.

Now, consider the time interval

$$J := \left\{T_{g,(v,w)} + 1, \ldots, T_{g,(v,w)} + \left\lfloor \tfrac{6\lambda g n^3}{1-p_{\mathrm{R}}} \right\rfloor \right\}.$$

If $\sum_{t\in J} I_t \geq g$, then the event $T_{(v,w)} \leq T_{g,(v,w)} + \frac{6\lambda g n^3}{1-p_{\mathrm{R}}}$ holds with certainty. Thus,

$$\Pr\left[T_{(v,w)} \leq T_{g,(v,w)} + \tfrac{6\lambda g n^3}{1-p_{\mathrm{R}}}\right] \geq \Pr\left[\sum_{t\in J} I_t \geq g\right].$$

By Theorem 1.18, we can now derive the proposition from the Chernoff bound (Corollary 1.10). Let $\mu = \mathrm{E}\left[\sum_{t\in J} I_t\right]$. Then, for sufficiently large n,

$$\mu = \left\lfloor \tfrac{6\lambda}{1-p_{\mathrm{R}}}\, g\, n^3 \right\rfloor \cdot \tfrac{1-p_{\mathrm{R}}}{2\,e\,n^3} > \lambda g.$$

Hence, using the Chernoff bounds we get

$$\Pr\left[\sum_{t\in J} I_t < g\right] < \Pr\left[\sum_{t\in J} I_t < \tfrac{1}{\lambda}\mu\right] \leq \mathrm{e}^{-\frac{(\lambda-1)^2 g}{2\lambda}} \leq n^{1-\lambda}.$$

□

Next, we let W_r be the set of all pairs $(v, w) \in W$ such that there exists an optimal v-w-path on at most r edges. Note that $W_n = W$, since there exists an optimal v-w-path on at most n edges for all $(v, w) \in W$. Furthermore, for all $r \in \mathbb{R}_0^+$ let $T_r := \max_{(v,w)\in W_r} T_{(v,w)}$ be the random variable that describes the first point in time $t \in \mathbb{N}$ for which there is an optimal v-w-path for all $(v, w) \in W_r$ in $x^{(t)}$. Clearly, $T_1 = 0$.

Finally, for $r, g \in \mathbb{N}$ with $g \leq r$ let $T_{g,r} := \max_{(v,w)\in W_r} T_{g,(v,w)}$ be the random variable that describes the first point in time $t \in \mathbb{N}$ such that for all $(v, w) \in W_r$ there is an optimal x-y-path in $x^{(t)}$ where (x, y) is a g-approximation of (v, w). Thus, $T_{r,r} = 0$.

Proposition 3.33. *Let $r, g \in \mathbb{R}_0^+$ with $g \leq r/2$ and $(v, w) \in W_{3r/2} \setminus W_r$. Then, for $\lambda > 0$ and n sufficiently large,*

$$\Pr\left[T_{g,(v,w)} \leq T_r + \tfrac{4\,\lambda\, n^4 \,\ln n}{p_{\mathrm{R}}\, r\, g^2}\right] \geq 1 - n^{-\lambda}\,.$$

Proof. We may condition on the event that $T_{g,(v,w)} > T_r$ holds, otherwise the event $T_{g,(v,w)} \leq T_r + \frac{4\,\lambda\, n^4\, \ln n}{p_{\mathrm{R}}\, r\, g^2}$ holds with certainty. Thus, without loss of generality, suppose that the event $T_{g,(v,w)} > T_r$ holds.

Let $t > T_r$. By the definitions of W_r and $W_{3r/2}$, there exists an optimal v-w-path $P := (v = u_0, (u_0, u_1), \ldots, u_k = w)$ on $k \in \mathbb{N}$ edges such that

$$r < k \leq 3r/2\,.$$

Let $a, b, j \in \mathbb{N}$ with $a + k - b \leq g$, $j \leq a + r$, and $j \geq b - r$. Since P is optimal, the two paths $(v = u_a, \ldots, u_j)$ and $(u_j, \ldots, u_b = w)$ are optimal, too. Moreover, by definition of j both paths are of length at most r. Thus, since $t > T_r$, at time t the $\mathrm{EA_{APSP}}$ has found an optimal u_a-u_j-path and an optimal u_j-u_b-path. The concatenation of these two optimal paths results in an optimal u_a-u_b-path. Hence, the vertex pair (u_a, u_b) is a g-approximation of (v, w).

Let a, b, and j be given. Then, the probability that $\mathrm{EA_{APSP}}$ performs a recombination step using this particular concatenation at time t is at least p_{R}/n^4. There are at least $g^2/2$ ways to choose the pair (a, b) and at least $2r - k + 1 \geq r/2$ ways to choose j. Note that these bounds hold independently of the variations and recombinations performed by the $\mathrm{EA_{APSP}}$ at times $T_r + 1, \ldots, t - 1$. Thus, for all $t > T_r$,

$$\Pr\left[T_{g,(v,w)} = t \,\middle|\, T_{g,(v,w)} \geq t\right] \geq \frac{p_{\mathrm{R}}\, r\, g^2}{4n^4}\,.$$

Let $\Delta := \frac{4\,n^4}{p_R\, r\, g^2}$. Then, by the union bound,

$$\Pr\left[T_{g,(v,w)} \geq T_r + \lambda \Delta \ln n\right] \leq \left(1 - \tfrac{1}{\Delta}\right)^{\lambda \Delta \ln n} \leq e^{-\lambda \ln n} = n^{-\lambda}. \qquad \square$$

Using the previous two propositions, we prove Theorem 3.31.

Proof of Theorem 3.31. Let $\lambda > 3$ and let n be sufficiently large. Let $L := \lceil \log_{3/2} n \rceil$ and

$$r(i) := \left(\tfrac{3}{2}\right)^i n^{1/4} \ln^{1/4} n \qquad \text{and} \qquad g(i) := \left(\tfrac{3}{2}\right)^{-i/3} n^{1/4} \ln^{1/4} n$$

for all $i \in \{0, \ldots, L\}$. Then, $g(0) = r(0) = n^{1/4} \ln^{1/4} n$ and $g(n) \leq r(n)/2$ for all $n \geq 1$. Furthermore, we have

$$0 = T_{g(0),r(0)} \leq T_{r(0)} \leq T_{g(1),r(1)} \leq T_{r(1)} \leq \cdots \leq T_{g(L),r(L)} \leq T_{k(L)} = T\,.$$

For all $(v, w) \in W_{r(0)}$ let

$$A_{(v,w)}: \qquad T_{(v,w)} \leq T_{g(0),(v,w)} + \frac{6\,\lambda\, g(0)\, n^3}{1 - p_R} \quad \text{and}$$

$$B_{(v,w)}: \qquad T_{g(0),(v,w)} = 0 \qquad \text{(this event occurs with certainty)}\,.$$

For all $i \in \{1, \ldots, L\}$ and all $(v, w) \in W_{r(i)} \setminus W_{r(i-1)}$ let

$$A_{(v,w)}: \quad T_{(v,w)} \leq T_{g(i),(v,w)} + \frac{6\,\lambda\, g(i)\, n^3}{1 - p_R} \quad \text{and}$$

$$B_{(v,w)}: \quad T_{g(i),(v,w)} \leq T_{r(i-1)} + \frac{4\,\lambda\, n^4 \ln n}{p_R\, r(i-1)\, g(i)^2}\,.$$

Suppose the events $A_{(v,w)}$ and $B_{(v,w)}$ hold for all $(v, w) \in W$. Then, it holds for all $i \in \{0, \ldots, L\}$, that

$$T_{r(i)} - T_{g(i),r(i)} \leq \frac{6\,\lambda\, g(i)\, n^3}{1 - p_R} = \left(\tfrac{2}{3}\right)^{i/3} \cdot \frac{6\,\lambda\, p_R\, n^{3+1/4} \ln^{1/4} n}{p_R\,(1 - p_R)}\,,$$

and, for all $i \in \{1, \ldots, L\}$, that

$$T_{g(i),r(i)} - T_{r(i-1)} \leq \frac{4\,\lambda\, n^4 \ln n}{p_R\, r(i-1)\, g(i)^2} = \left(\tfrac{2}{3}\right)^{i/3} \cdot \frac{6\,\lambda\,(1 - p_R)\, n^{3+1/4} \ln n^{1/4}}{p_R\,(1 - p_R)}\,.$$

Thus, by the geometric series we have

$$T \leq \frac{48}{p_R\,(1-p_R)}\, \lambda\, n^{3+1/4} \ln n^{1/4}\,.$$

Applying the union bound together with Proposition 3.32 and Proposition 3.33 we derive that $A_{(v,w)}$ and $B_{(v,w)}$ hold for all $(v, w) \in W$ with probability at least $1 - 2n^{3-\lambda}$ which concludes the proof. $\square$

In Theorem 3.31, the recombination probability is restricted to the open interval $(0, 1)$. For $p_R = 0$, the EA_{APSP} performs variation only. It was shown in Doerr, Happ and Klein (2008), that in this case the expected optimization time increases to $\Omega(n^4)$ for the K_n where each edge has weight n except for a Hamilton path with edge-weights 1.

For $p_R = 1$, the EA_{APSP} performs recombination only. For this case, it is possible to show an upper bound of $O(|V|^4 \ln |V|)$ on the expected runtime of the EA_{APSP}. This is done by setting $p_R = 1$, $g = 0$ and $r = (\frac{3}{2})^i$ in Proposition 3.33 and the summing over all i in $1, \ldots, \lceil \log_{3/2} |V| \rceil$ like in the proof of the previous theorem.

If we take a closer look at Proposition 3.33, it becomes clear that the $|V|^4$ factor in this expected optimization time results from the lower bound on the probability to concatenate two specific paths in the population $x^{(t)}$ at time t. This bound drops to $|V|^3$, if we restrict the selection operator for recombination to pairs of paths such that the end vertex of the first path is the start vertex of the second. In this case the bound on the expected optimization time drops to $O(|V|^3 \ln |V|)$.

The classical problem-specific algorithm for the APSP is the Floyd-Warshall algorithm (Floyd (1962); Warshall (1962)). This algorithm solves the APSP problem in time $\Theta(|V|^3)$. It uses dynamic programming techniques based on concatenation. Basically, the EA_{APSP} using recombination mimics the FloydWarshall algorithms. Further studies of this capability of the recombination based evolutionary algorithms to simulate dynamic programming is studied in a general context by Doerr, Eremeev, Horoba, Neumann and Theile (2009).

3.12. Conclusion

We have analyzed the optimization time of the (1+1) EA for several of the most common polynomial time solvable problems in combinatorial optimization. For some of these problems, we found that the (1+1) EA simulates known problem-specific algorithms. For others, the (1+1) EA does not find an optimal solution in polynomial time with probability tending to one.

The collection of problems we examined is by no means exhaustive. For example, another combinatorial problem that is solvable in polynomial time is the sorting problem. The (1+1) EA solves this problem in polynomial time (Scharnow, Tinnefeld and Wegener (2004); Doerr and Happ (2008)).

On the other hand, the min-cut problem is a second example of a polynomial time solvable problem for which the (1+1) EA has exponential expected optimization time (Neumann, Reichel and Skutella (2008)).

There also exist a number of runtime result for the (1+1) EA on NP-hard problems. He and Yao (2001) studied the the subset-sum problem, Storch (2006, 2007) the maximum clique problem, Neumann (2007) the multi-objective minimum spanning tree problem, and Horoba (2009) the multi-objective shortest path problem.

Again, not all NP-hard problems are equally accessible to the (1+1) EA. For the special case of the partition problem, Witt (2005) showed that the (1+1) EA can find a 4/3-approximation in quadratic time. He also showed that multiple runs of the (1+1) EA will result in a PRAS.

In contrast to this, Friedrich, Hebbinghaus, Neumann, He and Witt (2007) showed that with probability tending to one, the (1+1) EA for the vertex-cover and set-cover problems does not find a constant factor approximation in polynomial time.

However, under certain conditions this inapproximability can be dealt with by approaches based on multi-objective optimization (Friedrich, Hebbinghaus, Neumann, He and Witt (2007)), multiple runs (Oliveto, He and Yao (2007)), the use of diversity mechanisms (Oliveto, He and Yao (2008)), or hybridization (Friedrich, He, Hebbinghaus, Neumann and Witt (2009)).

For a multi-objective approach, Kratsch and Neumann (2009) conducted a bidimensional analysis using the optimal value as fixed parameter.

For most results in this chapter, there exists also similar analyses of RLS in the respective references. The reason for this is that often the analysis of RLS often guides that of the (1+1) EA. In fact, in most proofs of upper runtime bounds of the (1+1) EA in this chapter we conditioned on one of the two events that the (1+1) EA performs either one particular local variation or two. This means, we basically condition on the event that the (1+1) EA behaves exactly like RLS. Conversely, all upper bounds on the expected optimization times of the (1+1) EA proven in this chapter asymptotically apply to RLS, too, and can be shown using the same proof ideas.

Acknowledgments

I like to thank Thomas Jansen and Pascal Schweitzer for the careful proof-reading and the helpful comments.

References

Baswana, S., Biswas, S., Doerr, B., Friedrich, T., Kurur, P. P. and Neumann, F. (2009). Computing single source shortest paths using single-objective fitness, in *FOGA '09: Proceedings of the 10th ACM workshop on Foundations of Genetic Algorithms* (ACM), pp. 59–66.

Cormen, T. H., Leiserson, C. E., Rivest, R. L. and Stein, C. (2001). *Introduction to algorithms*, 2nd edn. (MIT Press).

Diestel, R. (2005). *Graph theory*, 3rd edn. (Springer).

Dijkstra, E. W. (1959). A note on two problems in connexion with graphs, *Numerische Mathematik* **1**, 1, pp. 269–271.

Doerr, B., Eremeev, A., Horoba, C., Neumann, F. and Theile, M. (2009). Evolutionary algorithms and dynamic programming, in *GECCO '09: Proceedings of the 11th annual Genetic and Evolutionary Computation Conference* (ACM), pp. 771–778.

Doerr, B. and Happ, E. (2008). Directed trees: A powerful representation for sorting and ordering problems, in *CEC '08: Proceedings of the IEEE Congress on Evolutionary Computation* (IEEE), pp. 3606–3613.

Doerr, B., Happ, E. and Klein, C. (2007a). A tight analysis of the (1+1)-EA for the single source shortest path problem, in *CEC '07: Proceedings of the IEEE Congress on Evolutionary Computation* (IEEE), pp. 1890–1895.

Doerr, B., Happ, E. and Klein, C. (2008). Crossover can provably be useful in evolutionary computation, in *GECCO '08: Proceedings of the 10th annual Genetic and Evolutionary Computation Conference* (ACM), pp. 539–546.

Doerr, B., Hebbinghaus, N. and Neumann, F. (2007b). Speeding up evolutionary algorithms through asymmetric mutation operators, *Evolutionary Computation* **15**, 4, pp. 401–410.

Doerr, B. and Johannsen, D. (2007). Adjacency list matchings — an ideal genotype for cycle covers, in *GECCO '07: Proceedings of the 9th annual Genetic and Evolutionary Computation Conference* (ACM), pp. 1203–1210.

Doerr, B. and Johannsen, D. (2010). Edge-based representation beats vertex-based representation in shortest path problems, in *GECCO '10: Proceedings of the 11th annual Genetic and Evolutionary Computation Conference* (ACM), pp. 759–766.

Doerr, B., Klein, C. and Storch, T. (2007c). Faster evolutionary algorithms by superior graph representation, in *FOCI '07: Proceedings of the IEEE Symposium on Foundations of Computational Intelligence* (IEEE), pp. 245–250.

Doerr, B. and Theile, M. (2009). Improved analysis methods for crossover-based algorithms, in *GECCO '09: Proceedings of the 11th annual Genetic and Evolutionary Computation Conference* (ACM), pp. 247–254.

Droste, S., Jansen, T. and Wegener, I. (2002). On the analysis of the (1+1) evolutionary algorithm, *Theoretical Computer Science* **276**, 1-2, pp. 51–81.

Euler, L. (1741). Solutio problematis ad geometriam situs pertinentis, *Commentarii academiae scientiarum Petropolitanae* **8**, pp. 128–140.

Fischer, S. and Wegener, I. (2004). The Ising model on the ring: mutation versus recombination, in *GECCO '04: Proceedings of the 6th annual Genetic and Evolutionary Computation Conference, Part I, Lecture Notes in Computer Science*, Vol. 3102 (Springer), pp. 1113–1124.

Floyd, R. W. (1962). Algorithm 97: shortest path, *Commuications of the ACM* **5**, 6, p. 345.

Friedrich, T., He, J., Hebbinghaus, N., Neumann, F. and Witt, C. (2009). Analyses of simple hybrid algorithms for the vertex cover problem, *Evolutionary Computation* **17**, 1, pp. 3–19.

Friedrich, T., Hebbinghaus, N., Neumann, F., He, J. and Witt, C. (2007). Approximating covering problems by randomized search heuristics using multiobjective models, in *GECCO '07: Proceedings of the 9th annual Genetic and Evolutionary Computation Conference* (ACM), pp. 797–804.

Giel, O. and Wegener, I. (2003). Evolutionary algorithms and the maximum matching problem, in *STACS '03: Proceedings of the 20th annual Symposium on Theoretical Aspects of Computer Science, Lecture Notes in Computer Science*, Vol. 2607 (Springer), pp. 415–426.

Giel, O. and Wegener, I. (2006). Maximum cardinality matchings on trees by randomized local search, in *GECCO '06: Proceedings of the 8th annual Genetic and Evolutionary Computation Conference* (ACM), pp. 539–546.

He, J. and Yao, X. (2001). Drift analysis and average time complexity of evolutionary algorithms, *Artificial Intelligence* **127**, 1, pp. 57–85.

Hierholzer, C. (1873). Über die Möglichkeit, einen Linenzug ohne Wiederholung und ohne Unterbrechung zu umfahren, *Mathematische Annalen* **6**, pp. 30–32.

Horoba, C. (2009). Analysis of a simple evolutionary algorithm for the multiobjective shortest path problem, in *FOGA '09: Proceedings of the 10th ACM workshop on Foundations of Genetic Algorithms* (ACM), pp. 113–120.

Horoba, C. and Sudholt, D. (2009). Running time analysis of ACO systems for shortest path problems, in *SLS '09: Proceedings of the 2nd International workshop on Engineering Stochastic Local Search Algorithms, Lecture Notes in Computer Science*, Vol. 5752 (Springer), pp. 76–91.

Jansen, T. and Sudholt, D. (2010). Analysis of an asymmetric mutation operator, *Evolutionary Computation* **18**, 1, pp. 1–26.

Jarník (1930). O jistém problému minimálním, *Práca Moravské Přírodovědecké Společnosti* **6**, pp. 57–63.

Kratsch, S. and Neumann, F. (2009). Fixed-parameter evolutionary algorithms and the vertex cover problem, in *GECCO '09: Proceedings of the 11th annual Genetic and Evolutionary Computation Conference* (ACM), pp. 293–300.

Kruskal, J. B. (1956). On the shortest spanning subtree of a graph and the traveling salesman problem, *Proceedings of the AMS* **7**, 1, pp. 48–50.

Mehlhorn, K. and Sanders, P. (2009). *Algorithms and data structures: the basic toolbox*, 1st edn. (Springer).

Micali, S. and Vazirani, V. V. (1980). An $O(\sqrt{|V|}\,|E|)$ algorithm for finding maximum matching in general graphs, in *FOCS '80: Proceedings of the 21st annual IEEE Syposium on Foundations of Computer Science* (IEEE), pp. 17–27.

Neumann, F. (2007). Expected runtimes of a simple evolutionary algorithm for the multi-objective minimum spanning tree problem, *European Journal of Operational Research* **181**, 3, pp. 1620–1629.

Neumann, F. (2008). Expected runtimes of evolutionary algorithms for the Eulerian cycle problem, *Computers and Operation Research* **35**, 9, pp. 2750–2759.

Neumann, F., Reichel, J. and Skutella, M. (2008). Computing minimum cuts by randomized search heuristics, in *GECCO '08: Proceedings of the 10th annual Genetic and Evolutionary Computation Conference* (ACM), pp. 779–786.

Neumann, F. and Wegener, I. (2005). Minimum spanning trees made easier via multi-objective optimization, in *GECCO '05: Proceedings of the 7th annual Genetic and Evolutionary Computation Conference* (ACM), pp. 763–769.

Neumann, F. and Wegener, I. (2007). Randomized local search, evolutionary algorithms, and the minimum spanning tree problem, *Theoretical Compututer Science* **378**, 1, pp. 32–40.

Oliveto, P. S., He, J. and Yao, X. (2007). Evolutionary algorithms and the vertex cover problem, in *CEC '07: Proceedings of the IEEE Congress on Evolutionary Computation* (IEEE), pp. 1870–1877.

Oliveto, P. S., He, J. and Yao, X. (2008). Analysis of population-based evolutionary algorithms for the vertex cover problem, in *CEC '08: Proceedings of the IEEE Congress on Evolutionary Computation* (IEEE), pp. 1563–1570.

Prim, R. C. (1957). Shortest connection networks and some generalizations, *Bell System Technology Journal* **36**, pp. 1389–1401.

Raidl, G. R. and Julstrom, B. A. (2003). Edge sets: an effective evolutionary coding of spanning trees, *IEEE Transactions on Evolutionary Computation* **7**, 3, pp. 225–239.

Reichel, J. and Skutella, M. (2007). Evolutionary algorithms and matroid optimization problems, in *GECCO '07: Proceedings of the 9th annual Genetic and Evolutionary Computation Conference* (ACM), pp. 947–954.

Reichel, J. and Skutella, M. (2009). On the size of weights in randomized search heuristics, in *FOGA '09: Proceedings of the 10th ACM workshop on Foundations of Genetic Algorithms* (ACM), pp. 21–28.

Sasaki, G. H. and Hajek, B. (1988). The time complexity of maximum matching by simulated annealing, *Journal of the ACM* **35**, 2, pp. 387–403.

Scharnow, J., Tinnefeld, K. and Wegener, I. (2004). The analysis of evolutionary algorithms on sorting and shortest paths problems, *Journal of Mathematical Modelling and Algorithms* **3**, 4, pp. 349–366.

Storch, T. (2006). How randomized search heuristics find maximum cliques in planar graphs, in *GECCO '06: Proceedings of the 8th annual Genetic and Evolutionary Computation Conference* (ACM), pp. 567–574.

Storch, T. (2007). Finding large cliques in sparse semi-random graphs by simple randomized search heuristics, *Theoretical Computer Science* **386**, 1–2, pp. 114–131.

Warshall, S. (1962). A theorem on Boolean matrices, *Journal of the ACM* **9**, 1, pp. 11–12.

Watson, R. A. and Jansen, T. (2007). A building-block royal road where crossover is provably essential, in *GECCO '07: Proceedings of the 9th annual Genetic and Evolutionary Computation Conference* (ACM), pp. 1452–1459.

Witt, C. (2005). Worst-case and average-case approximations by simple randomized search heuristics, in *STACS '05: Proceedings of the 22nd annual Symposium on Theoretical Aspects of Computer Science*, *Lecture Notes in Computer Science*, Vol. 3404 (Springer), pp. 44–56.

Chapter 4

Theoretical Aspects of Evolutionary Multiobjective Optimization

Dimo Brockhoff

INRIA Saclay — Île-de-France
Bat 490, Université Paris-Sud, 91405 Orsay Cedex, France
dimo.brockhoff@lri.fr,
http://www.lri.fr/~brockho/

Evolutionary multiobjective optimization (EMO), the optimization of problems with multiple objectives by means of evolutionary computation methods, has become one of the main approaches to tackle real-world problems in recent years. Although theory in EMO is less established than for single-objective randomized search heuristics or the classical field of deterministic multiobjective optimization, several important theoretical results have been accomplished in recent years. This chapter gives a broad overview over those theoretical studies obtained in the field while focusing on the topics performance assessment, hypervolume-based search, and rigorous runtime analyses and convergence results.

Contents

4.1. Introduction

Optimization problems in practice often involve the simultaneous optimization of 2 or more conflicting objectives. Evolutionary multiobjective optimization (EMO) techniques are well suited for tackling those multiobjective optimization problems because they are able to generate a set of solutions that represent the inherent trade-offs between the objectives. When the first multiobjective evolutionary algorithms (MOEAs) have been proposed in the mid-1980s, MOEAs have been seen as single-objective evolutionary algorithms where only the selection scheme needed to be tailored towards multiobjective optimization. In the meantime, EMO has become an independent research field with its specific research questions—and its own theoretical foundations. Although theory in EMO is less established than in the field of single-objective evolutionary algorithms or the classical field of deterministic multiobjective optimization, several important theoretical studies have been conducted in recent years which opened up a better understanding of the underlying principles and resulted in the proposition of better algorithms in practice.

Besides a brief introduction about the basic principles of EMO (Sections 4.1.1 and 4.1.2), the main goal of this chapter is to give a general overview of theoretical studies published in the field and to present some of the theoretical results and their proofs in more detail[a]. Due to space limitations, we only focus on three main aspects of previous and current research here: (i) performance assessment with quality indicators (Section 4.2), (ii) hypervolume-based search (Section 4.3), and (iii) rigorous runtime analyses and convergence properties of MOEAs (Section 4.4). Although the restriction to these aspects is due to space limitations, their selection reflects the importance of these topics to the theoretical development in recent years.

[a] The selection of the papers presented here is made as broad and objective as possible although such an overview can never be exhaustive. In particular, the author tried to collect all studies containing theoretical results on EMO published at the major conferences in the field, i.e., FOGA (1999–2009), EMO (2003–2009), GECCO, and CEC (2005–2009) as well as in all journal volumes of the IEEE Transactions on Evolutionary Computing and the Evolutionary Computation Journal. Furthermore, the EMOO web page `http://www.lania.mx/~ccoello/EMOO/` built the basis of the selection. However, due to page limitations, a few of the references found could not be cited here such that the interested reader is referred to the supplementary technical report (Brockhoff, 2009).

4.1.1. *Multiobjective Optimization*

In the following, we assume without loss of generality that k objective functions $f_i : X \to \mathbb{R}$, $1 \leq i \leq k$, mapping a solution x in decision space X to its objective vector $f(x) = (f_1(x), \ldots, f_k(x))$ in the objective space $\mathbb{R}^k$, have to be minimized simultaneously. This yields a major difference to single-objective optimization tasks: in most problems with more than one objective function, no solution exists that minimizes all objective functions simultaneously. Instead of finding or approximating the best objective function value, we consider in the following to find or approximate the set of so-called *Pareto-optimal solutions* representing the best trade-offs between the objectives. To this end, we define the Pareto dominance relation as follows. A solution $x \in X$ is said to *dominate* another solution $y \in X$ if and only if $f_i(x) \leq f_i(y)$ for all $1 \leq i \leq k$ and there exists an $i \in \{1, \ldots, k\}$ such that $f_i(x) < f_i(y)$. We also write $x \prec y$. A solution $x^* \in X$ is then called *Pareto-optimal* if and only if there is no other solution in X that dominates x^*. In the same way, the *weak dominance relation* $\preceq$ can be defined. A solution $x \in X$ weakly dominates a solution $y \in X$ ($x \preceq y$) if and only if $f_i(x) \leq f_i(y)$ for all $1 \leq i \leq k$. Two solutions that are mutually weakly dominating each other are called *indifferent* whereas they are called *non-dominated* if and only if none is weakly dominating the other. Both dominance relations can be generalized to relations between sets of solutions. For example, a solution set $A \subseteq X$ weakly dominates a solution set $B \subseteq X$ if and only if for all $b \in B$ there exists an $a \in A$ such that $a \preceq b$. Specific sets of solutions are the so-called *Pareto set approximations*, which are solution sets of pairwisely non-dominated solutions. More general definitions of the Pareto dominance concepts, e.g., via cones, exist but due to space limitations we refer the interesting reader to text books like (Ehrgott, 2005).

When comparing the differences between single-objective and multiobjective optimization, one viewpoint is the order-relation based view. In single-objective optimization, every solution is mapped to a real value and solutions can always be pairwisely compared via the less or equal relation $\leq$ on $\mathbb{R}$, i.e., the total order $\leq \subseteq \mathbb{R} \times \mathbb{R}$ induces via f an order on the search space X that is a total preorder, i.e., a reflexive, transitive, and total relation. In a multiobjective scenario, the $\leq$ relation is generalized to objective vectors, i.e., $\leq \subseteq \mathbb{R}^k \times \mathbb{R}^k$. Here, the totality is not given, e.g., due to vectors $a, b \in \mathbb{R}^k$ where $f_1(a) < f_1(b)$ but $f_2(a) > f_2(b)$—the relation $\leq$ is only a partial order, i.e., reflexive, antisymmetric, and transitive. Hence, also the induced weak Pareto dominance relation $\preceq \subseteq X \times X$ is not total. In terms of order relations, the search for all Pareto-optimal solutions is equivalent to the search for all minimal elements of the order relation $\prec$. This set of minimal elements, also denoted as the *Pareto set*, is well-defined for a finite search space but also in the case of continuous optimization, the existence of a non-empty Pareto set can be proven if some assumptions on the problem are given, see for example (Henig, 1982) or (Miettinen, 1999, p. 35). The image of the Pareto set under f is called *Pareto front* .

In practice, finding or approximating the Pareto set is not enough—usually a *decision maker* (DM) decides which of the non-dominated solutions found by an optimization algorithm is chosen in the end. Depending on when the DM is involved in the optimization, we distinguish between *a priori*, *a posteriori* and *interactive* methods (Miettinen, 1999). Until recently, research in the field of EMO focused on a posteriori methods—assuming that the DM is involved only after the optimization. Then, the optimization usually aims at a good approximation of the Pareto set that both maps to a region close to the Pareto front and that is diverse in objective space to provide a good set of alternatives to the DM, cf. (Deb, 2001, p. 24). Due to the absence of a DM in most of the EMO research, also theoretical EMO studies assumed the a posteriori scenario as we do here.

A recently proposed view on EMO should also be mentioned here. As the goal in an a posteriori scenario is to find a *set* of solutions, multiobjective problems can be seen as set problems: the search space Ψ is then the set of all Pareto set approximations, i.e., a set of sets, and a set preference relation on Ψ is leading an optimization algorithm towards the minimal elements of this relation on sets. The advantage of this viewpoint is the simple integration of user preferences into the search (Zitzler *et al.*, 2008).

4.1.2. *Multiobjective Evolutionary Algorithms — A Very Brief Chronology*

Since the first usage of evolutionary algorithms for multiobjective optimization in the mid-1980s (see for example Coello Coello *et al.*, 2007), researchers have proposed many evolutionary algorithms that are tailored towards the simultaneous optimization of several objectives. Among the well-established ones, NSGA-II (Deb *et al.*, 2002) and SPEA2 (Zitzler *et al.*, 2002) have to be mentioned here. Both use the Pareto dominance concept as the main selection criterion in an elitist manner where non-dominated solutions are favored over dominated ones. In addition, a second selection criterion establishes diversity among the solutions. However, experimental studies have shown that both algorithms do not scale well if the number of objectives increases and that—although non-dominated solutions are preferred over dominated ones—over time, previously dominated solutions can enter the population again, which results in an oscillating distance to the Pareto front. Several attempts have been made to circumvent this behavior of which the indicator-based algorithms have been shown to produce improved results in practice (Zitzler and Künzli, 2004; Beume *et al.*, 2007). Besides the various approaches to improve the selection criterion of EMO algorithms, only minor attention has been paid to the variation operators so far where the multiobjective version of the CMA-ES is the most noticeable approach (Igel *et al.*, 2007).

Along with the development of new algorithms, progress could also be seen with respect to theoretical results in recent years. Whereas theoretical investigations considered fundamental design issues and the development of more effective and faster algorithms for subroutines of MOEAs in the beginning, nowadays theoretical studies tackle many aspects of EMO. Performance assessment, hypervolume indicator based optimization and the analysis of runtime and approximation properties of simple MOEAs on combinatorial optimization problems are only a few but the most active of them. For a more comprehensive view of the entire field of EMO, in addition to the theoretical results presented in the next sections, we refer to the text books by Deb (2001) and Coello Coello *et al.* (2007).

4.2. Performance Assessment with the Attainment Function and Quality Indicators

With the huge amount of different MOEAs used in practice, it is required to be able to compare the performance of them with respect to both certain test functions and on real-world applications. Unlike in single-objective optimization, where the outcome of an algorithm run is usually the real-valued best function value found so far, standard statistical methods are not applicable in the case of multiobjective optimization where the outcome of an evolutionary multiobjective optimizer is not directly describable by a single real-valued random variable but by a random *set* of real-valued *vectors*. This results in several difficulties in comparison to single-objective optimization: (i) not all resulting sets of objective vectors are comparable; (ii) the size of the sets can vary between different algorithms and even between different runs; (iii) standard statistical approaches cannot be applied directly, e.g., the mean of the sets of objective vectors generated by different runs of an algorithm might lie beyond the Pareto front if the front is concave; (iv) moreover, the comparison of evolutionary multiobjective optimizers' performance always needs to take into account the preferences of a decision maker. To tackle the mentioned difficulties, the attainment function approach and the idea of quality indicators have been proposed which we present from a theoretical point of view here.

4.2.1. *The Attainment Function*

The attainment function approach can be seen as the generalization of the cumulative distribution function $F_{\mathcal{X}}(z) = \Pr(\mathcal{X} \leq z)$ of a real-valued random variable $\mathcal{X}$ with $z \in \mathbb{R}$ to the multiobjective domain. This generalization is necessary since the outcome of an evolutionary multiobjective optimizer cannot be modeled as a single real-valued random variable $\mathcal{X}$ for which the above distribution function allows to define common statistical measures such as the mean or the variance. Instead, the outcome of an evolutionary multiobjective algorithm needs to be modeled by a random *set* $\mathcal{A} = \{A_i \mid A_i \in \mathbb{R}^k, 1 \leq i \leq M\}$ of non-dominated objective vectors the size M of which is also a random variable. The assumption that $\mathcal{A}$ does not contain any dominated solution, i.e., that $A_i \not\leq A_j$ for any $i \neq j$ is not crucial here but simplifies the notations. The attainment function of such a set of non-dominated objective vectors $\mathcal{A}$, first proposed by Fonseca and Fleming (1996) and later on further investigated and generalized by Grunert da Fonseca *et al.* (2001) and Fonseca *et al.* (2005), is defined by the function $\alpha_{\mathcal{A}} : \mathbb{R}^k \to [0, 1]$ with

$$\alpha_{\mathcal{A}}(z) = \Pr(A_1 \leq z \vee A_2 \leq z \vee \ldots \vee A_M \leq z) =: \Pr(\mathcal{A} \preceq \{z\}) ,$$

where $\leq \subseteq \mathbb{R}^k \times \mathbb{R}^k$ is the less or equal relation on the set of all objective vectors and $z \in \mathbb{R}^k$. For an arbitrary objective vector z, the attainment function corresponds to the probability that at least one of the objective vectors produced by an algorithm run dominates z. It is a generalization of the multivariate cumulative distribution function $F_A(z) = \Pr(A \leq z)$ of a vector $A \in \mathbb{R}^k$ with $z \in \mathbb{R}^k$ to which it reduces if only $M = 1$ objective vector per run is produced by the algorithm. In case of only one objective it reduces further to the standard cumulative distribution function $F_{\mathcal{X}}(z) = \Pr(\mathcal{X} \leq z)$ of a real-valued random variable $\mathcal{X}$ where $z \in \mathbb{R}$.

In practice, the attainment function can be estimated by the *empirical attainment function*

$$\alpha_n(z) = \frac{1}{n}\sum_{i=1}^{n} \mathbf{I}\{\mathcal{A}_i \leq z\} ,$$

where $\mathbf{I}\{\cdot\}$ is the indicator function, giving 1 if and only if the argument is true and where the random sets $\mathcal{A}_1, \ldots, \mathcal{A}_n$ correspond to the outcomes of n independent runs of the optimizer. Figure 4.1 shows an example for three sets of two-dimensional objective vectors. Statistical tests can then be run on $\alpha_n(z)$ to reject a null hypothesis like "algorithms Alg_1 and Alg_2 are performing equally" (Fonseca and Fleming, 1996; Knowles, 2002).

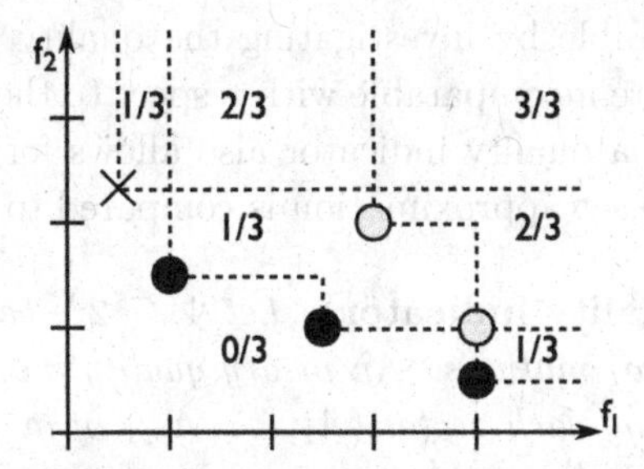

Fig. 4.1. The empirical attainment function values are shown for each part of the two-dimensional objective space for $n = 3$ sets of objective vectors with 1 (cross), 2 (empty circle), and 3 solutions (filled circle). For example, the area on the top right is dominated by all three sets, whereas the objective space close to the cross is either dominated by one (1/3) or no set (0/3).

Besides the definition and interpretation of the attainment function, Grunert da Fonseca *et al.* (2001) also pointed out an interesting relation to random closed set theory: Instead of the set $\mathcal{A} = \{A_i \mid A_i \in \mathbb{R}^k, 1 \leq i \leq M\}$, the outcome of an algorithm can also be described by the region $\mathcal{Y}$ in objective space that is weakly dominated by the solutions associated with $\mathcal{A}$ or in other words that is *attained* by $\mathcal{A}$:

$$\mathcal{Y} = \{y \in \mathbb{R}^k \mid A_1 \leq y \vee A_2 \leq y \vee \ldots \vee A_M \leq y\} .$$

With this alternative representation, the attainment function can be written as $\alpha_{\mathcal{A}}(z) = \Pr(z \in \mathcal{Y})$ for any $z \in \mathbb{R}^k$ and represents the expected value of the binary random field $\mathbf{I}\{\mathcal{Y} \cap \{z\} \neq \emptyset\} = \mathbf{I}\{\mathcal{A} \leq z\}$, see (Grunert da Fonseca *et al.*, 2001) for details. In other words, the attainment function represents the first order moment of the *location* of the objective vectors produced by an algorithm outcome $\mathcal{A}$. The generalization of the attainment function to higher order moments has been investigated as well (Fonseca *et al.*, 2005) but its applicability in practice failed up to now—mainly due to its high dimensionality. Nevertheless, the results on the attainment function belong to the main fundamental theoretical studies in the field of evolutionary multiobjective optimization.

4.2.2. *Quality Indicators*

Quality indicators are functions that map one or several Pareto set approximations to a real value—allowing performance assessment of multiobjective optimizers in a real-valued domain. Since with quality indicators only real numbers have to be compared, two Pareto set approximations can be made always comparable by investigating the quality indicator values of them—even if they are incomparable with respect to the Pareto dominance relation[b]. Moreover, a quality indicator also allows for statements of *how much better* a Pareto set approximation is compared to another one.

Definition 4.1 (Quality indicator). *Let $\Psi \subseteq 2^X$ be the set of all possible Pareto set approximations. An m-ary quality indicator is a function $I : \Psi^m \to \mathbb{R}$, assigning each vector $(A_1, \ldots, A_m)$ of m Pareto set approximations a real value $I(A_1, \ldots, A_m)$.*

An example of a unary ($m = 1$) quality indicator is the general distance measure that assigns a Pareto set approximation A the average Euclidean distance in objective space between all objective vectors $f(a)$ with $a \in A$ and the Pareto front. We should remark here, that the general distance measure is only applicable if the Pareto front is known which is usually not the case in practice. An example of a unary quality indicator where this knowledge about the Pareto front is not needed is the hypervolume indicator or $\mathcal{S}$-metric (Zitzler and Thiele, 1998, see also Section 4.3). It assigns a set A the hypervolume of the objective space dominated by A but which itself is dominating a specified reference set R. The ε-indicator of Zitzler *et al.* (2003) is one example of a binary quality indicator. The ε-indicator value $I_\varepsilon(A, B) = \inf_{\varepsilon \in \mathbb{R}}\{\forall b \in B : \exists a \in A : \forall 1 \leq i \leq k : f_i(a) \leq \varepsilon \cdot f_i(b)\}$ for two Pareto set approximations A, B and k objectives can be interpreted as the smallest value by which all objective vectors of solutions in A have to be

[b]Although all quality indicators allow to compare two Pareto set approximations, not all quality indicators induce an order relation that is meaningful, e.g., that assigns solution sets close to the Pareto set a better value than dominated solution sets. We will discuss this relation to the Pareto dominance relation in more detail later on in this section.

divided such that B is weakly dominated—if A already weakly dominates B, $I_\varepsilon(A,B) \leq 1$ holds whereas $I_\varepsilon(A,B) > 1$ indicates that $A \not\preceq B$.

With the help of quality indicators, the performance assessment of MOEAs usually follows the same theoretical framework:

Definition 4.2 (Comparison method (Zitzler *et al.*, 2003)). *Let $A, B \in \Psi$ be two Pareto set approximations, $\mathbf{I} = \{I_1, \ldots, I_l\}$ a vector of quality indicators and $E : \mathbb{R}^k \times \mathbb{R}^k \to \{false, true\}$ an interpretation function that maps two real vectors to a Boolean value. In case $\mathbf{I}$ contains only unary quality indicators, we define the* comparison method $C_{\mathbf{I},E}$ *as*

$$C_{\mathbf{I},E}(A,B) = E(\mathbf{I}(A), \mathbf{I}(B))$$

and in case $\mathbf{I}$ contains only binary indicators as

$$C_{\mathbf{I},E}(A,B) = E(\mathbf{I}(A,B), \mathbf{I}(B,A))$$

where $\mathbf{I}(A') = (I_1(A'), \ldots, I_l(A'))$ and $\mathbf{I}(A',B') = (I_1(A',B'), \ldots, I_l(A',B'))$ for $A', B' \in \Psi$.

Typical examples of using quality indicators fit to this formalization. For a single unary indicator I for example, $\mathbf{I}(A') = I(A')$ and $E(I(A), I(B))$ is usually defined as $I(A) > I(B)$ if we assume maximization of the indicator function. Thus, we can compare all types of Pareto set approximations by comparing their quality indicator values and interpreting a "true" of the interpretation function E as "A is better than B"—even if the solution sets are incomparable with respect to the dominance relation. In this case, the performance assessment of MOEAs can be done similar to the single-objective case and standard statistical approaches can be used to investigate the differences in indicator values between different algorithms.

However, there is an important fact that one should keep in mind when using quality indicators. Using two different indicators will, in general, result in two different decisions which algorithm is the best and a decision based on an indicator might not even reflect the weak Pareto dominance relation. Thus, one would like to use only those indicators that do not contradict decisions that can be made by the weak Pareto dominance relation itself as it represents the most general form of "outperformance": whenever a Pareto set approximation A is better than another set B with respect to the weak Pareto dominance relation, i.e., when $A \preceq B \wedge B \not\preceq A$, which we denote by $A \lhd B$, we do not want to get a different statement with the comparison method $C_{\mathbf{I},E}(A,B)$. In other words, we would like $C_{\mathbf{I},E}(A,B)$ to be a sufficient criterion for $A \lhd B$ such that $C_{\mathbf{I},E}(A,B)$ can state *that*

A is better than B. We say, the comparison method is compatible with the dominance relation $\lhd$ [c]. If $C_{\mathbf{I},E}(A,B)$ is in addition also a necessary condition for $A \lhd B$, i.e., if $C_{\mathbf{I},E}(A,B) \Leftrightarrow A \lhd B$, the comparison method can indicate *whether* A is better than B which is formalized as follows.

Definition 4.3 (Compatibility and completeness (Zitzler *et al.*, 2003)). Let $\blacktriangleleft$ be an arbitrary binary relation on the set Ψ of Pareto set approximations. The comparison method $C_{\mathbf{I},E}$ is denoted as $\blacktriangleleft$*-compatible* if either

$$\forall A,B \in \Psi : C_{\mathbf{I},E}(A,B) \Rightarrow A \blacktriangleleft B \quad \text{or} \quad \forall A,B \in \Psi : C_{\mathbf{I},E}(A,B) \Rightarrow B \blacktriangleleft A$$

and as $\blacktriangleleft$*-complete* if either

$$\forall A,B \in \Psi : A \blacktriangleleft B \Rightarrow C_{\mathbf{I},E}(A,B) \quad \text{or} \quad \forall A,B \in \Psi : B \blacktriangleleft A \Rightarrow C_{\mathbf{I},E}(A,B) \ .$$

Based on this definition, Zitzler *et al.* (2003) theoretically investigate the restrictions of using a set of only unary indicators, the main result of which we state here while referring to the paper for the proof.

Theorem 4.4 (Zitzler *et al.* (2003)). *For the case of an optimization problem with $k \geq 2$ objectives and objective space $\mathbb{R}^k$, there exists no comparison method $C_{\mathbf{I},E}$ that is based on a finite combination $\mathbf{I}$ of unary quality indicators that is $\lhd$-compatible and $\lhd$-complete at the same time.*

Although this result tells us that there is no comparison method based on unary indicators that yields the equivalence $C_{\mathbf{I},E}(A,B) \Leftrightarrow A \lhd B$, we can still hope for a unary indicator I that is $\not\lhd$-compatible and $\lhd$-complete such that whenever $I(A) > I(B)$ we know that B is not better than A ($C_{I,E}(A,B) \Rightarrow B \not\lhd A$) and on the other hand, the comparison method also detects that A is better than B whenever this is the case ($A \lhd B \Rightarrow C_{I,E}(A,B)$). In this case, we also say the indicator I is *compliant* with the weak Pareto dominance relation. The only unary indicators known so far that have this property are the unary hypervolume indicator I_H and its variants, further investigated in the next section. Many other indicators proposed in the literature, e.g., the mentioned generational distance measure, do not respect the Pareto dominance relation, i.e., they are neither $\blacktriangleleft$-compatible nor $\blacktriangleleft$-complete for any $\blacktriangleleft \in \{\lhd, \preceq, \prec\}$ (Zitzler *et al.*, 2003). However, a few quality indicators besides the hypervolume indicator

[c] Although our notation follows (Zitzler *et al.*, 2003) here, other studies define the compatibility in the same manner (Hansen and Jaszkiewicz, 1998; Knowles, 2002; Farhang-Mehr and Azarm, 2003) or relate it to so-called refinements of the weak Pareto dominance relation in set-based EMO (Zitzler *et al.*, 2008).

are known that are compatible or complete with respect to one of the dominance relations and which should be therefore preferred over indicators that do not respect any dominance relation. Only three of them do not need information about the Pareto front: the average best weight combination, the distance from reference set indicator, and the unary ε-indicator Zitzler *et al.* (2003). These three indicators are all $\ntriangleleft$-compatible and at the same time complete with respect to the strict Pareto dominance relation. Note also that the restriction of the previous theorem does not hold for binary indicators such that with several of them, e.g., the above mentioned binary ε-indicator, comparison methods that are at the same time $\lhd$-compatible and $\lhd$-complete can be constructed (Zitzler *et al.*, 2003).

4.2.3. *Future Research Directions*

With the help of quality indicators, it is possible to use techniques known from single-objective optimization also for multiobjective performance assessment. However, far the most studies nowadays fix the number of evaluations and consider the achieved quality indicator values. Instead, in single-objective optimization, the recent trend is to fix a target value (of the objective function, which would be the quality indicator function in the multiobjective case) and report the number of evaluations to reach it (Hansen *et al.*, 2009). Like that, the reported numbers have an absolute meaning: if an algorithm needs half as many function evaluations to reach a certain target than another one it is twice as fast, whereas reaching an objective function or quality indicator value that is twice as small does not mean the algorithm is twice as fast. Here, a rethinking needs to take place in the EMO field and theoretical investigations of this so-called *horizontal view*[d] of performance assessment might help here.

Another theoretical question is whether there exists a unary indicator that is both $\blacktriangleleft_1$-compatible and $\blacktriangleleft_2$-complete for a certain combination of set preference relations $\blacktriangleleft_1$ and $\blacktriangleleft_2$. For some combinations, Zitzler *et al.* (2003) gave already the answer, but for some combinations, the question is still open, compare Table III in (Zitzler *et al.*, 2003).

Last and in anticipation of the following section, the question which Pareto set approximations yield a maximal value of a given unary quality indicator needs to be investigated. Not only allows this to set appropriate target values in the mentioned horizontal view of performance assessment

[d]If the achieved objective function or quality indicator values for several algorithms are plotted over time, fixing a target value can be seen as comparing the runtimes of the algorithms at a horizontal line located at this fixed target value (Hansen *et al.*, 2009).

but also the bias of algorithms aiming at the optimization of a specific quality indicator can be studied as it is presented for the special case of the hypervolume indicator in the following.

4.3. Hypervolume-based Search

One of the latest areas of EMO where theoretical investigations have been made is the area of hypervolume-based search. The hypervolume indicator, initially proposed for performance assessment, has nowadays been used to guide the search in several MOEAs such as the ESP algorithm (Huband *et al.*, 2003), SMS-EMOA (Beume *et al.*, 2007), the multiobjective version of CMA-ES (Igel *et al.*, 2007), or HypE (Bader and Zitzler, 2008). A few other indicator based algorithms exist that use different indicators, e.g., IBEA (Zitzler and Künzli, 2004), but the fact that the hypervolume indicator is compliant with the Pareto dominance relation makes it the most frequently used indicator in practice.

As mentioned above, the hypervolume indicator is a unary quality indicator mapping a set of solutions to the hypervolume in objective space that is weakly dominated by the corresponding objective vectors of the solutions but at the same time is weakly dominating a so-called *reference point.* More formally, for a given set $A \in 2^X$ of solutions and a reference point $r \in \mathbb{R}^k$ in objective space, we define the set of objective vectors that are weakly dominated by A and that weakly dominate r according to Zitzler *et al.* (2008)[e] as $H(A,r) = \{h \mid \exists a \in A : f(a) \leq h \leq r\}$ and the corresponding hypervolume indicator $I_H(A,r)$ as the Lebesgue measure of this set

$$I_H(A,r) = \lambda(H(A,r)) = \int_{\mathbb{R}^n} \mathbf{1}_{H(A,r)}(z)dz$$

where $\mathbf{1}_{H(A,r)}(z)$ is the indicator function which equals 1 if and only if $z \in H(A,r)$ and 0 otherwise. Zitzler *et al.* (2007) generalized the hypervolume indicator to a weighted version, where a user-specified weight function on the objective space can be used to guide the search towards certain regions and/or points of high interest. With respect to a weight function $w : \mathbb{R}^k \to \mathbb{R}$, the weighted hypervolume indicator $I_H^w(A,r)$ for a solution set $A \subseteq X$ and a reference point $r \in \mathbb{R}^k$ is defined by

$$I_H^w(A,r) = \int_{\mathbb{R}^n} w(z)\mathbf{1}_{H(A,r)}(z)dz \ .$$

[e]The definition in (Zitzler *et al.*, 2008) is more general and includes a *set* of reference points. Here, we restrict the definition to the more common case of one reference point.

The main idea behind all hypervolume-based algorithms is the same, be it with respect to the original or the weighted version: applying the (weighted) hypervolume indicator to a set of solutions reformulates a multiobjective problem as a single-objective one and the aim is to find a solution set, the hypervolume indicator of which is as high as possible. In case of population-based algorithms where the considered number of solutions is finite and corresponds often to the population size μ, the goal is to find or approximate the set of μ solutions that maximizes the hypervolume indicator (Auger *et al.*, 2009c).

In the light of hypervolume-based search, two main lines of theoretical research can be identified. On the one hand, fast and effective algorithms to compute the hypervolume indicator are needed and theoretical research in this area focused on the computational complexity of the hypervolume computation and the search for fast algorithms. On the other hand, a theoretical characterization of how the μ solutions, optimizing the hypervolume indicator, are distributed on the Pareto front builds the basis for understanding hypervolume-based evolutionary algorithms in general and the comparison of algorithms with respect to the hypervolume indicator as performance criterion in particular. The characterization of such so called *optimal μ-distributions* is one of the more recent branches in theoretical evolutionary multiobjective optimization and is going to be presented together with the results on the computational complexity of the hypervolume computation in more detail.

4.3.1. *Computational Complexity of Computing the Hypervolume Indicator*

Right after Zitzler and Thiele (1998) proposed the hypervolume indicator for performance assessment, several algorithms to compute it exactly have been proposed and analyzed in the literature. The first such analysis has been performed by Knowles (2002) while proving the runtime of his algorithm to be $\mathcal{O}(n^k + 1)$ for n solutions and k objectives. Later on, While (2005) proved that a previously proposed algorithm has exponential runtime instead of the claimed polynomial one. While *et al.* (2006) also proposed the *hypervolume by slicing objectives* (HSO) algorithm and thoroughly proved the number of generated hypercubes to be $\binom{k+n-2}{n-1}$, and Fonseca *et al.* (2006) improved the calculation time further by proposing a dimension-sweep algorithm with a runtime of $\mathcal{O}(n^{k-2} \log n)$. The largest improvement in calculating the hypervolume exactly has been finally proposed by Beume and Rudolph (2006) by using a link between the hypervolume indicator calculation and Klee's measure problem which results in a runtime of $\mathcal{O}(n \log n + n^{k/2} \log n)$.

With Klee's measure problem, we associate the calculation of the size of the union of a set of n real-valued intervals. Generalized to an arbitrary dimension k, we ask for the computation of the union of a set of n axis-parallel hypercuboids. The relation to the hypervolume indicator calculation is straightforward, cf. (Beume and Rudolph, 2006): given a set $A \in 2^X$ of n solutions and the hypervolume's reference point $r \in \mathbb{R}^k$, one can build an instance for Klee's measure problem by using the objective vectors as lower bounds and the reference point as upper bound for n k-dimensional intervals or hypercuboids respectively. Therefore, known algorithms for Klee's measure problem can be directly used for the hypervolume indicator computation. Beume and Rudolph (2006) (and later on Beume (2009)) use a slightly modified version of one of such algorithms for Klee's measure problem with a runtime of $\mathcal{O}(n \log n + n^{k/2} \log n)$[f].

After the problem of the worst-case complexity of the hypervolume indicator computation remained open for years—only Beume *et al.* (2009a) proved a non-trivial lower bound of $\Omega(n \log n)$ for a fixed number of k objectives—Bringmann and Friedrich (2008) showed that the problem of computing the hypervolume indicator value for n solutions and k objectives is $\#\mathcal{P}$-hard. This result indicates that we cannot hope for an exact algorithm that is polynomial in the number of objectives unless $\mathcal{P} = \mathcal{NP}$[g].

The proof of the $\#\mathcal{P}$-hardness of the hypervolume indicator computation problem, denoted by `HYP` in the following, is a simple reduction from the $\#\mathcal{P}$-hard problem `#MON-CNF` that is defined as follows.

Definition 4.5. Given a monotone Boolean formula in conjunctive normal form, i.e., of the form $\texttt{CNF} = \bigwedge_{j=1}^{n} \bigvee_{i \in C_j} x_i$ where $C_j \subset \{1, \ldots, k\}$ are the clauses and the k variables x_i are only used in their non-negated forms. Then the problem `#MON-CNF` (`SATISFIABILITY PROBLEM FOR MONOTONE BOOLEAN FORMULAS IN CONJUNCTIVE NORMAL FORM`) asks for the number of satisfying variable assignments for `CNF`.

[f] In the meantime, a slightly faster algorithm for Klee's measure problem has been proposed by Chan (2010) which can be used to slightly reduce the runtime of the algorithm in (Beume and Rudolph, 2006; Beume, 2009) to $\mathcal{O}(n^{k/2} 2^{\mathcal{O}(\log^* n)})$ where $\log^* n$ is the iterated logarithm.

[g] A problem A is called $\#\mathcal{P}$-hard if all problems in $\#\mathcal{P}$ are Turing reducible to A where $\#\mathcal{P}$ is the set of all counting problems that can be solved by a *counting Turing machine*. For details, see (Valiant, 1979).

Theorem 4.6 (Bringmann and Friedrich (2008)). *HYP is #$\mathcal{P}$-hard.*

Proof. We show the #$\mathcal{P}$-hardness of HYP by a polynomial reduction from the #$\mathcal{P}$-hard problem #MON-CNF. To this end, let CNF $= \bigwedge_{j=1}^{n} \bigvee_{i \in C_j} x_i$ be the #MON-CNF instance where $C_j \subset \{1, \ldots, k\}$ are the clauses. We compute the number of satisfying variable assignments—instead for CNF—for the negated formula $\overline{\text{CNF}} = \bigvee_{j=1}^{n} \bigwedge_{i \in C_j} \neg x_i$ and return the number 2^k of all possible variable assignments minus the computed number of variable assignments for $\overline{\text{CNF}}$. For constructing the HYP instance, we introduce for each clause C_j a solution a^j with objective vector $(a_1^j, a_2^j, \ldots, a_k^j) \in \mathbb{R}^k$ where $a_i^j = 1$ if $i \in C_j$ and $a_i^j = 0$ otherwise for each $i \in \{1, \ldots, k\}$ and set the hypervolume's reference point to $(2, \ldots, 2) \in \mathbb{R}^k$. We observe that the hypervolume indicator value of the set of all solutions a^j, $1 \leq j \leq n$, can be written as the hypervolume of a union of boxes of the form

$$B_{(x_1, \ldots, x_k)} = [1 - x_1, 2 - x_1] \times [1 - x_2, 2 - x_2] \times \cdots \times [1 - x_k, 2 - x_k]$$

with $x_i \in \{0, 1\}$ for all $1 \leq i \leq k$. Moreover, $B_{(x_1, \ldots, x_k)}$ is a subset of the hypervolume dominated by all solutions a^j if and only if it is a subset of the hypervolume contribution $[a_1^j, 2] \times \cdots \times [a_k^j, 2]$ of at least one solution a^j if and only if we have $a_i^j \leq 1 - x_i$ for all $i \in \{1, \ldots, k\}$ if and only if $a_i^j = 0$ for all $i \in \{1, \ldots, k\}$ with $x_i \in 1$ if and only if $i \notin C_j$ for all i with $x_i = 1$ if and only if $(x_1, \ldots, x_k)$ satisfies $\bigwedge_{i \in C_j} \neg x_i$ for at least one $j \in \{1, \ldots, n\}$ if and only if $(x_1, \ldots, x_k)$ satisfies $\overline{\text{CNF}}$.

Since the hypervolume of a box $B_{x_1, \ldots, x_k}$ equals 1, we have therefore

$$I_H\Big(\bigcup_{1 \leq j \leq n} a^j, r\Big) = \left|\{(x_1, \ldots, x_k) \in \{0,1\}^k \mid (x_1, \ldots, x_k) \text{ satisfies } \overline{CNF}\}\right| .$$

Thus, a polynomial time algorithm for HYP would result in a polynomial time algorithm for #MON-CNF which proves the theorem. □

Along with this result, Bringmann and Friedrich (2009a) also investigated the computational complexity for related problems, such as finding the solution of a given set with the smallest hypervolume contribution. It turns out that the investigated problems are also hard to compute exactly. Furthermore, Bringmann and Friedrich (2008, 2009a) proposed efficient *approximation algorithms* for calculating the hypervolume indicator and the related problems and proved bounds on their approximation quality and runtime. Also an exact algorithm for computing the set of μ solutions out of $M \geq \mu$ that maximizes the hypervolume indicator has been proposed—reducing the runtime of all previously known algorithms by a factor of

$n^{\min\{M-\mu,d/2\}}$ to $\mathcal{O}(n^{k/2} \log n + n^{M-\mu})$ for $k > 3$ (Bringmann and Friedrich, 2009b). Moreover, further algorithms have been proposed that are useful in hypervolume-based search, e.g., a fast algorithm to compute the hypervolume contributions of each solution (Bradstreet *et al.*, 2008). We refrain from presenting more details here and refer the interested reader to the corresponding literature.

4.3.2. *Optimal μ-Distributions*

Studies on quality indicators have shown that the hypervolume indicator is compliant with the weak Pareto dominance relation (see Section 4.2.2). One implication is that maximizing the hypervolume indicator will result in a set of solutions the objective vectors of which (i) lie on the Pareto front and (ii) even cover the entire Pareto front if the number of solutions is larger or equal than the number of Pareto-optimal solutions (Fleischer, 2003). However, the maximal hypervolume indicator value might not be achievable with a MOEA since its population size is usually bounded above by a constant μ. Then, the question arises which set of μ solutions maximizes the hypervolume indicator, how such a set is distributed on the Pareto front and which indicator value can be achieved with μ solutions only. Moreover, the question arises whether hypervolume-based algorithms that aim at finding a set of μ solutions that maximizes the hypervolume indicator are really converging to the optimal achievable hypervolume value. In the light of this discussion, we denote a set of μ points that have the largest hypervolume indicator value among all sets of μ points as *optimal μ-distribution*[h].

Definition 4.7. A set $X^\mu \subseteq X$ of μ solutions that maximizes the hypervolume indicator, i.e., with

$$X^\mu = \underset{A \subseteq X, |A| = \mu}{\operatorname{argmax}} \{I_H(A, r)\} ,$$

is called *optimal μ-distribution.*

In case of a finite Pareto front, the existence of an optimal μ-distribution is trivial and for the case of a continuous Pareto front, Auger *et al.* (2009c) gave a simple existence proof based on the Extreme Value Theorem. Auger *et al.* (2009c) have also been the first to rigorously study where the points in an optimal μ-distribution are located on the front. This helps to understand the bias, the hypervolume (and especially the weighted hypervolume) is

[h]Note that optimal μ-distributions can be defined with respect to any quality indicator although we only consider optimal μ-distributions for the hypervolume indicator here.

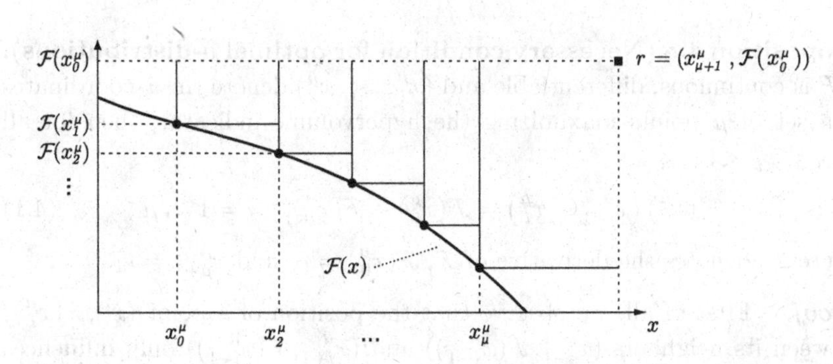

Fig. 4.2. Hypervolume (hatched area) for a set of objective vectors (filled circles) on a biobjective continuous Pareto front. The reference point is depicted by a filled square.

introducing as well as to investigate whether hypervolume-based algorithms converge towards the optimum of the formulated optimization problem, i.e., an optimal μ-distribution.

4.3.2.1. *Properties of Optimal μ-Distributions*

To state the main results about optimal μ-distributions, we introduce the basic notations from (Auger *et al.*, 2009c). All published results in this area only deal with biobjective problems, i.e., $k = 2$ in the remainder of this section. We furthermore assume that the Pareto front can be described by a continuous function $x \in [x_{\min}, x_{\max}] \mapsto \mathcal{F}(x)$ where x lies in the image of the decision space under the first objective f_1 and $\mathcal{F}(x)$ lies in the image of the decision space under f_2, see Figure 4.2. For simplicity, we neglect the decision space completely, i.e., identifying a solution with its objective vector. A solution on the Pareto front can then be—due to the restriction to biobjective problems—unambiguously identified with its first coordinate $x \in [x_{\min}, x_{\max}]$. The hypervolume indicator value of a set of μ solutions on the Pareto front can then be written as

$$I_H^\mu((x_1^\mu, \ldots, x_\mu^\mu), r) := \sum_{i=1}^{\mu} (x_{i+1}^\mu - x_i^\mu)(\mathcal{F}(x_0^\mu) - \mathcal{F}(x_i^\mu))$$

where we define $x_{\mu+1}^\mu := r_1$ and $\mathcal{F}(x_0^\mu) := r_2$ and denote the hypervolume's reference point as $r = (r_1, r_2)$, cf. Figure 4.2.

Based on this formalization, Auger *et al.* (2009c) proved a necessary condition on optimal μ-distributions in terms of a recurrence relation describing the positions of neighbored points:

Proposition 4.8 (Necessary condition for optimal μ-distributions). If $\mathcal{F}$ is continuous, differentiable and $(x_1^\mu, \ldots, x_\mu^\mu)$ denote the x-coordinates of a set of μ points maximizing the hypervolume indicator, then for all $x_{\min} < x_i^\mu < x_{\max}$

$$\mathcal{F}'(x_i^\mu)\left(x_{i+1}^\mu - x_i^\mu\right) = \mathcal{F}(x_i^\mu) - \mathcal{F}(x_{i-1}^\mu),\ i = 1 \ldots \mu \qquad (4.1)$$

where $\mathcal{F}'$ denotes the derivative of $\mathcal{F}$, $\mathcal{F}(x_0^\mu) = r_2$ and $x_{\mu+1}^\mu = r_1$.

Proof. First of all, we observe that the position of a point $(x_i^\mu, \mathcal{F}(x_i^\mu))$ between its neighbors $(x_{i-1}^\mu, \mathcal{F}(x_{i-1}^\mu))$ and $(x_{i+1}^\mu, \mathcal{F}(x_{i+1}^\mu))$ only influences the hypervolume with respect to the objective space that is solely dominated by $(x_i^\mu, \mathcal{F}(x_i^\mu))$. The volume of this hypervolume contribution of the point $(x_i^\mu, \mathcal{F}(x_i^\mu))$ equals

$$H_i = (x_{i+1}^\mu - x_i^\mu)(\mathcal{F}(x_{i-1}^\mu) - \mathcal{F}(x_i^\mu)) \ .$$

For an optimal μ-distribution, each of the contributions H_i has to be maximal with respect to x_i^μ since otherwise one would be able to increase I_H^μ by moving x_i^μ. Since the hypervolume contribution becomes zero for x_i^μ at the boundary of $[x_{i-1}^\mu, x_{i+1}^\mu]$, the maximum must lie in the interior of the domain. Therefore, the necessary condition holds that the derivative of H_i with respect to x_i^μ is zero or x_i^μ is an endpoint of $\mathcal{F}$, i.e., either $x_i^\mu = x_{\min}$ or $x_i^\mu = x_{\max}$. The derivative of H_i with respect to x_i^μ equals

$$H_i'(x_i^\mu) = -x_{i+1}^\mu \mathcal{F}'(x_i^\mu) - \mathcal{F}(x_{i-1}^\mu) + \mathcal{F}(x_i^\mu) + x_i^\mu \mathcal{F}'(x_i^\mu) \ .$$

Reorganizing the terms and setting $H_i'(x_i^\mu) = 0$, we obtain Equation 4.1. □

The recurrence relation of Theorem 4.1 can be directly used to show that saddle points, i.e., points where the derivative of the function equals zero, are never included in an optimal μ-distribution.

Corollary 4.9. *If $x_i^\mu, i = 2 \ldots \mu - 1$ is a point of a set of μ points maximizing the hypervolume indicator and x_i^μ is not an endpoint of the Pareto front, then $\mathcal{F}'(x_i^\mu) \neq 0$.*

Here, as well as for the following results, we refrain from showing the proof due to space restrictions and refer to the original publication instead. Besides these general results, Auger *et al.* (2009c) argue that the characterization of optimal μ-distributions is not an easy task and therefore pursue their study on the one hand by proving results for special front shapes, i.e., for linear fronts, and on the other hand by investigating the case where μ goes to infinity and the optimal distribution of μ points on the Pareto front converges to a density. For the case of linear fronts, Auger *et al.* (2009c) proved the uniqueness of the optimal μ-distribution and gave a formula describing the location of the corresponding μ points exactly.

Theorem 4.10 (Auger *et al.* (2009c)). *If the Pareto front is a (connected) line, optimal μ-distributions are such that the distance is the same between all neighbored solutions.*

This result covers earlier results on even more special cases by Emmerich *et al.* (2007) and Beume *et al.* (2009a) where the slope of the linear front equals 1 and where the extreme points of the front are always assumed to be included in the optimal μ-distribution. In the same manner, Friedrich *et al.* (2009) compute optimal μ-distributions analytically for Pareto fronts that can be described by a function $f(x) = c/x$ in an interval $[1, c]$ for a constant $c > 1$.

As to the results for μ going to infinity, i.e., results on the density $\delta(x)$ of points on the Pareto front at a point $(x, \mathcal{F}(x))$, we only state the most general result for the weighted hypervolume indicator informally due to space limitations. For a fixed integer μ, Auger *et al.* (2009b) consider a sequence of μ ordered points on the Pareto front, for which their x-coordinates are $x_1^\mu, \ldots, x_\mu^\mu$. Then, the authors assume that this sequence converges—when μ goes to ∞—to a density $\delta(x)$ where the density in a point x is formally defined as the limit of the number of points contained in a small interval $[x, x+h[$ normalized by the total number of points μ when both μ goes to ∞ and h to 0, i.e., $\delta(x) = \lim_{\substack{\mu\to\infty \\ h\to 0}} \left(\frac{1}{\mu h} \sum_{i=1}^{\mu} \mathbf{1}_{[x,x+h[}(x_i^\mu) \right)$. Then, one can prove that the density δ corresponding to an optimal μ-distribution equals

$$\delta(x) = \frac{\sqrt{-\mathcal{F}'(x) w(x, \mathcal{F}(x))}}{\int_0^{x_{\max}} \sqrt{-\mathcal{F}'(x) w(x, \mathcal{F}(x))} dx} .$$

When looking carefully at this formula for the density, we see that for a fixed weight w, the density of points only depends on the slope of the front and not on whether the front is convex or concave as in previous belief.

The results in (Auger *et al.*, 2009c,a) do not only allow for general statements of the hypervolume indicator like the one above but also to numerically approximate an optimal μ-distribution and the corresponding indicator value. This is highly useful to investigate the performance of so-called hypervolume-based algorithms that aim at optimizing the hypervolume indicator explicitly by using the indicator as the main selection criterion.

Nevertheless, it remains to investigate whether such hypervolume-based algorithms converge to an optimal μ-distribution.

4.3.2.2. *Hypervolume-Based Selection and the Convergence to Optimal μ-Distributions*

Optimally, having generated λ offspring from μ parents in a hypervolume-based algorithm, one would like to choose the next population as the set of μ solutions out of all $\mu + \lambda$ solutions that maximizes the hypervolume indicator. Like this, one can easily guarantee the convergence to an optimal μ-distribution if $\lambda \geq \mu$ and every point in the search space can be sampled with a positive probability[i]. This strategy, however, is computationally expensive as, in principle, $\binom{\mu+\lambda}{\mu}$ solution sets have to be considered. Although Bringmann and Friedrich (2009b) proposed a faster algorithm, its runtime is still exponential in k and a faster strategy is needed in practice.

Such a greedy, but not always optimal strategy is used in most known hypervolume-based algorithms: When a population P of $\mu+\lambda$ solutions has to be reduced to μ solutions, the solution s with the smallest hypervolume loss $d(s) := I_H(P, r) - I_H(P \setminus \{s\}, r)$ is deleted iteratively until the desired size is reached where the hypervolume loss is recalculated every time a solution is deleted. Algorithm 10 shows a general framework of such a hypervolume-based algorithm with greedy strategy in terms of the Simple Indicator Based Evolutionary Algorithm (SIBEA) (Zitzler *et al.*, 2007). To be even more efficient, a Pareto dominance based ranking is used in SIBEA and other hypervolume-based algorithms before the reduction, such that only pairwisely incomparable solutions are taken into account within P. Note that in general, other strategies to optimize the hypervolume indicator are known, e.g., the k-greedy strategy of (Zitzler *et al.*, 2008) or the one in HypE (Bader and Zitzler, 2008). However, practically relevant algorithms such as ESP, SMS-EMOA and MO-CMA-ES are using the described greedy strategy which explains the interest of theoretical studies into the greedy approach.

One of the basic questions for such hypervolume-based algorithms with greedy environmental selection, for which often $\lambda = 1$ is chosen to reduce the runtime further, is whether they converge to an optimal μ-distribution.

---

[i] Under these circumstances, the probability to sample μ new solutions with a better hypervolume indicator value than the current μ parents is always positive when no optimal μ-distribution is found.

Algorithm 10 Simple Indicator-Based Evolutionary Algorithm (SIBEA)

Given: population size μ; number of generations N

Step 1 (Initialization): Generate an initial set of decision vectors P of size μ; set the generation counter $m := 0$

Step 2 (Environmental Selection): Iterate the following three steps until the size of the population does no longer exceed μ:

(1) Rank the population using dominance depth (number of dominating solutions, (Zitzler *et al.*, 2004)) and determine the set of solutions $P' \subseteq P$ with the worst rank

(2) For each solution $x \in P'$ determine the loss of hypervolume $d(x) = I_H(P', r) - I_H(P' \setminus \{x\}, r)$ if it is removed from P'

(3) Remove the solution with the smallest loss $d(x)$ from the population P (ties are broken randomly)

Step 3 (Termination): If $m \geq N$ then output P and stop; otherwise $m := m+1$.

Step 4 (Mating): Randomly select λ elements from P as temporary mating pool Q. Apply variation operators such as recombination and mutation to Q which yields Q'. Set $P := P + Q'$ (multi-set union) and continue with Step 2.

Theorem 4.11 (Zitzler *et al.* (2008)). *Hypervolume-based evolutionary algorithms with a greedy environmental selection step as in Algorithm 10 do not converge to the optimal μ-distribution in general.*

The proof is based on an example with four solutions of a biobjective problem where the greedy heuristic is not able to find the optimal μ-distribution in finite time, see (Zitzler *et al.*, 2008). This result can be seen as a major drawback of current hypervolume-based evolutionary algorithms. However, algorithms that use the greedy strategy such as SMS-EMOA and MO-CMA-ES are known to work well for applications in practice which might indicate that the above example is quite artificial and rare. To further investigate this question, Beume *et al.* (2009b) generalized the example from (Zitzler *et al.*, 2008) to a continuous one showing that the non-convergence of the greedy strategy is not restricted to a single pathological example. Moreover, Bringmann and Friedrich (2009b) showed that the greedy strategy is not only unable to converge to an optimal μ-distribution in general but that the set of λ deleted solutions can even have a hypervolume indicator value that is arbitrarily far from the best achievable value.

4.3.3. *Future Research Directions*

Although the recent complexity results for the hypervolume indicator calculation reduced the chances of fast exact algorithms, enhancements of the current algorithms are desirable—using Monte Carlo sampling is a first step towards this goal (Bader and Zitzler, 2008; Bringmann and Friedrich, 2008, 2009a). Also the use of specialized data structures to store the hypervolume boxes might decrease the actual runtime of exact algorithms further as it was done in (Bringmann and Friedrich, 2009b).

With respect to optimal μ-distributions, the generalization of the biobjective results to problems with more than 2 objectives might give insights on the differences between problems with a few and many objectives. Although Friedrich *et al.* (2009) already investigated the connection between optimal μ-distributions and the approximation quality with respect to the multiplicative ε-indicator[j], the high number of practically relevant quality indicators leaves enough room for further studies. The transfer of the optimal μ-distribution concept to other quality indicators will also help to better understand the influence of quality indicators on the performance assessment. Furthermore, research about convergence properties and runtime analyses of indicator based evolutionary algorithms in general are needed to argue towards their application in practice.

Last, we would like to mention to investigate the influence of different selection strategies in hypervolume-based algorithms. It is known that theoretically, the often used greedy selection can yield arbitrary bad approximation sets and also the non-convergence results indicate that hypervolume-based algorithms might fail in general. However, further theoretical studies about *when* this happens and whether other strategies, such as HypE (Bader and Zitzler, 2008), can circumvent such a behavior need to be carried out.

4.4. Runtime Analyses and Convergence Properties

For the sake of completeness, we also want to state the main results in the most active branch of theoretical EMO studies: convergence and runtime analyses. Due to the limited space, we refrain from stating details but instead give an overview of what has been achieved in the last years.

4.4.1. *Convergence Results*

The first theoretical papers in the EMO field dealt with the convergence properties of evolutionary algorithms. Convergence of an algorithm A in

[j] In more detail, the authors investigated the differences between optimal μ-distributions for the hypervolume indicator and solution sets of μ points that yield an optimal approximation ratio, or in other words optimal μ-distributions for the multiplicative ε-indicator with the smallest possible ε which depends on μ.

the multiobjective case is given if the population of A contains the Pareto set with probability 1 if time goes to infinity. In other words, the distance between the population and the minimal elements of the Pareto dominance relation goes to zero if time goes to infinity. However, depending on whether the search space is discrete and finite or continuous, the used distance measures and therefore the mathematical formulation of convergence differs. We refer to the papers mentioned below for details.

Many convergence results of multiobjective evolutionary algorithms are due to Rudolph. He and his co-authors investigated algorithms for which at least one solution is Pareto-optimal after a finite number of steps on finite and discrete search spaces (Rudolph, 1998a; Rudolph and Agapie, 2000) as well as algorithms with a fixed population size on finite search spaces where all solutions in the population converge to the Pareto set (Rudolph, 2001a). Rudolph also provided convergence results of multiobjective optimizers in continuous domain (Rudolph, 1998b) and investigated noisy and interval-based MOEAs and their convergence (Rudolph, 2001b). Early results on the convergence of MOEAs are also provided by Hanne (1999). Here, the focus lies on different selection schemes, the possibilities of temporary fitness deterioration, and problems with unreachable solutions.

Later on, Yan *et al.* (2003) proved the convergence of a MOEA which has been specifically developed for solving the traveling salesperson problem. Also differential evolution approaches (Xue *et al.*, 2005) and other meta-heuristics (Villalobos-Arias *et al.*, 2005) for multiobjective optimization have been investigated. Recently, even the convergence properties of quantum-inspired algorithms have been investigated (Li *et al.*, 2007). The latest result is a convergence proof for an algorithm that approaches the Pareto set both from the feasible and the infeasible region (Hanne, 2007).

All mentioned theoretical studies on the convergence of multiobjective algorithms have in common that they use standard probability theory and/or the theory of Markov chains. As the first main result, we can state that one necessary condition for convergence in finite search spaces is—similar to single-objective optimization—the fact that the variation operators can produce any solution with positive probability (Rudolph, 1998a). A second interesting result states that a simple elitist strategy does not converge to the Pareto set if the population size is fixed to μ (Rudolph, 2001a). It is worth to mention that both NSGA-II and SPEA2 use a similar type of elitism and are known to diverge in practice if the number of objectives is large. Last, we mention that for continuous search spaces, single-objective results, e.g., regarding the choice of the step-size rule, cannot be fully transferred to the multiobjective case (Rudolph, 1998b).

In contrast to the convergence results mentioned above, Teytaud (2007) investigated the *convergence rate* of multiobjective evolutionary algorithms, i.e., the speed of convergence or in other words the (asymptotic) time until the distance between the population and the Pareto set reaches a certain precision[k]. More precisely, Teytaud showed a general lower bound for the computation time of comparison-based algorithms until a certain precision of the Haussdorff distance is found and an upper bound for a simple random search. The results imply that no comparison-based MOEA performs better than random search if the number of objectives is large.

4.4.2. *The First Runtime Analyses, Common Algorithms, and Special Proof Techniques*

In terms of runtime analyses of multiobjective evolutionary algorithms in discrete domain such as $X = \{0,1\}^n$, the goal is to prove bounds on the *expected runtime*, i.e., the expected number of objective function evaluations until the population of an algorithm contains the entire Pareto set or an approximation thereof, especially if the Pareto set is too large.

The main difference between runtime analyses of single- and multiobjective evolutionary algorithms is that in single-objective optimization only one search point with the smallest objective function value has to be found whereas in the multiobjective case almost always a set of solutions is sought. The investigated algorithms for evolutionary multiobjective optimization are therefore population-based. The simplest of such algorithms, comprising their single-objective counterparts randomized local search (RLS) and (1+1)EA, are the local and the global version of the Simple Evolutionary Multiobjective Optimizer. The local version is simply denoted by SEMO and is detailed for minimization in Algorithm 11 whereas Global SEMO refers to the global version. After sampling a first solution uniformly at random in the search space X, in each iteration of the algorithms, a parent x is drawn uniformly at random from the population P and an offspring x' is generated from x by either flipping one randomly chosen bit (SEMO) or each bit with probability $1/n$ (Global SEMO). If x' is not weakly dominated by any other solution in P, the new solution x' is added to P and

[k]In general, any distance between the population and the Pareto set or the Pareto front can be considered. In his work, Teytaud actually used the Haussdorff distance in objective space, i.e., a distance measure with respect to the Pareto front.

Algorithm 11 Simple Evolutionary Multiobjective Optimizer (SEMO, Laumanns *et al.* (2002b))

Choose an initial individual x uniformly at random from $X = \{0,1\}^n$
$P \leftarrow \{x\}$
loop
  Select one element x uniformly at random from P
  Create offspring x' by flipping a randomly chosen bit of x
  $P \leftarrow P \setminus \{z \in P \,|\, x' \prec z\}$
  if $\nexists z \in P$ such that $(z \prec x' \vee f(z) = f(x'))$ **then**
    $P \leftarrow P \cup \{x'\}$
  end if
end loop

all solutions dominated by x' are deleted from P. This guarantees that the algorithms converge with probability 1 to a set covering the entire Pareto front which, at the same time, implies that the population size is not bounded by a constant as in most of the algorithms used in practice.

SEMO was introduced by Laumanns *et al.* (2002b) to perform the first runtime analysis of a multiobjective evolutionary algorithm on a simple test problem called LeadingOnesTrailingZeros (LOTZ)[1]. Here, as well as in several other studies, the analyses of the algorithm is divided into two parts—the first phase covers the time until the first Pareto-optimal point is found and the second phase stops if the entire Pareto set is found.

Theorem 4.12 (Laumanns *et al.* (2002b)). *Let the biobjective maximization problem* LOTZ : $\{0,1\}^n \to \mathbb{N}^2$ *be defined as*

$$\mathrm{LOTZ}(x_1,\ldots,x_n) = \left(\sum_{i=1}^{n}\prod_{j=1}^{i} x_j, \sum_{i=1}^{n}\prod_{j=i}^{n}(1-x_j)\right).$$

The expected runtime until SEMO finds the entire Pareto set for LOTZ *is* $\Theta(n^3)$.

Proof. First of all, we investigate the problem itself in more detail; see also Figure 4.3. The decision space can be partitioned into $n+1$ sets $S_0 \ldots, S_n$ with $S_i = \{x \in X \,|\, f_1(x) + f_2(x) = i\}$, i.e., S_i contains all decision vectors of the form $1^a 0 *^{(n-i-2)} 1 0^b$ with $a+b=i$ for $i<n$ and $1^a 0^b$ with $a+b=n$ for the set S_n which coincides with the Pareto set of size $n+1$. The set S_{n-1} is empty. All solutions within a set S_i are incomparable whereas for

[1]Although in the remainder of the chapter we consider minimization problems only, the following result is stated in its original version of maximization.

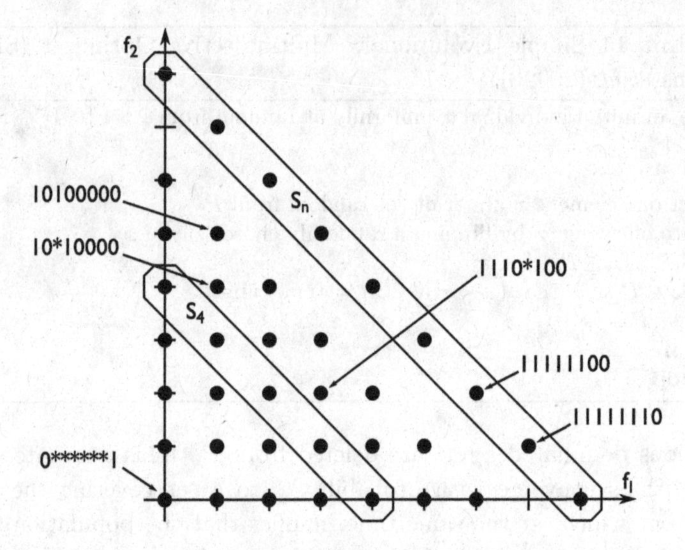

Fig. 4.3. Objective space and corresponding decision vectors for the problem LOTZ with $n = 8$ bits; a '$*$' can be either a one or a zero. Exemplary, the sets S_4 and S_n are shown.

each solution in set S_i with $i < n$ there is at least one solution in each set S_j with $i < j$ which dominates it.

For the runtime analyses, we first consider the time until the first Pareto-optimal point is found. During this first phase, the population consists of one individual only since the mutation of a single bit either produces a dominating or a dominated solution in another set S_i. Therefore, a new solution x' that is added to the population of SEMO will replace its parent x if and only if $x \in S_i$ and $x' \in S_j$ with $i < j$. As there is always a one-bit mutation with probability $1/n$ from the current solution in S_i to one in a higher set S_j, the waiting time to leave S_i is $\mathcal{O}(n)$. The time until S_n is reached is thus $\mathcal{O}(n^2)$ since there are at most $\mathcal{O}(n)$ such steps necessary.

Once a Pareto-optimal solution is found, we can derive tight bounds for the time until the entire Pareto set is found. We observe that the population is always contiguous on the Pareto front and that we can generate a new Pareto-optimal point by sampling a solution the objective vector of which is one of the outmost of the population and flipping a single bit. The probability to sample an outmost point is at least $1/i$ and at most $2/i$ if i Pareto-optimal solutions are found and the probability to sample a new Pareto-optimal point from there is at least $1/n$ and at most $2/n$. By summing up the expected runtimes $E(T_i)$ for finding the $(i{+}1)$th Pareto-optimal solution which is at least $i/2 \cdot n/2 = in/4$ and at most in, we bound the expected waiting time $E(T)$ until the entire Pareto set is found by $1/8(n^3 - n^2) \leq \sum_{i=1}^{n-1} ni/4 \leq E(T) = \sum_{i=1}^{n-1} E(T_i) \leq \sum_{i=1}^{n-1} ni \leq 1/2(n^3 - n^2)$. □

The proof techniques in this very first proof on the runtime of multiobjective evolutionary algorithms are seemingly the same than in runtime analyses of single-objective algorithms. The same holds for the first analyses of Global SEMO on LOTZ by Giel (2003) and the runtime analysis of SEMO on another simple biobjective test problem called multiobjective counting ones (MOCO, (Thierens, 2003)). Also in the first study comparing single-objective approaches and SEMO by Giel and Lehre (2006), common proof techniques of single-objective runtime analyses have been used to show that a real multiobjective approach like SEMO can improve the runtime drastically compared to single-objective evolutionary algorithms. However, a few proof techniques which are tailored towards multiobjective problems have been proposed which we will describe briefly in the following.

Two proof techniques are due to Laumanns *et al.* (2004b) who analyzed SEMO and two other simple algorithms on several simple test problems. The first of these proof techniques can be seen as a generalization of the fitness-based partitioning approach used frequently in single-objective runtime analyses, see (Wegener, 2003). The idea is to partition the decision space into l disjoint sets $S_1, \ldots, S_l$ where S_l coincides with the Pareto set. If for arbitrary two sets S_i and S_j with $i, j \neq l$ all solutions in S_j dominate all solutions in S_i, a lower bound on the probability to leave S_i gives an upper bound on the expected number of times mutations are performed on non-Pareto-optimal solutions. This so-called *decision space partitioning* approach can mainly be used to bound the time until the first Pareto-optimal solution is found as for example in (Laumanns *et al.*, 2004b,a).

To prove bounds on the time until all Pareto-optimal solutions are found if one of them is known already, a second technique called *general graph search* is helpful (Laumanns *et al.*, 2004b,a). Given a fully connected weighted graph $G = (V, E)$ with nodes V, edges $E = V \times V$, and weights $w : E \to \mathbb{R}$ for the edges, we identify the nodes with the Pareto-optimal solutions and the edges with mutations between them. The weights of the edges correspond to the mutation probabilities of the associated mutations. Instead of analyzing the full optimization process on the entire search space, the general graph search method restricts the analysis to the mutations between the Pareto-optimal solutions by analyzing a simplified graph search algorithm that starts by visiting a randomly chosen node, marks all visited nodes, and iteratively jumps from a—according to the evolutionary algorithm's mating selection operator—selected marked node to a neighbor that is chosen uniformly at random until all nodes are visited. Instead of investigating *all* mutation probabilities between Pareto-optimal solutions to do the runtime analysis, the following Lemma allows to bound the runtime until all Pareto-optimal solutions are found even if the graph's weights, i.e., the mutation probabilities, are known only on a *spanning tree* on G.

Lemma 4.13 (General graph search (Laumanns *et al.*, 2004a)). *If the edge weights of a spanning tree on $G = (V, E)$ can be lower bounded by p then the above described graph search algorithm has found all nodes and edges of G after $(c+1)\frac{|V|}{p}\ln|V|$ jumps with probability at least $1-|V|^{-c}$ and the expected number of jumps is bounded by $\mathcal{O}(\frac{|V|}{p}\log|V|)$.*

The general graph search approach is nicely applied to the runtime analyses of the Fair Evolutionary Multiobjective Optimizer (FEMO) with a global mutation operator on a 0-1 knapsack problem in (Laumanns *et al.*, 2004a). There, the considered spanning tree on the Pareto set of size $\Theta(n^2)$ is constructed by connecting search points with Hamming distance 2 which bounds the mutation probability by $\Omega(1/n^2)$ and results in a runtime bound of $\mathcal{O}(n^4 \log n)$. Note, that the way a marked node is selected by the graph search algorithm depends on the investigated evolutionary algorithm.

Another proof technique that is tailored towards the analysis of the diversity maintaining evolutionary multiobjective optimizer (DEMO) has been proposed by Horoba (2009). Similar to the fitness-based partitions method, Horoba (2009) proved a theorem that upper bounds the optimization time of DEMO until it finds an approximation of the Pareto set if three assumptions on the population and the mutation operator are fulfilled. Due to space limitations, we refer to (Horoba, 2009) for details.

4.4.3. *Other Runtime Analysis Results*

While the first runtime analyses of multiobjective evolutionary algorithms only investigated simple test functions, several practically relevant combinatorial optimization problems and other aspects of evolutionary multiobjective optimization have been investigated in terms of rigorous runtime analyses in the meantime. Due to space limitations, we only mention the studies and their results briefly here.

Combinatorial Optimization Problems: Similar to the development in single-objective optimization, most of the recent studies on the runtime analyses of MOEAs investigate combinatorial optimization problems. Table 4.1 on page 130 presents an overview of the main results, the used algorithms and the used proof techniques for all publications on multiobjective runtime analyses that can be found in the literature[m].

Hypervolume-based Search: Although many algorithms aiming at optimizing the hypervolume indicator are known in practice, only one study analyzed the runtime of SIBEA on two simple test problems so far (Brockhoff *et al.*, 2008). The results show[n] that optimizing the hypervolume indicator by always generating $\lambda = 1$ new offspring per generation might be sufficient for some problems although the general convergence of such algorithms cannot be guaranteed, cf. Section 4.3.2.2.

Multiobjectivization: Multiobjectivization, i.e., the addition of objectives to a problem or the decomposition of a single objective into two or more objectives, has been originally proposed from a practical point of view to speed-up the performance of MOEAs by providing additional information that might guide the search towards the Pareto set. However, several studies questioned this approach and argued that increasing the number of objectives will in general increase the difficulty of the problem for MOEAs due to the increased number of incomparable solution pairs. The recent runtime analyses by Brockhoff *et al.* (2009) for the addition of objectives and by Handl *et al.* (2008) for the decomposition of objectives show that multiobjectivization can provable reduce the runtime of Global SEMO—depending on the objective function that is added or how the decomposition is performed. Furthermore, Brockhoff *et al.* (2009) showed that two equally difficult single-objective problems can be solved faster simultaneously when combined to a biobjective problem.

EMO Operators and Mechanisms: The impact of single operators or of a specific concept of EMO algorithms on the runtime has been the focus of other recent studies where different algorithms that only differ in a single

[m]For a description of the proof techniques that are not described in this chapter, we refer to (Wegener, 2003) and (Neumann and Wegener, 2007).

[n]Note that there is some ongoing discussion among the authors whether the proof for the large Pareto front is fully correct.

Table 4.1. Overview of runtime analyses for combinatorial multiobjective optimization problems in the literature. The abbreviations for the proof techniques are coupon collector theorem (CCT), variants of decision space partitioning (DSP), expected multiplicative weight decrease (EMWD), general graph search (GGS), potential functions (PF), and typical runs (TR).

Problem	Publications	Main result	Proof techniques
Knapsack problems (KP)	(Laumanns *et al.*, 2004a)	runtime of Global SEMO and Global FEMO on a biobjective KP	DSP,TR, GGS
	(Kumar and Banerjee, 2006)	runtime of REMO on LOTZ and a quadratic function; runtime of REMO until $(1+\varepsilon)$-approximation for 0-1-KP is found	DSP,TR
Minimum spanning trees (MST)	(Neumann, 2007)	runtime of Global SEMO on two multiobjective formulations of MST	DSP
	(Neumann and Wegener, 2006)	multiobjective formulation results in lower runtime for SEMO and Global SEMO than for RLS/(1+1)EA	DSP,PF
Cutting problems	(Neumann *et al.*, 2008)	runtime of RLS and (1+1)EA on minimum cut problem; runtime of Global SEMO and DEMO on a biobjective version (min. costs and flow)	TR,EMWD
	(Neumann and Reichel, 2008)	runtime until Global SEMO and DEMO find an approximation of the front for a multiobjective formulation of a minimum multicut problem	DSP,PF, EMWD
Shortest path problems (SP)	(Scharnow *et al.*, 2004)	runtime of a specialized (1+1)EA on multiobjectivized version of single source SP with $n-1$ objectives	DSP
	(Horoba, 2009)	DEMO is FPRAS for single source multiobjective SP with k weights per edge; new fitness based partition approach	DSP
covering problems and plateau functions	(Friedrich *et al.*, 2007a)	runtime and approximation results for RLS and (1+1)EA with lexicographic ordering and SEMO and Global SEMO on a biobjective vertex cover problem and on a biobjective set cover problem	CCT,PF, TR,EMWD
	(Friedrich *et al.*, 2007b)	runtime of (1+1)EA with lexicographic order is polynomial and for Global SEMO exponential on a plateau function and on a biobjective SETCOVER instance	TR,DSP, EMWD
	(Kratsch and Neumann, 2009)	parameterized complexity results of Global SEMO on two biobjective formulations of vertex cover	DSP,PF
Graph bisectioning problem	(Greiner, 2009)	runtime analyses on two instances of the graph bisectioning problem where either (1+1)EA or Global SEMO (with the number of free nodes as second objective) is faster	DSP,PF, TR

operator or mechanism have been analyzed theoretically. Horoba and Neumann (2008, 2009), for example, investigated the diversity mechanisms in Global SEMO, in an algorithm called Global DEMO$_\varepsilon$ (Diversity Evolutionary Multi-objective Optimizer) and in a simplified version of SPEA2. The rigorous analyses of the time until an ε-approximation of the Pareto front is found indicate that there is at least one problem for which every investigated diversity mechanism fails to find a good approximation in polynomial time whereas the other two algorithms are fast. In another study, (Friedrich *et al.*, 2008) investigated the concept of fairness in the same manner: the analyses of two different versions of FEMO on two test functions show that each of the algorithms can outperform the other with a runtime gap that is exponential in the bitstring length.

4.4.4. *Future Research Directions*

As we have pointed out, various convergence studies and runtime analyses of MOEAs have been performed in recent years. However, the theoretical understanding of why and when MOEAs are efficient and how general parameters should be chosen lies way behind the knowledge in current research on single-objective algorithms. The first approximation results on hard problems and the comparison of different operators are first steps towards a better understanding of the algorithms' principles and work in this direction should be continued, e.g., by deriving approximation results for other problems or by comparing algorithms with different selection schemes. Hypervolume-based search is another area where theoretical studies are needed. Further studies about when hypervolume-based algorithms are beneficial and when obstructive are needed as well as detailed analyses about how basic parameters such as the number of offsprings should be chosen or how other selection schemes, such as the one in HypE (Bader and Zitzler, 2008), are working. Moreover, other indicator-based algorithms could be analyzed in the near future which might result in a better understanding of the underlying principles of quality indicators in general.

4.5. Other Areas of Interest

Although the title of this chapter gives the impression that it covers all theoretical aspects of evolutionary multiobjective optimization, the limited space does not allow for an exhaustive discussion of all theoretical studies in this area. Besides the mentioned topics, theoretical research has also focused on archiving (Laumanns *et al.*, 2002a; Knowles and Corne, 2003, 2004; Laumanns, 2007; Schütze *et al.*, 2007, 2008), resulted in runtime improvements for some operators (Jensen, 2003), or transferred no free lunch results to the multiobjective case (Corne and Knowles, 2003).

4.6. Summary

The number of theoretical studies about evolutionary multiobjective optimization has been growing quickly in the recent years. Where in the exhaustive overview of publications about evolutionary multiobjective optimization by Coello Coello *et al.* (2007), only 20 publications on theoretical aspects of evolutionary multiobjective algorithms are cited, both in 2008 and in 2009, more than 10 papers with theoretical investigations in the field of evolutionary multiobjective optimization can be reported. This chapter presents an extensive overview over these developments in the theory of evolutionary multiobjective optimization where the focus lies on quality indicators, hypervolume-based search, and runtime analyses. The detailed list of references and the identification of open research questions makes it a good starting point to further advance our fundamental understanding of the underlying principles of evolutionary multiobjective optimization.

Acknowledgments

This work has been supported by the French national research agency (ANR) within the SYSCOMM project ANR-08-SYSC-017. In addition, the author would like to thank his former employer ETH Zurich for the support during the literature research and Nicola Beume for her careful proofreading and useful comments.

References

Auger, A., Bader, J., Brockhoff, D. and Zitzler, E. (2009a). Articulating User Preferences in Many-Objective Problems by Sampling the Weighted Hypervolume, in *Genetic and Evolutionary Computation Conference (GECCO 2009)* (ACM), pp. 555–562.

Auger, A., Bader, J., Brockhoff, D. and Zitzler, E. (2009b). Investigating and Exploiting the Bias of the Weighted Hypervolume to Articulate User Preferences, in *Genetic and Evolutionary Computation Conference (GECCO 2009)* (ACM), pp. 563–570.

Auger, A., Bader, J., Brockhoff, D. and Zitzler, E. (2009c). Theory of the Hypervolume Indicator: Optimal μ-Distributions and the Choice of the Reference Point, in *Foundations of Genetic Algorithms (FOGA 2009)* (ACM), pp. 87–102.

Bader, J. and Zitzler, E. (2008). HypE: An Algorithm for Fast Hypervolume-Based Many-Objective Optimization, TIK Report 286, Computer Engineering and Networks Laboratory (TIK), ETH Zurich.

Beume, N. (2009). $\mathcal{S}$-Metric Calculation by Considering Dominated Hypervolume as Klee's Measure Problem, *Evolutionary Computation* **17**, 4.

Beume, N., Fonseca, C. M., Lopez-Ibanez, M., Paquete, L. and Vahrenhold, J. (2009a). On the Complexity of Computing the Hypervolume Indicator, *IEEE Trans. Evol. Comput.* **13**, 5, pp. 1075–1082.

Beume, N., Naujoks, B. and Emmerich, M. (2007). SMS-EMOA: Multiobjective Selection Based on Dominated Hypervolume, *European J. Oper. Res.* **181**, pp. 1653–1669.

Beume, N., Naujoks, B., Preuss, M., Rudolph, G. and Wagner, T. (2009b). Effects of 1-Greedy $\mathcal{S}$-Metric-Selection on Innumerably Large Pareto Fronts, in *Evolutionary Multi-Criterion Optimization (EMO 2009)* (Springer), pp. 21–35.

Beume, N. and Rudolph, G. (2006). Faster S-Metric Calculation by Considering Dominated Hypervolume as Klee's Measure Problem, Tech. Rep. CI-216/06, Universität Dortmund.

Bradstreet, L., While, L. and Barone, L. (2008). A Fast Incremental Hypervolume Algorithm, *IEEE Trans. Evol. Comput.* **12**, 6, pp. 714–723.

Bringmann, K. and Friedrich, T. (2008). Approximating the Volume of Unions and Intersections of High-Dimensional Geometric Objects, in *Intl. Symp. on Algorithms and Computation (ISAAC 2008)* (Springer), pp. 436–447.

Bringmann, K. and Friedrich, T. (2009a). Approximating the Least Hypervolume Contributor: NP-hard in General, But Fast in Practice, in *Evolutionary Multi-Criterion Optimization (EMO 2009)* (Springer), pp. 6–20.

Bringmann, K. and Friedrich, T. (2009b). Don't Be Greedy When Calculating Hypervolume Contributions, in *Foundations of Genetic Algorithms (FOGA 2009)* (ACM), pp. 103–112.

Brockhoff, D. (2009). Theoretical Aspects of Evolutionary Multiobjective Optimization—A Review, Rapport de Recherche RR-7030, INRIA Saclay—Île-de-France.

Brockhoff, D., Friedrich, T., Hebbinghaus, N., Klein, C., Neumann, F. and Zitzler, E. (2009). On the Effects of Adding Objectives to Plateau Functions, *IEEE Trans. Evol. Comput.* **13**, 3, pp. 591–603.

Brockhoff, D., Friedrich, T. and Neumann, F. (2008). Analyzing Hypervolume Indicator Based Algorithms, in *Parallel Problem Solving From Nature (PPSN X)* (Springer), pp. 651–660.

Chan, T. (2010). A (Slightly) Faster Algorithm for Klee's Measure Problem, *Computational Geometry* **43**, pp. 243–250.

Coello Coello, C. A., Lamont, G. B. and Van Veldhuizen, D. A. (2007). *Evolutionary Algorithms for Solving Multi-Objective Problems* (Springer).

Corne, D. and Knowles, J. (2003). No Free Lunch and Free Leftovers Theorems for Multiobjective Optimisation Problems, in *Evolutionary Multi-Criterion Optimization (EMO 2003)* (Springer), pp. 327–341.

Deb, K. (2001). *Multi-Objective Optimization Using Evolutionary Algorithms* (Wiley).

Deb, K., Pratap, A., Agarwal, S. and Meyarivan, T. (2002). A Fast and Elitist Multiobjective Genetic Algorithm: NSGA-II, *IEEE Trans. Evol. Comput.* **6**, 2, pp. 182–197.

Ehrgott, M. (2005). *Multicriteria Optimization*, 2nd edn. (Springer).

Emmerich, M., Deutz, A. and Beume, N. (2007). Gradient-Based/Evolutionary Relay Hybrid for Computing Pareto Front Approximations Maximizing the S-Metric, in *Hybrid Metaheuristics* (Springer), pp. 140–156.

Farhang-Mehr, A. and Azarm, S. (2003). Minimal Sets of Quality Metrics, in *Evolutionary Multi-Criterion Optimization (EMO 2003)* (Springer), pp. 405–417.

Fleischer, M. (2003). The Measure of Pareto Optima. Applications to Multi-Objective Metaheuristics, in *Evolutionary Multi-Criterion Optimization (EMO 2003)* (Springer), pp. 519–533.

Fonseca, C. M. and Fleming, P. J. (1996). On the Performance Assessment and Comparison of Stochastic Multiobjective Optimizers, in *Parallel Problem Solving from Nature (PPSN IV)* (Springer), pp. 584–593.

Fonseca, C. M., Grunert da Fonseca, V. and Paquete, L. (2005). Exploring the Performance of Stochastic Multiobjective Optimisers with the Second-Order Attainment Function, in *Evolutionary Multi-Criterion Optimization (EMO 2005)* (Springer), pp. 250–264.

Fonseca, C. M., Paquete, L. and López-Ibáñez, M. (2006). An Improved Dimension-Sweep Algorithm for the Hypervolume Indicator, in *Congress on Evolutionary Computation (CEC 2006)* (IEEE Press), pp. 1157–1163.

Friedrich, T., He, J., Hebbinghaus, N., Neumann, F. and Witt, C. (2007a). Approximating Covering Problems by Randomized Search Heuristics Using Multi-Objective Models, in *Genetic and Evolutionary Computation Conference (GECCO 2007)* (ACM), pp. 797–804.

Friedrich, T., Hebbinghaus, N. and Neumann, F. (2007b). Plateaus Can Be Harder in Multi-Objective Optimization, in *Congress on Evolutionary Computation (CEC 2007)* (IEEE Press), pp. 2622–2629.

Friedrich, T., Horoba, C. and Neumann, F. (2008). Runtime Analyses for Using Fairness in Evolutionary Multi-Objective Optimization, in *Parallel Problem Solving from Nature (PPSN X)* (Springer), pp. 671–680.

Friedrich, T., Horoba, C. and Neumann, F. (2009). Multiplicative Approximations and the Hypervolume Indicator, in *Genetic and Evolutionary Computation Conference (GECCO 2009)* (ACM), pp. 571–578.

Giel, O. (2003). Expected Runtimes of a Simple Multi-Objective Evolutionary Algorithm, in *Congress on Evolutionary Computation (CEC 2003)*, IEEE Press, pp. 1918–1925.

Giel, O. and Lehre, P. K. (2006). On The Effect of Populations in Evolutionary Multi-Objective Optimization, in *Genetic and Evolutionary Computation Conference (GECCO 2006)* (ACM), pp. 651–658.

Greiner, G. (2009). Single- and Multi-Objective Evolutionary Algorithms for Graph Bisectioning, in *Foundations of Genetic Algorithms (FOGA 2009)* (ACM), pp. 29–38.

Grunert da Fonseca, V., Fonseca, C. M. and Hall, A. O. (2001). Inferential Performance Assessment of Stochastic Optimisers and the Attainment Function, in *Evolutionary Multi-Criterion Optimization (EMO 2001)* (Springer), pp. 213–225.

Handl, J., Lovell, S. C. and Knowles, J. (2008). Multiobjectivization by Decomposition of Scalar Cost Functions, in *Parallel Problem Solving From Nature (PPSN X)* (Springer), pp. 31–40.

Hanne, T. (1999). On the Convergence of Multiobjective Evolutionary Algorithms, *European J. Oper. Res.* **117**, 3, pp. 553–564.

Hanne, T. (2007). A Primal-Dual Multiobjective Evolutionary Algorithm for Approximating the Efficient Set, in *Congress on Evolutionary Computation (CEC 2007)* (IEEE Press), pp. 3127–3134.

Hansen, M. P. and Jaszkiewicz, A. (1998). Evaluating the Quality of Approximations of the Non-Dominated Set, IMM Technical Report IMM-REP-1998-7, Institute of Mathematical Modeling, Technical University of Denmark.

Hansen, N., Auger, A., Finck, S. and Ros, R. (2009). Real-Parameter Black-Box Optimization Benchmarking 2009: Experimental Setup, INRIA Research Report RR-6828, INRIA Saclay—Ile-de-France.

Henig, M. I. (1982). Existence and Characterization of Efficient Decision With Respect to Cones, *Mathematical Programming* **23**, pp. 111–116.

Horoba, C. (2009). Analysis of a Simple Evolutionary Algorithm for the Multiobjective Shortest Path Problem, in *Foundations of Genetic Algorithms (FOGA 2009)* (ACM), pp. 113–120.

Horoba, C. and Neumann, F. (2008). Benefits and Drawbacks for the Use of Epsilon-Dominance in Evolutionary Multi-Objective Optimization, in *Conference on Genetic and Evolutionary Computation (GECCO 2008)* (ACM), pp. 641–648.

Horoba, C. and Neumann, F. (2009). Additive Approximations of Pareto-Optimal Sets by Evolutionary Multi-Objective Algorithms, in *Foundations of Genetic Algorithms (FOGA 2009)* (ACM), pp. 79–86.

Huband, S., Hingston, P., White, L. and Barone, L. (2003). An Evolution Strategy with Probabilistic Mutation for Multi-Objective Optimisation, in *Congress on Evolutionary Computation (CEC 2003)* (IEEE Press), pp. 2284–2291.

Igel, C., Hansen, N. and Roth, S. (2007). Covariance Matrix Adaptation for Multi-Objective Optimization, *Evolutionary Computation* **15**, 1, pp. 1–28.

Jensen, M. T. (2003). Reducing the Run-Time Complexity of Multiobjective EAs: The NSGA-II and Other Algorithms, *IEEE Trans. Evol. Comput.* **7**, 5, pp. 503–515.

Knowles, J. and Corne, D. (2003). Properties of an Adaptive Archiving Algorithm for Storing Nondominated Vectors, *IEEE Trans. Evol. Comput.* **7**, 2, pp. 100–116.

Knowles, J. and Corne, D. (2004). Bounded Pareto Archiving: Theory and Practice, in *Metaheuristics for Multiobjective Optimisation* (Springer), pp. 39–64.

Knowles, J. D. (2002). *Local-Search and Hybrid Evolutionary Algorithms for Pareto Optimization*, Ph.D. thesis, University of Reading.

Kratsch, S. and Neumann, F. (2009). Fixed-Parameter Evolutionary Algorithms and the Vertex Cover Problem, in *Genetic and Evolutionary Computation Conference (GECCO 2009)* (ACM), pp. 293–300.

Kumar, R. and Banerjee, N. (2006). Analysis of a Multiobjective Evolutionary Algorithm on the 0-1 Knapsack Problem, *Theoretical Computer Science* **358**, 1, pp. 104–120.

Laumanns, M. (2007). Stochastic Convergence of Random Search to Fixed Size Pareto Set Approximations, arXiv.org:0711.2949.

Laumanns, M., Thiele, L., Deb, K. and Zitzler, E. (2002a). Combining Convergence and Diversity in Evolutionary Multiobjective Optimization, *Evolutionary Computation* **10**, 3, pp. 263–282.

Laumanns, M., Thiele, L. and Zitzler, E. (2004a). Running Time Analysis of Evolutionary Algorithms on a Simplified Multiobjective Knapsack Problem, *Natural Computing* **3**, 1, pp. 37–51.

Laumanns, M., Thiele, L. and Zitzler, E. (2004b). Running Time Analysis of Multiobjective Evolutionary Algorithms on Pseudo-Boolean Functions, *IEEE Trans. Evol. Comput.* **8**, 2, pp. 170–182.

Laumanns, M., Thiele, L., Zitzler, E., Welzl, E. and Deb, K. (2002b). Running Time Analysis of Multi-Objective Evolutionary Algorithms on a Simple Discrete Optimization Problem, in *Parallel Problem Solving From Nature (PPSN VII)* (Springer), pp. 44–53.

Li, Z., Li, Z. and Rudolph, G. (2007). On the Convergence Properties of Quantum-Inspired Multi-Objective Evolutionary Algorithms, in *Conference on Intelligent Computing (ICIC 2007)* (Springer), pp. 245–255.

Miettinen, K. (1999). *Nonlinear Multiobjective Optimization* (Kluwer).

Neumann, F. (2007). Expected Runtimes of a Simple Evolutionary Algorithm for the Multi-Objective Minimum Spanning Tree Problem, *European J. Oper. Res.* **181**, 3, pp. 1620–1629.

Neumann, F. and Reichel, J. (2008). Approximating Minimum Multicuts by Evolutionary Multi-Objective Algorithms, in *Parallel Problem Solving From Nature (PPSN X)* (Springer), pp. 72–81.

Neumann, F., Reichel, J. and Skutella, M. (2008). Computing Minimum Cuts by Randomized Search Heuristics, in *Genetic and Evolutionary Computation Conference (GECCO 2008)* (ACM), pp. 779–786.

Neumann, F. and Wegener, I. (2006). Minimum Spanning Trees Made Easier Via Multi-Objective Optimization, *Natural Computing* **5**, 3, pp. 305–319.

Neumann, F. and Wegener, I. (2007). Randomized Local Search, Evolutionary Algorithms, and the Minimum Spanning Tree Problem, *Theoretical Computer Science* **378**, 1, pp. 32–40.

Rudolph, G. (1998a). Evolutionary Search for Minimal Elements in Partially Ordered Finite Sets, in *Conference on Evolutionary Programming VII* (Springer), pp. 345–353.

Rudolph, G. (1998b). On a Multi-Objective Evolutionary Algorithm and Its Convergence to the Pareto Set, in *IEEE International Conference on Evolutionary Computation* (IEEE Press), pp. 511–516.

Rudolph, G. (2001a). Evolutionary Search under Partially Ordered Fitness Sets, in *International NAISO Congress on Information Science Innovations (ISI 2001)* (ICSC Academic Press), pp. 818–822.

Rudolph, G. (2001b). Some Theoretical Properties of Evolutionary Algorithms under Partially Ordered Fitness Values, in *Evolutionary Algorithms Workshop (EAW-2001)*, pp. 9–22.

Rudolph, G. and Agapie, A. (2000). Convergence Properties of Some Multi-Objective Evolutionary Algorithms, in *Congress on Evolutionary Computation (CEC 2000)*, Vol. 2 (IEEE Press), pp. 1010–1016.

Scharnow, J., Tinnefeld, K. and Wegener, I. (2004). The Analysis of Evolutionary Algorithms on Sorting and Shortest Paths Problems, *Journal of Mathematical Modelling and Algorithms* **3**, 4, pp. 349–366.

Schütze, O., Coello Coello, C. A., Tantar, E. and Talbi, E. G. (2008). Computing Finite Size Representations of the Set of Approximate Solutions of an MOP with Stochastic Search Algorithms, in *Genetic and Evolutionary Computation Conference (GECCO 2008)* (ACM), pp. 713–720.

Schütze, O., Laumanns, M., Tantar, E., Coello Coello, C. A. and Talbi, E.-G. (2007). Convergence of Stochastic Search Algorithms to Gap-Free Pareto Front Approximations, in *Genetic and Evolutionary Computation Conference (GECCO 2007)* (ACM), pp. 892–899.

Teytaud, O. (2007). On the Hardness of Offline Multi-Objective Optimization, *Evolutionary Computation* **15**, 4, pp. 475–491.

Thierens, D. (2003). Convergence Time Analysis for the Multi-Objective Counting Ones Problem, in *Evolutionary Multi-Criterion Optimization (EMO 2003)* (Springer), pp. 355–364.

Valiant, L. G. (1979). The Complexity of Computing the Permanent, *Theoretical Computer Science* **8**, pp. 189–201.

Villalobos-Arias, M., Coello Coello, C. A. and Hernández-Lerma, O. (2005). Asymptotic Convergence of Some Metaheuristics Used for Multiobjective Optimization, in *Foundations of Genetic Algorithms (FOGA 2005)* (Springer), pp. 95–111.

Wegener, I. (2003). Methods for the Analysis of Evolutionary Algorithms on Pseudo-Boolean Functions, in *Evolutionary Optimization*, International Series in Operations Research & Management Science (Springer), pp. 349–369.

While, L. (2005). A New Analysis of the LebMeasure Algorithm for Calculating Hypervolume, in *Evolutionary Multi-Criterion Optimization (EMO 2005)* (Springer), pp. 326–340.

While, L., Hingston, P., Barone, L. and Huband, S. (2006). A Faster Algorithm for Calculating Hypervolume, *IEEE Trans. Evol. Comput.* **10**, 1, pp. 29–38.

Xue, F., Sanderson, A. C. and Graves, R. J. (2005). Multi-Objective Differential Evolution—Algorithm, Convergence Analysis, and Applications, in *Congress on Evolutionary Computation (CEC 2005)* (IEEE Press), pp. 743–750.

Yan, Z., Zhang, L., Kang, L. and Lin, G. (2003). A New MOEA for Multi-Objective TSP and Its Convergence Property Analysis, in *Evolutionary Multi-Criterion Optimization (EMO 2003)* (Springer), pp. 342–354.

Zitzler, E., Brockhoff, D. and Thiele, L. (2007). The Hypervolume Indicator Revisited: On the Design of Pareto-compliant Indicators Via Weighted Integration, in *Evolutionary Multi-Criterion Optimization (EMO 2007)* (Springer), pp. 862–876.

Zitzler, E. and Künzli, S. (2004). Indicator-Based Selection in Multiobjective Search, in *Parallel Problem Solving from Nature (PPSN VIII)* (Springer), pp. 832–842.

Zitzler, E., Laumanns, M. and Bleuler, S. (2004). A Tutorial on Evolutionary Multiobjective Optimization, in *Metaheuristics for Multiobjective Optimisation, Lecture Notes in Economics and Mathematical Systems*, Vol. 535 (Springer), pp. 3–37.

Zitzler, E., Laumanns, M. and Thiele, L. (2002). SPEA2: Improving the Strength Pareto Evolutionary Algorithm for Multiobjective Optimization, in *Evolutionary Methods for Design, Optimisation and Control with Application to Industrial Problems (EUROGEN 2001)*, pp. 95–100.

Zitzler, E. and Thiele, L. (1998). Multiobjective Optimization Using Evolutionary Algorithms—A Comparative Case Study, in *Parallel Problem Solving from Nature (PPSN V)*, pp. 292–301.

Zitzler, E., Thiele, L. and Bader, J. (2008). On Set-Based Multiobjective Optimization (Revised Version), TIK Report 300, Computer Engineering and Networks Laboratory (TIK), ETH Zurich.

Zitzler, E., Thiele, L., Laumanns, M., Fonseca, C. M. and Grunert da Fonseca, V. (2003). Performance Assessment of Multiobjective Optimizers: An Analysis and Review, *IEEE Trans. Evol. Comput.* **7**, 2, pp. 117–132.

Chapter 5

Memetic Evolutionary Algorithms

Dirk Sudholt
International Computer Science Institute
Berkeley, CA 94704, USA
sudholt@icsi.berkeley.edu

Memetic Evolutionary Algorithms include local search techniques into the random search process of evolutionary algorithms. This kind of hybridization has become very popular in recent years and there are many experimental studies where memetic evolutionary algorithms outperform plain evolutionary algorithms. This chapter reviews the first runtime analyses for memetic algorithms. In the first part, we deal with one of the most important issues in the design of memetic evolutionary algorithms: how to properly balance global and local search. We construct functions that are very sensitive to the parametrization in a sense that only small changes to "ideal" parameter values can turn a polynomial runtime into a superpolynomial or even exponential one, with high probability.

In the second part, we consider memetic evolutionary algorithms with variable-depth search, also known as Kernighan-Lin. It performs a chained sequence of local steps and returns the best solution found in the sequence. As it naturally moves away from its starting point, it can help an evolutionary algorithm to escape from local optima that are very hard to overcome otherwise. We make this precise for simply-structured instances for two combinatorial problems: MINCUT and MAXSAT. While a simple memetic algorithm with variable-depth search finds global optima in polynomial expected time, plain evolutionary algorithms, iterated local search with a simple local search, and simulated annealing need exponential time. The results also demonstrate the benefits of hybridization as the hybrid drastically outperforms its single components.

Contents

5.1. Introduction

Over the past years, it has become increasingly popular to include local search techniques into the random search process of evolutionary algorithms. These hybrid algorithms are known by various names such as memetic (evolutionary) algorithms (MAs), evolutionary local search, genetic local search, global-local search hybrids, large-step Markov chains, and others. The term *memetic algorithms* was coined by Pablo Moscato who related local search to a process of cultural evolution; the word "memetic" is derived from biologist Richard Dawkins's notion of "memes" as the cultural and social equivalent of "genes." We fix this name in the following.

Memetic algorithms apply local search to newly created offspring to quickly find high-fitness individuals and to discover promising regions of the search space. Researchers consistently report very good results for memetic algorithms on practical problems and there is much experimental evidence where this approach outperforms common evolutionary algorithms. See Moscato (1999) for a survey or the book by Hart *et al.* (2004). The benefit of local search is manifold. Firstly, we can expect high-fitness solutions to be found more quickly due to the increased amount of "greediness." Moreover, there may be low-fitness offspring located in the basin of attraction of a high-fitness local optimum. In a standard evolutionary algorithm, such solutions are likely to get lost immediately in the selection process. In a memetic approach local search may improve upon such solutions and reach a local optimum with high fitness. This effect is particularly visible in constraint optimization problems where often infeasible solutions are penalized with respect to their fitness and the penalty decreases towards feasible regions. If mutation or crossover create an infeasible offspring, local search can work as repair mechanism and find a feasible solution. Finally, there is a possibility to include problem-specific knowledge into the search as local search may be adopted to the problem at hand. Such an approach is often possible since local search strategies are typically easy to design, even in cases where no global problem-specific strategy is known.

In recent years, many different kinds and variants of memetic algorithms have emerged. Krasnogor and Smith (2005) present a taxonomy of memetic algorithms and discuss several design issues. One trend is to employ multiple local search operators. These algorithms are sometimes called *multimeme algorithms* (Neri *et al.*, 2007); each local search operator is called a "meme." The choice of memes is often made adaptively or even self-adaptively, see the survey by Ong, Lim, Zhu and Wong (2006). Furthermore, memetic approaches have been proposed for optimization in continuous spaces (Hart, 2003) and for other algorithmic paradigms such as estimation of distribution algorithms (Aickelin *et al.*, 2007) and ant colony optimization (Dorigo and Stützle, 2004). A first rigorous analysis of a "memetic" ant colony optimizer was recently presented by Neumann *et al.* (2008).

Hybridization of search heuristics poses new challenges to both theory and practice. For theory it is hard to keep track with the state-of-the-art as the algorithms become more and more elaborate. It is widely acknowledged that a solid theoretical foundation of these algorithms is needed. Many operators commonly used in such a hybrid are fairly simple and easy to implement; this holds for standard genetic operators such as mutation, crossover, and selection as well as for many local search strategies. However, the dynamics of an algorithm and the interplay of these components is very hard to tackle analytically. Therefore, studies on memetic algorithms are mostly empirical (e.g., Hart *et al.*, 2004) or rely on non-rigorous arguments (e.g., Sinha *et al.*, 2004).

The first rigorous theoretical results on the optimization time of memetic algorithms were presented in Sudholt (2006b,a). These results are reviewed in Section 5.3; our presentation is based on selected improved results published in Sudholt (2009). We consider the parametrization of memetic algorithms and the balance between global and local search. In applications the available computational resources have to be spread among global and local search, but finding a good balance is not always easy. If the effect of local search is too weak, we fall back to standard evolutionary algorithms. If the effect of local search is too strong, the algorithm may quickly get stuck in local optima of bad quality. Moreover, the algorithm is likely to rediscover the same local optimum over and over again, wasting computational effort. Lastly, too much local search quickly leads to a loss of diversity within the population.

We consider a simple memetic algorithm that captures basic working principles of memetic algorithms—the interplay of genetic operators like mutation and selection with local search. We then present artificial examples where good parameter settings are very hard to find. Moreover, only small changes to such a good parametrization lead to a phase transition from polynomial optimization times to superpolynomial or even exponential times, with high probability. Our results demonstrate that parametrizing memetic evolutionary algorithms can be extremely hard. Moreover, they rule out simple and effective design guidelines for the considered algorithm that do not depend on the problem at hand.

Section 5.4 covers the use of memetic algorithms for combinatorial problems. Our motivation is to give an explanation why memetic algorithms are so effective for many practical problems. To this end, we present instances for problems from combinatorial optimization where a memetic algorithm with so-called variable-depth search drastically outperforms many other algorithms like the (1+1) EA, memetic algorithms with a simple local search operator and simulated annealing. The negative results even hold if we allow a broad range of parameter settings for these algorithms. In addition, we prove that in one case the memetic algorithm is more powerful than its single components. A preliminary version of these results appeared in Sudholt (2008b).

A section with preliminaries (Section 5.2) and conclusions and remarks for future work in Section 5.5 complete the outline of the chapter.

5.2. Preliminaries

We consider the setting of pseudo-Boolean optimization where a function $f\colon \{0,1\}^n \to \mathbb{R}$ is to be maximized. Let $\mathrm{H}(x, y)$ denote the Hamming distance between x and y and $\mathrm{H}(x, Y) := \min_{y \in Y} \mathrm{H}(x, y)$ for a non-empty set $Y \subseteq \{0,1\}^n$. We denote by $N(x)$ the open Hamming neighborhood of x and by $N^*(x)$ the closed Hamming neighborhood that also includes x itself. For a set $X \subseteq \{0,1\}^n$ we denote $N(X) := \bigcup_{x \in X} N(x)$ and $N^*(X) := N(X) \cup X$.

In the sequel, all asymptotic statements refer to the problem dimension n. When estimating probabilities, we say that an event E occurs *with high probability* if $\Pr(E) \geq 1 - n^{-\varepsilon}$ for some constant $\varepsilon > 0$. We say E occurs *with overwhelming probability* if $\Pr(E) = 1 - 2^{-\Omega(n^{\varepsilon})}$.

In memetic algorithms local search is often called with a fixed frequency, the *local search frequency*, and then run for a fixed number of iterations, the *local search depth*. This is reflected in the following $(\mu+\lambda)$ Memetic Algorithm, shortly $(\mu+\lambda)$ MA, defined in Sudholt (2009). We first describe two operators used by the $(\mu+\lambda)$ MA. One is the standard mutation operator that flips each bit independently with a fixed mutation probability p_m. Unless noted otherwise, we consider the default value $p_m := 1/n$.

The local search operator employed in the $(\mu+\lambda)$ MA is defined as follows. This local search is generic in that no pivot rule is specified (i.e., it is not specified how to choose between several local improvements). It generalizes several concrete local search strategies such as first ascent or steepest ascent. The local search depth is denoted by δ.

Operator 12 Local search(y)

---

for δ iterations **do**
    if there is a $z \in N(y)$ with $f(z) > f(y)$ **then** $y := z$
    else stop and **return** y.
return y.

The $(\mu+\lambda)$ MA operates with a population, a multiset of size μ, and creates λ offspring in each generation. This is done by choosing randomly a parent, then mutating it, and, every τ generations, additionally applying local search to the result of the mutation.

Algorithm 13 $(\mu+\lambda)$ Memetic algorithm

Let $t := 0$.
Initialize P_0 with μ individuals chosen uniformly at random.
repeat
    $P'_t := \emptyset$.
    Do λ times:
        Choose $x \in P_t$ uniformly at random.
        Create y by flipping each bit in x independently with prob. p_m.
        if $t \bmod \tau = 0$ **then** $y :=$ local search(y).
        $P'_t := P'_t \cup \{y\}$.
    Create P_{t+1} by selecting the best μ individuals from $P_t \cup P'_t$.
        (Break ties in favor of P'_t.)
    $t := t + 1$.

Note that the $(\mu+\lambda)$ MA does not accept worsenings at the end of a generation as the best individuals from the multiset $P_t \cup P'_t$ are chosen for the next generation. Such a selection strategy is called *elitist selection*. Several well-known randomized search heuristics can be identified as special cases of the $(\mu+\lambda)$ MA. The $(\mu+\lambda)$ MA without local search, i.e., $\delta = 0$ or $\tau = \infty$, is known as $(\mu+\lambda)$ EA. The (1+1) MA with $\tau = 1$ represents an iterated local search algorithm (Lourenço *et al.*, 2002).

The time until a global optimum is found—the optimization time—is defined by the number of f-evaluations until a global optimum is evaluated. Local search constitutes an inner loop within the main loop, hence the number of f-evaluations within local search must be accounted for, too. Whatever pivot rule is used, the number of f-evaluations in one local search call is trivially bounded by δn (assuming each neighbor is evaluated at most once). During t generations the number of f-evaluations is bounded by $t\lambda(1 + \delta n/\tau)$. This implies that if λ and δ are polynomial then the optimization time is polynomial if and only if the number of generations until an optimum is found is polynomial.

The analysis of memetic algorithms is, in general, more challenging than the analysis of plain evolutionary algorithms. One reason is that iterative algorithms often become harder to tackle analytically when more operators are added to the search process. This holds in particular as local search is called due to a dynamic schedule, specified by the local search frequency. In addition, local search itself is a rather complex operator. It constitutes a search process on its own, with a separate access to the objective function. In contrast to mutation, a call of local search can yield a very large jump if it ends with a search point that is far away from its starting point. While it may be hard to predict the outcome of local search, predicting the behavior of the global process might be even harder when local search is started after mutation, i.e., with a random starting point. In the sequel we will face these difficulties and shed some light on how the interplay of mutation and local search can be analyzed.

5.3. The Impact of Parametrization

The importance of the parametrization of memetic algorithms has already been recognized by Hart (1994) who investigated empirically the impact of the local search frequency and the local search depth on three artificial test functions. Land (1998) extended this study to combinatorial optimization, referring to the balance between global search and local search as the local/global ratio. Ishibuchi *et al.* (2003) considered a hybrid algorithm where local search is called with a fixed probability and investigated the impact of this parameter on a flowshop scheduling problem. Another line of research was to adapt the parametrization of a memetic algorithm to the problem at hand according to an analysis of the problem structure (Merz, 2004; Watson *et al.*, 2003). Finally, there were theoretical investigations of simplified models of memetic algorithms explaining how to balance global and local search (Sinha *et al.*, 2004). However, as their arguments were non-rigorous, their work could not lead to formal proofs.

We address the topic of parametrization from a rigorous theoretical perspective. Section 5.3.1 deals with the impact of the local search depth on the $(\mu+\lambda)$ MA while in Section 5.3.2 we focus on the impact of the local search frequency.

5.3.1. *The Impact of the Local Search Depth*

Section 4 in Sudholt (2009) investigates the impact of the local search depth on the performance of the $(\mu+\lambda)$ MA. A class of functions is defined where there is only a small critical window for values of the local search depth δ in which an efficient optimization is possible. More precisely, there is an ideal value $D = D(n)$ for the local search depth in a sense that the choice $\delta = D$ guarantees an efficient optimization with high probability, while even small deviations of δ from D lead to superpolynomial runtimes. The value of D can be chosen almost arbitrarily given a fixed value of n.

The construction is heavily based on so-called *long k-paths* (Rudolph, 1997). A long k-path $\mathcal{P}_n^k$ is a self-avoiding path $\mathrm{P}_0, \mathrm{P}_1, \dots, \mathrm{P}_\ell$ of Hamming neighbors with properties described in the following lemma. For a formal definition of long k-paths and a proof for the lemma we refer to Sudholt (2009).

Lemma 5.1.

(1) The number of bit strings in $\mathcal{P}_n^k$ equals $k \cdot 2^{n/k} - k + 1$. All points on the path are different.

(2) Let $\mathrm{P}_s \in \mathcal{P}_n^k$ and $\mathrm{P}_{s+i} \in \mathcal{P}_n^k$ with $s, i \in \mathbb{N}_0$. If $i < k$ then $H(\mathrm{P}_s, \mathrm{P}_{s+i}) = i$, otherwise $H(\mathrm{P}_s, \mathrm{P}_{s+i}) \geq k$.

The second condition implies that, in order to mutate P_s into some successor on the path that is by at least k positions ahead, at least k bits have to flip simultaneously. If $k = \Theta(\sqrt{n})$ then this probability is at most

$e^{-\Omega(\sqrt{n}\log n)}$ and hence exponentially small. This implies that finding a shortcut on the path by mutation is extremely unlikely.

Let P_i be the i-th point of the long k-path $\mathcal{P}_n^k$ for $k := \sqrt{n}$. For path points $P_i, P_j \in \mathcal{P}_n^k$ we define the *path distance* between P_i and P_j as $|i-j|$, i.e., the absolute index difference. Note that according to Lemma 5.1 the path distance coincides with the Hamming distance if $|i-j| < k$. Let $m = \log^3 n$ and choose some value $D = D(n)$ such that $D \geq m$. We now identify n subsequent sections on the long k-path. The first section just contains P_0. For each $0 \leq i < n$ there is a section containing $\{P_{i(D+m)+1}, \ldots, P_{(i+1)(D+m)}\}$. The fitness of points on the long k-path is defined such that on each section the fitness is strictly increasing and the section ends with a local optimum, also called peak. This construction is sketched in Figure 5.1, reproduced from Sudholt (2009). Each peak has a path distance of $D+m$ to the peak of the next section. The fitness of a point P_i on the long k-path is specified by the following height function. Define $q_i := \lceil i/(D+m) \rceil$ and

$$\text{height}(i) := \left(1 + \frac{2m}{D}\right)^{q_i} \cdot \left(i - q_i \cdot (D+m) + \frac{D}{2} + m\right).$$

This function is linearly increasing for every section. It can be easily verified that the last point of a section represents a local optimum and that $\text{height}(i(D+m)) = \text{height}(i(D+m)+D)$, i.e., $P_{i(D+m)+D}$ is the first suc-

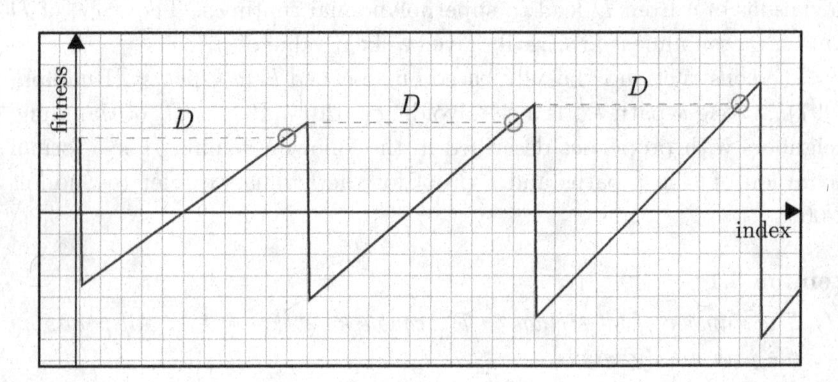

Fig. 5.1. Sketch of the function f_D. The x-axis shows the index on the long k-path. The y-axis shows the fitness. The thick solid line shows the fitness of the points on the long k-path. Encircled path points are close to a target region with respect to Hamming distance. The long k-path can be separated into n subsequent sections with increasing fitness, each one ending with a local optimum. For the sake of clarity, only the first three out of n sections are shown.

cessor of $i(D+m)$ with a competitive fitness value. The points $\mathrm{P}_{i(D+m)+D}$ are called *competitive points*. The idea is that if the local search depth is significantly smaller than D then local search stops with a search point that is worse than its parent and gets removed in the selection step.

On the other hand, the functions are designed in such a way that a too large local search depth also yields an inefficient behavior. We define regions of target points representing global optima somewhere on the way to a peak such that with a large local search depth the algorithm runs past the target region. We thereby exploit that mutation and local search employ different neighborhoods: the target is placed with Hamming distance 2 to the path and Hamming distance at most $m/2$ to all competitive points. The latter condition, along with $m \leq k$, immediately implies that the Hamming distance from every peak to every target point is at least $m/2$. All other search points, in particular those in between the long k-path and the target, are assigned a very low fitness. This barrier prevents local search from traversing between the path and the target. Let $S := \{\mathrm{P}_i \mid 0 \leq i \leq n(D+m)\}$ be the union of all sections and C be the set of all competitive points, then $T := \{x \mid \mathrm{H}(x, S) = 2 \wedge \mathrm{H}(x, C) \leq m/2\}$. Finally, the fitness function f_D is defined as follows.

Definition 5.2. Let $M := \max_{i=0}^{n(D+m)}\{|\operatorname{height}(i)|\} + 1$ and

$$f_D(x) := \begin{cases} \operatorname{height}(i) & \text{if } x = \mathrm{P}_i \in S \\ +M & \text{if } x \in T, \\ -M & \text{otherwise.} \end{cases}$$

Obviously, all points in T with fitness M are global optima. All points with fitness $-M$ form a plateau of equal fitness. Hence, if mutation creates an offspring y with $\mathrm{H}(y, S \cup T) > 1$, then y is surrounded by equally bad neighbors and local search stops with y. If the $(\mu{+}\lambda)$ MA starts with an initial population of larger fitness, such an offspring is rejected immediately.

Randomized search heuristics are usually initialized uniformly at random. The following theorem, however, considers a deterministic initialization where all individuals in the population occupy the first point on the path, P_0. This modification is not essential as the results can be adapted to hold for random initialization with a more complicated construction for f_D (Sudholt, 2006b). The same holds for the main result from Section 5.3.2.

Theorem 5.3. *Let $D \geq 2\log^3 n$, $\lambda = O(\mu)$, and $\mu, \delta, \tau = \text{poly}(n)$. Initialize the $(\mu+\lambda)$ MA with μ copies of P_0, then the following holds with high probability:*

- *if $\delta = D$, then the $(\mu+\lambda)$ MA optimizes f_D in polynomial time and*
- *if $|\delta - D| \geq \log^3 n$, then the $(\mu+\lambda)$ MA needs superpolynomial time on f_D.*

We present the main proof ideas and refer to Sudholt (2009) for a complete formal proof. All peaks have Hamming distance at least $m/2$ to all global optima. Recall that $m = \log^3 n$. The probability of flipping at least $m/2$ bits in a single mutation is bounded by

$$\binom{n}{m/2} \cdot \left(\frac{1}{n}\right)^{m/2} \leq \frac{1}{(m/2)!} = 2^{-\Omega(m \log m)}.$$

By the union bound, with high probability even within $2^{cm \log m}$ mutations the probability that $m/2$ bits flip simultaneously in at least one mutation is still superpolynomially small if the constant $c > 0$ is small enough. In the following, we assume that less than $m/2$ bits flip in the first $2^{cm \log m}$ mutations and keep in mind the superpolynomially small error probability.

For the case $\delta \leq D - m$ we observe that by construction, P_D is the first successor of P_0 such that $f_D(\mathrm{P}_D) \geq f_D(\mathrm{P}_0)$. Since all initial search points have fitness $f_D(\mathrm{P}_0)$, along with elitist selection, the only way to alter the current population is to create an offspring P_i with $i \geq D$ by mutation and/or local search. As $\delta \leq D - m$, it is necessary for mutation to create some $y \in N^*(\{\mathrm{P}_{m+1}, \ldots, \mathrm{P}_{D+m}\})$. However, all these points have Hamming distance at least m from P_0, which completes the first case.

In case $\delta = D$ the local search depth is large enough such that the sections can be optimized one after another, unless the target is found. If the best individual is not a local optimum, the fitness can be easily improved by 1-bit mutations or by local search. If the best individual is a local optimum, it can be efficiently mutated into a point on the next section such that local search climbs the section and ends with an offspring with larger fitness. The probability for such a mutation is bounded below by $1/(en^2\mu)$. Using the reciprocal as an estimate for the expected waiting time, it can be shown that the expected number of generations until a new section is optimized is $O(n^2\mu\tau/\lambda)$ using $\lambda = O(\mu)$. Roughly speaking, the factor τ/λ reflects the average number of local search calls in one generation.

It remains to be shown that the target is found efficiently. Unless $m/2 - O(1)$ bits flip in one mutation, a new section can only be climbed by local search. We can also say that local search ends with a point in $S_i := \{P_{i(D+m)+D}, \ldots, P_{i(D+m)+D+m/2-2}\}$ for some $0 \leq i \leq n-1$. Moreover, a search point with fitness larger than the best fitness in S_i can only be created if a search point in S_i is chosen as parent. If this happens, the parent is mutated and the probability of creating a target point is $\Omega(1)$ as for each point in S_i almost every point with Hamming distance 2 belongs to the target set by definition of T. Therefore, the probability that the target is not found in all n sections is exponentially small and the expected number of generations is $O(n^2\mu\tau/\lambda)$. As the computational effort in one generation is polynomial, Markov's inequality implies a polynomial bound on the optimization time that holds with high probability.

In case $\delta \geq D+m$ the next section can only be climbed by local search, but every call of local search ends with a peak. As every peak has Hamming distance at least $m/2$ to every target point, the result follows for $\mu = 1$. In case of larger populations, it is possible that an individual on a peak creates a predecessor in its section by mutation. Although the offspring is worse than its parent, it is accepted if the population still contains worse individuals. So, the long k-path might be climbed down and the Hamming distance to the set of target points can decrease.

However, this only holds as long as the population contains worse individuals. Fix a local optimum x and call an individual y *fit* if $f_D(y) \geq f_D(x)$. If a fit individual is chosen as parent, a fit offspring is created with probability $\Omega(1)$ since a clone is created with probability $(1-1/n)^n$. The more fit individuals there are in the population, the more likely it is to create even more fit individuals. The expected time until the population only contains fit individuals is $O(\mu/\lambda \cdot \log \mu)$ if $\lambda = O(\mu)$. By standard arguments this takeover happens within a period of $t = O(\mu m/\lambda)$ generations, with high probability. Hence, for each section there are typically only t generations where offspring may climb down the long k-path. Using family tree arguments (Sudholt, 2009, Section 3)—originally introduced by Witt (2006) for the (μ+1) EA—one can show that the probability of climbing down the path by a distance of at least $m/2$ during t generations is $2^{-\Omega(m)}$ and hence superpolynomially small. We conclude that with high probability the population does not reach the target until the population only contains copies of the last peak and then $m/2$ bits have to flip in order to reach the target.

5.3.2. *The Impact of the Local Search Frequency*

The precise choice of the local search depth is crucial for the performance of the algorithm. Similar results are obtained for the local search frequency.

Two functions called Race^{con} and $\text{Race}^{\text{uncon}}$ are defined according to n, δ, and τ. It is proved that the (1+1) MA is efficient on Race^{con}, but inefficient on $\text{Race}^{\text{uncon}}$. Now, if the local search frequency is halved, the (1+1) MA suddenly becomes inefficient on Race^{con}, but efficient on $\text{Race}^{\text{uncon}}$.

The functions Race^{con} and $\text{Race}^{\text{uncon}}$, called *race functions*, are constructed in similar ways. All bit strings are partitioned into their left and right halves, which form two subspaces $\{0,1\}^{n/2}$ within the original space $\{0,1\}^n$ for even n. Each subspace contains a part of a long path. Except for special cases, the fitness is the (weighted) sum of the positions on the two paths. This way, climbing either path is rewarded and the (1+1) MA is encouraged to climb both paths in parallel.

The difference between the two paths is that they are adapted to the two neighborhoods used by mutation and local search, respectively. In the left half we have a connected path consisting of the first $\Theta(n^4 \cdot \delta/\tau)$ points of a long k-path, with k redefined to $\sqrt{n}/2$. The right half contains a similar path with path distance only $\Theta(n)$ to the end of the path, but only every third point of the path is present. Instead of a connected path, we have a sequence of isolated peaks where the closest peaks have Hamming distance 3. As the peaks form a path of peaks, we speak of an *unconnected path*. While the unconnected path cannot be climbed by local search, mutation can jump from peak to peak as a mutation of 3 specific bits has probability at least $1/(en^3)$. To conclude, local search is well suited to climb the connected path while mutation is well suited to climb the unconnected path.

Now, the main idea is as follows: if the local search frequency is high, we expect the (1+1) MA to optimize the connected path prior to the unconnected path. Contrarily, if the local search frequency is low, the (1+1) MA is likely to optimize the unconnected path prior to the connected one. Which path is optimized first can make a large performance difference. In the special cases where the end of any path is reached, we define separate fitness values for Race^{con} and $\text{Race}^{\text{uncon}}$. For Race^{con}, if the connected path is optimized first (i.e., wins the race), a global optimum is found. However, if the unconnected path wins the race, Race^{con} turns into a deceptive function that gives hints to move away from all global optima. In this situation, the expected time to reach a global optimum is exponential. For $\text{Race}^{\text{uncon}}$, the (1+1) MA gets trapped in the same way if the connected path wins and a global optimum is found in case the unconnected path wins.

We formalize the above-mentioned race functions. Given an appropriate constant $\varepsilon > 0$, the connected path has length $\ell = \frac{1-\varepsilon}{e} \cdot$

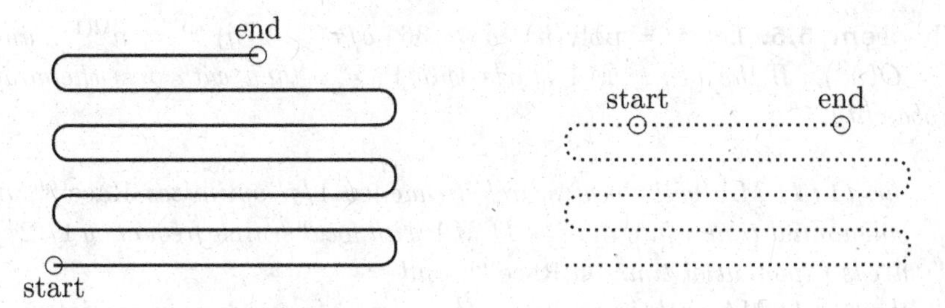

Fig. 5.2. Sketch of the connected path (left) and the unconnected path (right), including start and end points.

$\left(n^3 + \frac{2n^4}{\tau} \cdot (\delta - 3)\right) - n^{3/4} = \Theta(n^4 \cdot \delta/\tau)$ if $\delta/\tau = \Omega(1/n)$, hence proportional to the average number of local search iterations per generation. The unconnected path has length $r = n^5 + \frac{3(1+\sqrt{2})}{2e} \cdot n = n^5 + \Theta(n)$. For technical reasons the (1+1) MA starts with the n^5-th point on the path. A sketch of the two paths and their starting points is given in Figure 5.2.

Definition 5.4. Let $n = 2k^2$ for some $k \in \mathbb{N}$, k a multiple of 3, and let $\mathcal{P}^k_{n/2} = (\mathrm{P}_0, \mathrm{P}_1, \dots)$ be the long k-path of dimension $n/2$. For $w \in \mathcal{P}^k_{n/2}$ let $p(w) = i$ if $w = \mathrm{P}_i$. Denote $x = x'x''$ with $x', x'' \in \{0,1\}^{n/2}$ and call x *well-formed* if and only if $x', x'' \in \mathcal{P}^k_{n/2}$ and $p(x'')/3 \in \mathbb{N}_0$.

$$\mathrm{Race}^{\mathrm{con}}(x) := \begin{cases} n \cdot p(x') + p(x'') & \text{if } x \text{ well-f., } p(x') < \ell, \text{ and } p(x'') < r, \\ 2^n - p(x') & \text{if } x \text{ well-f., } p(x') < \ell, \text{ and } p(x'') \geq r, \\ 3^n & \text{if } x \text{ well-f. and } p(x') \geq \ell, \\ -1 & \text{otherwise.} \end{cases}$$

$$\mathrm{Race}^{\mathrm{uncon}}(x) := \begin{cases} n \cdot p(x') + p(x'') & \text{if } x \text{ well-f., } p(x') < \ell, \text{ and } p(x'') < r, \\ 2^n - p(x'') & \text{if } x \text{ well-f., } p(x') \geq \ell, \text{ and } p(x'') < r, \\ 3^n & \text{if } x \text{ well-f. and } p(x'') \geq r, \\ -1 & \text{otherwise.} \end{cases}$$

Now we state the main result of this section. The preconditions $\delta \geq 36$, $\delta/\tau \geq 2/n$, and $\tau = O(n^3)$ ensure that "enough" iterations of local search are performed during a polynomial number of generations. The reason is that local search must be a visible component in the algorithm for the different local search frequencies to take effect. The condition $\tau = n^{\Omega(1)}$ is required for technical reasons.

Theorem 5.5. *Let* $\delta = \text{poly}(n)$*,* $\delta \geq 36$*,* $\delta/\tau \geq 2/n$*,* $\tau = n^{\Omega(1)}$*, and* $\tau = O(n^3)$*. If the (1+1) MA starts with* $\mathrm{P}_0\,\mathrm{P}_{n^5}$ *then with overwhelming probability*

- *the (1+1) MA with local search frequency* $1/\tau$ *optimizes* Race^{con} *in polynomial time while the (1+1) MA with local search frequency* $1/(2\tau)$ *needs exponential time on* Race^{con} *and*
- *the (1+1) MA with local search frequency* $1/\tau$ *needs exponential time on* $\text{Race}^{\text{uncon}}$ *while the (1+1) MA with local search frequency* $1/(2\tau)$ *optimizes* $\text{Race}^{\text{uncon}}$ *in polynomial time.*

The basic proof idea is to find highly concentrated estimates for the progress of the (1+1) MA for the two paths, in terms of index differences. If $x_t = x'_t x''_t$ is the current population in generation t, the progress in generation t on the connected (unconnected) path is defined as $p(x'_{t+1}) - p(x'_t)$ ($p(x''_{t+1}) - p(x''_t)$). The progress in a set of generations is the sum of the progress values for all considered generations. Note that the progress in one generation is 0 if the offspring is rejected in the selection step. The total progress is estimated separately for generations with and without local search, respectively. We consider a time period of $\Theta(n^4)$ generations, partitioned into $T_{\text{mut}} = \Theta(n^4)$ generations without local search and T_{ls} generations with local search. Note that $T_{\text{ls}} = \Omega(n)$ and $T_{\text{ls}} = O(n^{4-\Theta(1)})$ by assumptions on τ.

The task of estimating the progress on the two paths is complicated by the fact that selection is based on both paths and hence the random progress variables for both paths are not independent. Assuming that no shortcut is taken for any long k-path, the progress on the connected path dominates the progress on the unconnected path in one generation as the position on the connected path is weighted with the factor n. The (1+1) MA may step back on the unconnected path if it advances on the connected path at the same time, but the probability for this event is by a factor of $O(1/n)$ smaller than the probability that the (1+1) MA does not change its position on the connected path. For both paths the expected progress is dominated by the probabilities of hitting the next well-formed path point. Using multiple applications of Chernoff bounds for appropriately defined random variables, the progress on the connected path in $T_{\text{mut}} = \Theta(n^4)$ generations is with overwhelming probability between $(1-\varepsilon)\cdot T_{\text{mut}}/(en)$ and $(1+\varepsilon)\cdot T_{\text{mut}}/(en)$ and the progress on the unconnected path is with overwhelming probability between $(1-\varepsilon)\cdot 3T_{\text{mut}}/(en^3)$ and $(1+\varepsilon)\cdot 3T_{\text{mut}}/(en^3)$.

For progress estimations in generations with local search we need to take a more detailed look at the interplay of mutation and local search. Consider a well-formed parent x, a mutation y of x and the outcome z of local search applied to y. It is obvious that, if y has Hamming distance larger than 1 to all well-formed search points, local search stops with $z = y$, resulting in an offspring with fitness -1 that is rejected by selection. If y has Hamming distance 1 to the set of all well-formed search points, y has exactly one well-formed Hamming neighbor due to the structure of long paths and the fact that the Hamming distance between two peaks is larger than 2. Hence, there is a unique "wrong" bit in x' or x'' that is flipped by the first iteration of local search. The remaining $\delta - 1$ iterations then climb the connected path. Lastly, if y is well-formed, then all δ iterations climb the connected path.

We conclude that either $\delta - 1$ or δ iterations of local search climb the connected path. The progress in one generation of local search is given by the combined effects of mutation and local search if the outcome of local search is accepted. The most typical situation for an accepted offspring is that mutation creates a search point with Hamming distance at most 1 to its parent, which happens with probability close to $2/e$. The reason is that for every $i \geq 2$ there are only $O(n)$ points with Hamming distance i that can lead to acceptance and the probability of hitting any of these points by mutation is $O(1/n)$. It can be shown that in $T_{\text{ls}} = n^{\Theta(1)}$ generations with local search, neglecting small order terms, the total progress on the connected path is in between $(1-\varepsilon) \cdot 2(\delta - 2)T_{\text{ls}}/e$ and $(1+\varepsilon) \cdot 2(\delta + 2)T_{\text{ls}}/e$, with overwhelming probability.

For the unconnected path, one iteration of local search can have a large effect. The probability of reaching the next peak on the unconnected path by a direct mutation is approximately $1/(en^3)$. However, in a generation with local search, the same peak is also reached if mutation hits one of its Hamming neighbors as then the first iteration of local search will climb the peak. Three of these Hamming neighbors have only Hamming distance 2 to x, yielding a probability of approximately $3/(en^2)$ of reaching the next peak. We also have a probability of around $3/(en^2)$ of reaching the previous peak if $x'' \neq \mathrm{P}_0$ (this condition is guaranteed through $O(n^4)$ generations by initialization, unless the (1+1) MA takes a shortcut). Together, the expected progress on the unconnected path is close to 0.

A stochastic process with expected progress 0 is known as a *martingale*. Several concentration results are known in martingale theory, reflecting the fact that a martingale typically does not deviate too far from its initial position. See for example the book by Williams (1991). We also remark that somewhat inverse results have been used recently for the analysis of randomized search heuristics (Jansen and Sudholt, 2010; Neumann *et al.*, 2009). These analyses use lower bounds on the probability that a martingale deviates from its initial position, according to lower bounds on the variance. Here, we make use of a concentration result, the method of martingale differences, which is also known as Azuma's inequality, see Theorem 1.15 in Chapter 1. A straightforward application yields that the absolute total progress on the unconnected path in T_{ls} generations with local search is bounded by $6(T_{\mathrm{ls}}/n^2)^{1/2+\varepsilon} + n^{3/4}$ with overwhelming probability.

The main result now follows from putting together progress bounds for appropriate time periods. First, we consider $T := n^4$ generations with local search frequency $1/\tau$. Summing up the corresponding progress bounds, the (1+1) MA reaches the end of the connected path in this period before it finds a successor of P_{r-k} on the unconnected path, with overwhelming probability. This implies that on $\mathrm{Race}^{\mathrm{con}}$, a global optimum is found. On $\mathrm{Race}^{\mathrm{uncon}}$, however, the objective is turned to minimizing the position on the unconnected path. The Hamming distance to each point $x' \, \mathrm{P}_{r+i}$ for $i \geq 0$ is at least k and all points with smaller Hamming distance have worse fitness. The only way to reach a global optimum is a direct jump flipping at least k bits. The probability for such an event is at most $1/(k!) = 2^{-\Omega(n^{1/2} \log n)}$. Moreover, the probability of finding the optimum within $2^{cn^{1/2} \log n}$ generations is still exponentially small if $c > 0$ is small enough, meaning that the (1+1) MA needs exponential time with overwhelming probability.

The argumentation for the (1+1) MA with local search frequency $1/(2\tau)$ is similar. We now consider a period of $T := \sqrt{2}n^4$ generations of the (1+1) MA with local search frequency $1/(2\tau)$. Compared to the previous setting, the number of generations with local search is roughly by a factor of $\sqrt{2}$ smaller and the number of generations without local search is roughly by a factor of $\sqrt{2}$ larger. This is both a disadvantage for the connected path and an advantage for the unconnected path. With overwhelming probability the unconnected path wins the race before a successor of $\mathrm{P}_{\ell-k}$ is found on the connected path. In this case the (1+1) MA needs polynomial time on $\mathrm{Race}^{\mathrm{uncon}}$ and exponential time on $\mathrm{Race}^{\mathrm{con}}$.

5.4. Memetic Algorithms in Combinatorial Settings

Section 5.3 has shown that memetic algorithms can outperform their equivalents without local search (i.e., $\delta = 0$) on artificial functions. In this section, we ask whether there are combinatorial settings where memetic algorithms outperform plain evolutionary algorithms. Thereby, we review parts of Sudholt (2008b) where a more sophisticated local search has been investigated: variable-depth search.

Variable-depth search (VDS) is well known for the TSP as *Lin-Kernighan* (Lin and Kernighan, 1973) and for GRAPH BISECTION as *Kernighan-Lin* (Kernighan and Lin, 1970). The idea is to perform a chained sequence of local moves. The next local move is chosen greedily. If there is a move that increases fitness, then a move with maximal fitness gain is chosen. Otherwise, a move with minimal loss in fitness is selected. To prevent the algorithm from looping, certain parts of the search space are made "tabu." For binary search spaces this means that if VDS flips some bit, then this bit cannot be flipped again during the run of VDS. The output of VDS is then a best solution encountered during the sequence of local moves. In the following procedure S denotes the sequence of solutions encountered during VDS and L is a set of indices for all bits that have been locked.

Operator 14 Variable-depth search(y)

$S, L := \emptyset$.
while $V_y := \{z \in N(y) \mid \forall i\colon (y_i \neq z_i \Rightarrow i \notin L)\} \neq \emptyset$ **do**
  Choose $z \in V_y$ with maximal f-value uniformly at random.
  $S := S \cup \{z\}$.
  $L := L \cup \{i \mid y_i \neq z_i\}$.
  $y := z$.
return $z \in S$ with maximal f-value chosen uniformly at random.

Note that one run of VDS takes at most n iterations since in every iteration of the loop at least one index is added to L.

We refer to *iterated VDS* as the (1+1) MA calling VDS in every generation. Sudholt (2008b) compares iterated VDS against several other trajectory-based algorithms, that is, algorithms with population size 1. Simply structured examples of combinatorial settings are presented where iterated VDS outperforms the (1+1) EA, simulated annealing, and a (1+1) MA with a simple local search. Thus, also the choice of the local search operator can be essential to design effective memetic algorithms.

The mentioned simple local search represents the generic local search operator (Operator 12) with a pivot rule that chooses a Hamming neighbor with larger fitness uniformly at random. To avoid confusion with the algorithm known as randomized local search, we call it *standard local search.* To make sure that the bad performance of the (1+1) MA with standard local search does not result from a wrong parametrization, we consider arbitrary values for the local search depth and the local search frequency; in fact, the results even hold for arbitrary criteria when to use local search. In addition, we also consider larger mutation probabilities $p_m \gg 1/n$. This accounts for the fact that iterated local search algorithms often use a stronger mutation, referred to as *perturbation.*

We also give a definition of simulated annealing, formulated for maximization. Simulated annealing chooses a random Hamming neighbor and then decides whether to accept it as new search point. It always accepts better solutions, but it also allows worse solutions to be accepted, depending on the magnitude of the fitness decrease and a parameter called *temperature.* The larger the temperature, the more likely it is to accept worse solutions. Usually, simulated annealing starts with a high temperature and decreases the temperature over time. This way, simulated annealing can explore the search space in the beginning and then gradually turns into a hill climber focusing on exploitation. A strategy to turn down the temperature is called a *cooling schedule.*

Algorithm 15 Simulated annealing

Let $t := 0$.
Choose x uniformly at random.
repeat
  Choose $y \in N(x)$ uniformly at random.
  Set $x := y$ with probability $\min\{1, \exp((f(y) - f(x))/T(t))\}$.
  $t := t + 1$.

We investigate instances of problems from combinatorial optimization, namely MINCUT and MAXSAT. Sudholt (2008b) and the extended version (Sudholt, 2008a, Section 4.2) also contain results for the KNAPSACK problem and some negative results are generalized to many different neighborhoods used by local search. We restrict ourselves to the Hamming neighborhood for the sake of simplicity. Section 5.4.1 introduces a characterization of local optima that are hard to overcome, along with lower bounds for the (1+1) MA and simulated annealing after having reached or gotten close to such local optima. In Sections 5.4.2 and 5.4.3 we then consider instances for the problems MINCUT and MAXSAT, respectively.

5.4.1. *Characterizing Difficult Local Optima*

A local optimum is particularly difficult for a search heuristic if the fitness decreases with every local move leading away from it. Combinatorial fitness landscapes often contain multiple local optima that are close to one another. Therefore, we consider a set of local optima and focus on the Hamming distance to such a set. If every distance-increasing local move decreases the fitness up to a Hamming distance of α, the set is called *α-difficult*. The reader is referred to Figure 5.4 in Section 5.4.3 for an illustrative example.

Definition 5.6. A non-empty set $S^* \subseteq \{0,1\}^n$ is called *α-difficult* for $\alpha = \alpha(n)$ with respect to the function f if $y \in N(x)$ and $\mathrm{H}(x, S^*) < \mathrm{H}(y, S^*) \leq \alpha$ implies $f(x) > f(y)$ for every $x, y \in \{0,1\}^n$ and S^* does not contain global optima.

The definition implies that all points with Hamming distance less than α to S^* have worse fitness than every point in S^*. This immediately leads to lower bounds for the (1+1) MA once S^* has been reached.

Lemma 5.7. *Let S^* be α-difficult. If the (1+1) MA using standard local search and mutation probability $p_m \leq (1-\varepsilon)\cdot\alpha/n$ for some $\varepsilon > 0$ reaches S^*, the remaining time until a global optimum is found is at least $2^{c\alpha}$ with probability $1 - 2^{-\Omega(\alpha)}$ for some constant $c > 0$.*

An α-difficult local optimum is challenging for memetic algorithms and simulated annealing as both have difficulties with large "valleys" in the fitness landscape. This similarity between evolutionary algorithms and simulated annealing has already been recognized by Jansen and Wegener (2007). If the temperature is low enough, an application of the Drift Theorem (Theorem 2.12) from Chapter 2 implies the following result.

Lemma 5.8. *Let S^* be α-difficult and let $|f(x) - f(y)| \geq \Delta = \Delta(\alpha)$ if $y \in N(x)$ and x, y have Hamming distance at most α to S^*. If simulated annealing with temperature $T(t) \leq \Delta/(\ln(4n/\alpha))$ reaches some x with $H(x, S^*) \leq (1-\varepsilon)\alpha$ for some constant $0 < \varepsilon \leq 1$, the remaining optimization time is at least $2^{c\alpha}$ with probability $1 - 2^{-\Omega(\alpha)}$ for some constant $c > 0$.*

5.4.2. *Mincut*

Given an undirected graph $G = (V, E)$, $V = \{v_1, \ldots, v_n\}$, the problem MINCUT asks for a partition of vertices into two non-empty subsets V_0

and V_1 such that the number of edges between V_0 and V_1 is minimized. Sudholt (2008b) considered a binary encoding where a bit x_i indicates that $v_i \in V_{x_i}$. The following fitness function counts the number of non-cut edges and penalizes infeasible solutions where V_0 or V_1 is empty.

$$f(x) := \begin{cases} \sum_{\{u,v\} \in E} (x_u x_v + (1 - x_u)(1 - x_v)) & \text{if } x \notin \{0^n, 1^n\}, \\ -1 & \text{if } x \in \{0^n, 1^n\}. \end{cases}$$

The instance $G = (V, E)$ with $V = \{u_1, \dots, u_{n/2}, v_1, \dots, v_{n/2}\}$ and $E = \{\{u_i, u_j\}, \{v_i, v_j\} \mid 1 \leq i < j \leq n/2\}$ consists of two cliques of size $n/2$, each. It is obvious that all solutions where both cliques form different sets are global optima. As empty sets V_0 or V_1 are penalized, all Hamming neighbors of infeasible solutions form local optima (see Figure 5.3). Even stronger, the set S^* of partitions with $|V_0| = 1$ or $|V_0| = n - 1$ is α-difficult for $\alpha = \lfloor n/4 - 1 \rfloor$. Infeasible solutions do not have neighbors with a larger distance from S^*. For all other points every local move shifting a vertex to the smaller set of the partition increases the cut size, unless it contains at least half of a clique.

During a typical run of the (1+1) EA or the (1+1) MA with standard local search it is equally likely for both cliques to move towards different

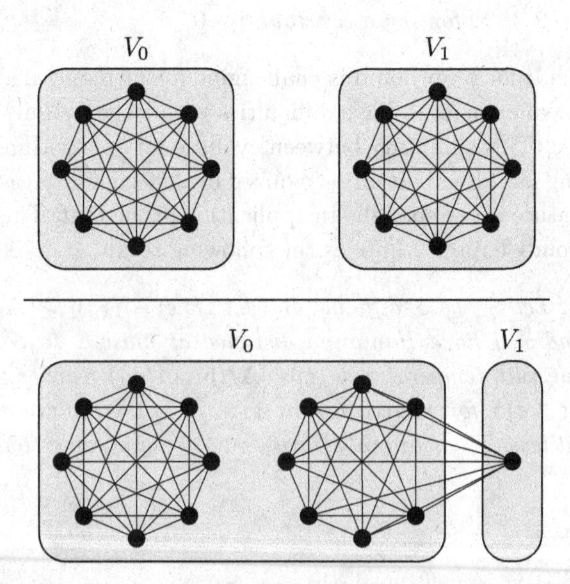

Fig. 5.3. A global optimum (top) and a local optimum (bottom) for the MINCUT instance with $n = 16$.

sides of the cut or move towards the same side. We prove that the latter case leads to exponential optimization times and that this happens with probability close to 1/2. Recall that the (1+1) EA is a special case of the (1+1) MA, hence it suffices to consider the (1+1) MA.

Theorem 5.9. *Consider the (1+1) MA with standard local search and mutation probability $p_m \leq 1/5$. The optimization time for the* MINCUT *instance is at least 2^{cn} with probability $1/2 - O(1/\sqrt{n})$ for some constant $c > 0$.*

Proof. Let S^* be the set of partitions with $|V_0| = 1$ or $|V_0| = n - 1$. If n is large enough, we have $1/5 \leq (1 - \varepsilon) \cdot \alpha/n$ for an appropriate $\varepsilon > 0$, hence the claim follows from Lemma 5.7 if we can prove that the algorithm reaches S^* with probability $1/2 - O(1/\sqrt{n})$.

Let S_0 contain all four (feasible and infeasible) solutions with no cut edge and let $S_1 = N(S_0)$. We will exploit that under certain conditions the fitness function appears perfectly symmetric to the algorithm; in fact, this symmetry can be guaranteed until the first point from S_0 is evaluated. We claim that S_1 is found before S_0 with probability $1 - O(1/\sqrt{n})$. By symmetry, this will then yield a (conditional) probability of 1/2 that a local optimum is reached.

Initialization creates a search point in S_0 with probability 2^{-n+2}. Assume that the algorithm does not start in $S_0 \cup S_1$. Consider a generation with current solution $x \notin (S_0 \cup S_1)$ and the next offspring creation. If $S_0 \cup S_1$ is reached during local search, the claim is trivial. Hence we focus on mutation only and fix a search point $z \in S_0$. If x has Hamming distance k to z, we have k solutions in S_1 with Hamming distance $k-1$ to x and $n-k$ solutions in S_1 with Hamming distance $k+1$. The probability of reaching a specific $y \in N(z)$ differs from the probability $\Pr(z)$ of reaching z in just one bit position. More precisely, the probabilities differ by factors $p_m/(1-p_m)$ or $(1 - p_m)/p_m$, depending on whether this bit has to be flipped or not. Let $\Pr(N(z))$ denote the probability of reaching $N(z)$, then

$$\begin{aligned}
\Pr(N(z)) &\geq k \cdot \Pr(z) \cdot \frac{1-p_m}{p_m} + (n-k) \cdot \Pr(z) \cdot \frac{p_m}{1-p_m} \\
&= \Pr(z) \left(\frac{k(1-p_m)^2 + (n-k)p_m^2}{p_m(1-p_m)} \right) \\
&= \Pr(z) \left(\frac{k(1-2p_m) + np_m^2}{p_m(1-p_m)} \right).
\end{aligned}$$

This term is increasing with k, hence in the worst case $k = 2$. Along with $p_m \leq 1/5$, $\Pr(N(z)) \geq \Pr(z)\left(\frac{1+np_m^2}{p_m}\right)$. The term in brackets is $1/p_m + np_m \geq \sqrt{n}$, hence $\Pr(N(z)) \geq \Pr(z) \cdot \sqrt{n}$. Since this holds for all $z \in S_0$ and the neighborhoods for all points in S_0 are disjoint, $p_1 \geq p_0 \cdot \sqrt{n}$ follows and S_1 is found before S_0 with probability $1 - O(1/\sqrt{n})$.

As long as no local optimum is found, the fitness is indifferent to the question whether the majority of the u-vertices is in V_0 or in V_1. The same holds independently for the v-vertices. Hence, if the first point in S_1 is created, given that S_0 has not been found yet, all points in S_1 have the same probability of being found. Half of these search points are local optima, hence the probability that S^* is found equals 1/2. By the union bound, the probability of reaching S^* is at least $1/2 - O(1/\sqrt{n}) - 2^{-n+2} = 1/2 - O(1/\sqrt{n})$. □

Reusing ideas from the proof of Theorem 5.9, we show that also simulated annealing fails with probability close to 1/2.

Theorem 5.10. *Simulated annealing with an arbitrary cooling schedule where $T(t)$ is monotone decreasing needs at least 2^{cn} steps for the* MINCUT *instance with probability $1/2 - 2^{-\Omega(n)}$ for some constant $c > 0$.*

Proof. We divide a run into two phases: the first phase ends when the temperature first drops to $n/12$ and then the second phase starts. Let T_1 be the number of generations in Phase 1 and let S_0 and S_1 be defined as in the proof of Theorem 5.9. We first prove that in Phase 1 no solution in S_0 will be evaluated in exponential time, with high probability.

Consider a search point x with $\mathrm{H}(x, S_0) = k \leq n/(4e^6)$. The probability of increasing the Hamming distance to S_0 is $p^+ \geq \frac{n-k}{n} \cdot e^{-n/(2T)} \geq \frac{1}{2} \cdot e^{-6}$ as the worst fitness decrease equals $n/2$. The probability of decreasing the Hamming distance to S_0 equals $p^- = \frac{k}{n} \leq \frac{1}{4} \cdot e^{-6}$. Together, the conditional probability of decreasing the Hamming distance is bounded by 1/3, provided that the Hamming distance is changed.

The probability of initializing simulated annealing with a search point x such that $\mathrm{H}(x, S_0) \leq n/(4e^6)$ is $2^{-\Omega(n)}$. Assuming to start at a larger distance and applying the Drift Theorem 2.12 to $\mathrm{H}(x, S_0)$ in the interval $[0, n/(4e^6)]$, the probability that S_0 is reached within the first $\min\{T_1, 2^{cn}\}$ steps is $2^{-\Omega(n)}$.

This concludes the proof if $T_1 \geq 2^{cn}$. Otherwise, we consider Phase 2 and assume that S_0 has not been reached in Phase 1. As simulated annealing only makes local moves, S_1 is reached before S_0 and the probability that a local optimum is found equals 1/2. Given a fixed Hamming distance $k \leq n/12$ from S_0, the minimal fitness difference Δ between two neighbors, as defined in Lemma 5.8, is attained when only one clique is cut. In that case the cut clique has k vertices on one side of the partition and when adding a $(k+1)$-st vertex, the fitness decreases by $n/2 - 2k - 1 \geq n/3$ if $k+1 \leq n/12$. Applying Lemma 5.8 with $\alpha = n/12$ and $\Delta(\alpha) = n/3$ proves the claim for $T \leq \Delta/(\ln(4n/\alpha)) = 1/(3\ln(48))$ and hence for $T \leq n/12$. □

The local optima of the MINCUT instance are extremely hard for standard evolutionary algorithms, memetic algorithms, and simulated annealing. Contrarily, iterated VDS easily escapes from this local optimum.

Theorem 5.11. *The expected number of generations of iterated VDS with mutation probability $p_m = 1/n$ on the* MINCUT *instance is $O(1)$.*

Proof. Observe that the most fitness-increasing local moves always lead towards local or global optima and these points can be reached by flipping each bit at most once. Hence, the first VDS reaches a global or local optimum. Assume that a local optimum is reached. With probability at least $(1-1/n)^n \geq 1/4$ for $n \geq 2$ the following mutation does not flip any bit and VDS is started from a local optimum. W. l. o. g. this local optimum is $V_0 = \{u_1\}$. The first step of VDS is to add some other u-vertex, say u_2, to V_0. But then the fitness is again increased by removing u_1, so we get $V_0 = \{u_2\}$. Next, all other u-vertices are moved to V_0 leading to a solution $V_0 = \{u_2, \ldots, u_{n/2}\}$. As now all u-vertices are frozen, the algorithm proceeds by moving all v-vertices one after another, ending with the solution $V_0 = V \setminus \{u_1\}$. Note that all four solutions that have been stated explicitly have $n/2 - 1$ cut edges and all other intermediate solutions are worse. The result of VDS is chosen uniformly among these four solutions. Three of these solutions are local optima, but $V_0 = \{u_2, \ldots, u_{n/2}\}$ is neighbored to a global optimum. If this solution is returned, there is a probability of at least $(1-1/n)^n \geq 1/4$ that the next mutation does not flip any bit and then VDS deterministically moves u_1 to create a global optimum. Otherwise, we again have a local optimum and repeat the argumentation. The expected waiting time until a local optimum is turned into a global one in two subsequent generations is at most $(1/4)^3 = \Omega(1)$, hence the expected number of generations is $O(1)$. □

5.4.3. *Maxsat*

Given n Boolean variables $x_1, \ldots, x_n$ a literal is either a variable or a negated variable. A clause is a disjunction of literals. A clause is satisfied with respect to an assignment x to the variables if the clause evaluates to true. Given a set C of clauses, the problem MAXSAT asks for an assignment of the variables maximizing the number of satisfied clauses.

Already Droste, Jansen and Wegener (2002) have investigated the MAXSAT problem in the context of evolutionary algorithms. As fitness function they chose the number of satisfied clauses. The instance they considered contains n clauses of length 1 (unit clauses) as well as all possible clauses with 3 different variables and two negated literals.

$$\forall i \neq j \neq k \neq i\colon (x_i \vee \overline{x_j} \vee \overline{x_k}) \in C$$
$$(x_1), (x_2), \ldots, (x_n) \in C$$

The many negated literals give strong hints to set many variables to 0. Note that all variables are symmetric, hence the number of satisfied clauses only depends on the number of bits set to 1 in x, denoted by $|x|_1$. If $|x|_1 = i$, then i unit clauses are satisfied. Among the other $\binom{n}{3}$ clauses there are $n \cdot \binom{i}{2}$ clauses where the last two literals evaluate to false. Moreover, there are $(n - i)$ choices for the first variable such that the first literal also evaluates to false. Hence, $(n - i) \cdot \binom{i}{2}$ clauses of length 3 are unsatisfied and the fitness is given by $f(x) = \binom{n}{3} - (n - |x|_1) \cdot \binom{|x|_1}{2} + |x|_1$. A sketch is shown in Figure 5.4.

It is easy to see that $|x|_1 = n$ implies a global optimum with fitness $\binom{n}{3} + n$. The search point 0^n has fitness $\binom{n}{3}$ and all x with $|x|_1 = 1$ have fitness $\binom{n}{3} + 1$ as $\binom{1}{2} = 0$. Assuming $n \geq 6$ and n a multiple of 3, the

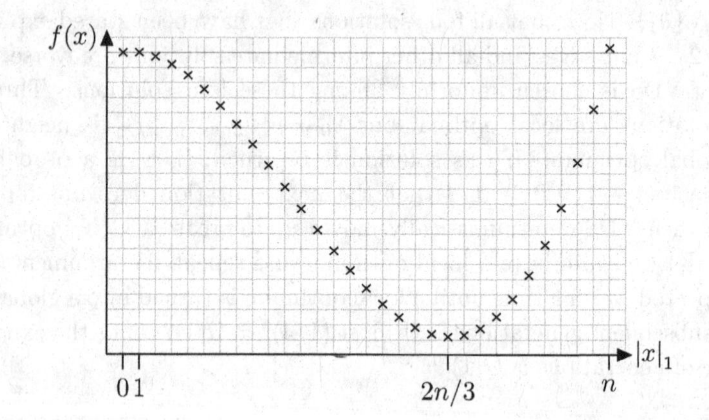

Fig. 5.4. Sketch of the fitness landscape according to the MAXSAT instance with $n = 30$. The set $S^* = \{x \mid |x|_1 = 1\}$ is $(2n/3 - 1)$-difficult as every local move leading away from S^* decreases fitness.

fitness decreases with $|x|_1$ in the interval $[1, 2n/3]$. It follows that the set $S^* = \{x \mid |x|_1 = 1\}$ is $(2n/3 - 1)$-difficult. Random initialization with overwhelming probability creates a search point with $2n/3 - \Omega(n)$ 1-bits. The following theorem is an easy implication from Lemma 5.7.

Theorem 5.12. *Consider the (1+1) MA with standard local search and mutation probability* $p_m \leq 1/2$. *The optimization time for the* MAXSAT *instance is at least* 2^{cn} *with probability* $1 - 2^{-\Omega(n)}$ *for some constant* $c > 0$.

A similar result can be proven for simulated annealing using an arbitrary cooling schedule. If $3n/5 \leq |x|_1 < 2n/3$ holds for the current search point, then the probability of decreasing the number of 1-bits in the new search point is larger than the probability of increasing it. In addition, a decrease is always accepted. An application of the Drift Theorem yields the following result.

Theorem 5.13. *The optimization time of simulated annealing on the* MAXSAT *instance is at least* 2^{cn} *with probability* $1 - 2^{-\Omega(n)}$ *for every cooling schedule and some constant* $c > 0$.

Again, we ask ourselves what iterated VDS can do. Interestingly, iterated VDS without mutation is not effective for the MAXSAT instance.

Theorem 5.14. *Iterated VDS without mutation (i.e.,* $p_m = 0$*) finds the global optimum of the* MAXSAT *instance only with probability* $2^{-\Theta(n)}$.

Proof. The lower bound on the success probability follows trivially from the fact that random initialization creates the global optimum with probability 2^{-n}. For the upper bound, Chernoff bounds yield that the probability of starting with x such that $2 \leq |x|_1 < 2n/3$ is $1 - 2^{-\Omega(n)}$. In this case the fitness can only be increased by flipping a single 1-bit. Since this bit afterwards cannot flip back to 1, VDS returns a local optimum with a single 1-bit. Having reached such a local optimum, the least fitness decrease is obtained by flipping the unique 1-bit to 0. However, this implies that 1^n cannot be reached. As all other search points have worse fitness, VDS again returns a local optimum with a single 1-bit. □

However, with mutation the global optimum can be reached efficiently.

Theorem 5.15. *The expected number of generations of iterated VDS with mutation probability* $p_m = 1/n$ *on the* MAXSAT *instance is* $O(n)$.

Proof. The first VDS creates a local or global optimum as one such point can be reached by a fitness-increasing path of Hamming neighbors where each bit is flipped at most once in total. If a local optimum with a single 1-bit is reached, then mutation creates 0^n with probability $1/n \cdot (1-1/n)^{n-1} \geq 1/(en)$ and the following VDS reaches 1^n with probability 1. The expected number of generations for this event is at most en. □

The MAXSAT example therefore demonstrates that a hybrid algorithm can outperform its single components on a combinatorial problem.

5.5. Conclusions and Future Work

We have reviewed the first rigorous theoretical analyses of memetic evolutionary algorithms. Similar to plain evolutionary algorithms, the investigation of memetic algorithms started with simple algorithms on artificial problems. Section 5.3 has dealt with the fundamental task of finding a proper balance between evolutionary and local search. Hierarchy results for the choice of the local search depth and the local search frequency, respectively, show that small changes of the parametrization can have a tremendous impact on the performance of the $(\mu+\lambda)$ MA. In particular, they rule out parameter settings for the $(\mu+\lambda)$ MA that work well for every problem.

The investigations from Section 5.4 have demonstrated that memetic algorithms with Kernighan-Lin's variable-depth search can be successful where many other trajectory-based algorithms fail. The MAXSAT instance has shown that hybridization can be essential as the hybrid algorithm clearly outperforms its single components. The problem instances have a very clear and simple structure and are well suited for teaching purposes.

The analyses of memetic algorithms have led to insights into their dynamic behavior. In many proofs a detailed observation of the interplay of mutation, local search, and selection has been essential. There are several directions for future work, towards more complex memetic algorithms and towards more realistic problems. An obvious next step is to take into account the effect of crossover and to perform analyses for memetic algorithms with crossover, mutation, selection, and local search. Many other memetic approaches may be analyzed as well. This includes non-binary search spaces, multimeme algorithms, or memetic hybridizations in the context of other optimization paradigms. A first rigorous study of ant colony optimization with local search has been presented by Neumann *et al.* (2008). Another challenging goal is to analyze adaptive or even self-adaptive memetic algorithms.

While we have demonstrated that iterated VDS can outperform common algorithms, this does not imply that iterated VDS is always superior. One can certainly find problem instances where the effect is reversed. It is an important open problem to characterize problems where memetic algorithms are superior to plain evolutionary algorithms. More thorough investigations on problems from combinatorial optimization like, e.g., vertex coloring, TSP, and scheduling problems are needed in order to advance the theoretical understanding of hybrid algorithms and to explain the broad empirical success of these metaheuristics from a theoretical perspective.

Acknowledgment

The author was partially supported by a postdoctoral fellowship from the German Academic Exchange Service.

References

Aickelin, U., Burke, E. K. and Li, J. (2007). An estimation of distribution algorithm with intelligent local search for rule-based nurse rostering, *Journal of the Operational Research Society* **58**, pp. 1574–1585.

Dorigo, M. and Stützle, T. (2004). *Ant Colony Optimization* (MIT Press).

Droste, S., Jansen, T. and Wegener, I. (2002). Optimization with randomized search heuristics—the (A)NFL theorem, realistic scenarios, and difficult functions, *Theoretical Computer Science* **287**, 1, pp. 131–144.

Hart, W. E. (1994). *Adaptive Global Optimization with Local Search*, Ph.D. thesis, University of California, San Diego, CA.

Hart, W. E. (2003). Locally-adaptive and memetic evolutionary pattern search algorithms, *Evolutionary Computation* **11**, 1, pp. 29–51.

Hart, W. E., Krasnogor, N. and Smith, J. E. (eds.) (2004). *Recent Advances in Memetic Algorithms, Studies in Fuzziness and Soft Computing*, Vol. 166 (Springer).

Ishibuchi, H., Yoshida, T. and Murata, T. (2003). Balance between genetic search and local search in memetic algorithms for multiobjective permutation flowshop scheduling, *IEEE Transactions on Evolutionary Computation* **7**, 2, pp. 204–223.

Jansen, T. and Sudholt, D. (2010). Analysis of an asymmetric mutation operator, *Evolutionary Computation* **18**, 1, pp. 1–26.

Jansen, T. and Wegener, I. (2007). A comparison of simulated annealing with a simple evolutionary algorithm on pseudo-Boolean functions of unitation, *Theoretical Computer Science* **386**, 1-2, pp. 73–93.

Kernighan, B. and Lin, S. (1970). An efficient heuristic procedure for partitioning graphs, *The Bell System Tech Journal* **49**, 2, pp. 291–307.

Krasnogor, N. and Smith, J. (2005). A tutorial for competent memetic algorithms: model, taxonomy, and design issues, *IEEE Transactions on Evolutionary Computation* **9**, 5, pp. 474–488.

Land, M. W. S. (1998). *Evolutionary Algorithms with Local Search for Combinatorial Optimization*, Ph.D. thesis, University of California, San Diego, CA.

Lin, S. and Kernighan, B. W. (1973). An effective heuristic algorithm for the traveling salesman problem, *Operations Research* **21**, pp. 498–516.

Lourenço, H. R., Martin, O. and Stützle, T. (2002). Iterated local search, in *Handbook of Metaheuristics, International Series in Operations Research & Management Science*, Vol. 57 (Kluwer Academic Publishers, Norwell, MA), pp. 321–353.

Merz, P. (2004). Advanced fitness landscape analysis and the performance of memetic algorithms, *Evolutionary Computation* **12**, 3, pp. 303–326.

Moscato, P. (1999). Memetic algorithms: a short introduction, in D. Corne, M. Dorigo and F. Glover (eds.), *New Ideas in Optimization* (McGraw-Hill), pp. 219–234.

Neri, F., Toivanen, J., Cascella, G. L. and Ong, Y.-S. (2007). An adaptive multimeme algorithm for designing HIV multidrug therapies, *IEEE/ACM Transactions on Computational Biology and Bioinformatics* **4**, 2, pp. 264–278.

Neumann, F., Sudholt, D. and Witt, C. (2008). Rigorous analyses for the combination of ant colony optimization and local search, in *Proceedings of the Sixth International Conference on Ant Colony Optimization and Swarm Intelligence (ANTS '08)*, *LNCS*, Vol. 5217 (Springer), pp. 132–143.

Neumann, F., Sudholt, D. and Witt, C. (2009). Analysis of different MMAS ACO algorithms on unimodal functions and plateaus, *Swarm Intelligence* **3**, 1, pp. 35–68.

Ong, Y.-S., Lim, M.-H., Zhu, N. and Wong, K.-W. (2006). Classification of adaptive memetic algorithms: a comparative study, *IEEE Transactions on Systems, Man, and Cybernetics, Part B* **36**, 1, pp. 141–152.

Rudolph, G. (1997). How mutation and selection solve long-path problems in polynomial expected time, *Evolutionary Computation* **4**, 2, pp. 195–205.

Sinha, A., Chen, Y. and Goldberg, D. E. (2004). Designing efficient genetic and evolutionary algorithm hybrids, in [Hart *et al.* (2004)], pp. 259–288.

Sudholt, D. (2006a). Local search in evolutionary algorithms: the impact of the local search frequency, in *Proceedings of the 17th International Symposium on Algorithms and Computation (ISAAC '06)*, *LNCS*, Vol. 4288 (Springer), pp. 359–368.

Sudholt, D. (2006b). On the analysis of the (1+1) memetic algorithm, in *Proceedings of the Genetic and Evolutionary Computation Conference (GECCO '06)* (ACM Press), pp. 493–500.

Sudholt, D. (2008a). *Computational Complexity of Evolutionary Algorithms, Hybridizations, and Swarm Intelligence*, Ph.D. thesis, Technische Universität Dortmund, URL `http://hdl.handle.net/2003/25954`.

Sudholt, D. (2008b). Memetic algorithms with variable-depth search to overcome local optima, in *Proceedings of the Genetic and Evolutionary Computation Conference (GECCO '08)* (ACM Press), pp. 787–794.

Sudholt, D. (2009). The impact of parametrization in memetic evolutionary algorithms, *Theoretical Computer Science* **410**, 26, pp. 2511–2528.

Watson, J.-P., Howe, A. E. and Whitley, L. D. (2003). An analysis of iterated local search for job-shop scheduling, in *Fifth Metaheuristics International Conference (MIC '03)*.

Williams, D. (1991). *Probability with Martingales* (Cambridge University Press).

Witt, C. (2006). Runtime analysis of the (μ+1) EA on simple pseudo-Boolean functions, *Evolutionary Computation* **14**, 1, pp. 65–86.

Chapter 6

Simulated Annealing

Thomas Jansen

University College Cork
Department of Computer Science
Cork, Ireland
t.jansen@cs.ucc.ie

Simulated annealing is a randomized search heuristic that is inspired by the annealing process in metallurgy. Like randomized local search it solves optimization problems by randomly moving from one candidate solution to a neighboring one. Unlike randomized local search it accepts such a move with a positive probability even if the new candidate solution is worse. The probability of accepting such a worsening move depends on the difference in function value and the current temperature that is defined by a cooling schedule. Using a constant temperature without any annealing yields the Metropolis algorithm.

The most important theoretical results concerning simulated annealing are given. The emphasis is on results concerning the performance by means of analyzing the optimization time. Concentrating mainly on combinatorial optimization problems strengths and weaknesses of simulated annealing are highlighted. Thorough analytical comparisons with the Metropolis algorithm and a simple and structurally similar evolutionary algorithm, the (1+1) EA, are presented.

Contents

6.1. Introduction

Simulated annealing is a general randomized search heuristic that is structurally simple and that has been applied successfully in many different contexts. It may be described as a nature-inspired search heuristic since it draws its motivation from the annealing process in metallurgy. In this process, heating and controlled cooling are applied to some material such that the atoms engage in a kind of random walk on their states ending up in a configuration with minimal energy. Capturing this idea in an algorithm yields a random process over a space of configurations where the probability of actually moving to a new configuration is determined by the difference in energy and the current 'temperature.'

Different equivalent algorithms are around that all implement this idea. In this chapter we define simulated annealing to be the algorithm described below. The algorithm is designed in a form that is suitable for the minimization of an objective function $f\colon S \to \mathbb{R}$ where S is some finite *search space.* Typical examples of such search spaces are the set of fixed-length bit strings $\{0,1\}^n$ or the set of all permutations over $\{1,2,\ldots,n\}$, S_n. For the search space S some notion of a *neighborhood* N needs to be defined. A neighborhood is a relation $N \subseteq S \times S$. We use the notation of a function $N\colon S \to \mathcal{P}(S)$ to refer to the neighborhood of a search point $s \in S$, given as $N(s) = \{s' \in S \mid (s,s') \in N\}$. Often the neighborhood N is irreflexive, i.e., $s \notin N(s)$ for all $s \in S$. We refer to $|N| := \max\{|N(s)| \mid s \in S\}$ as the *size* of the neighborhood. In many cases $|N| = |N(s)|$ holds for all $s \in S$. It is safe to assume that the search space S is very large, otherwise minimization of f can trivially be achieved by enumeration of f. One considers simulated annealing to be efficient if it is able to locate a global maximum of f with sufficiently large probability within a number of steps that is polynomially bounded in $\log(|S|)$.

The current temperature $T(t)$ is determined by a function $T\colon \mathbb{N}_0 \to \mathbb{R}^+$ that is called *cooling schedule.* If the search point $y \in N(x_t)$ has some function value $f(y)$ that is not larger than $f(x_t)$, it will replace x_t as current search point. If its function value is larger it may still be accepted. This happens with probability $e^{-(f(y)-f(x_t))/T(t)} > 0$. Clearly, this probability decreases with increasing difference $f(y) - f(x_t)$ and decreasing current temperature $T(t)$. If the current temperature approaches zero such moves become impossible. If the new search point $y \in N(x_t)$ replaces the current search point x_t we say that the move to y is *accepted.* Otherwise we say that

Algorithm 16 Simulated annealing

1. Set $t := 0$. Choose $x_t \in S$ uniformly at random.
2. Repeat
3. Choose $y \in N(x_t)$ uniformly at random.
4. With probability $\min\left\{1, e^{-(f(y)-f(x-t))/T(t)}\right\}$ set $x_{t+1} := y$ else set $x_{t+1} := x_t$.
5. $t := t + 1$
6 Until some stopping criterion is fulfilled.

the move to y is *rejected*. Due to its natural inspiration one often considers only non-increasing cooling schedules. Note that this is not necessary from an algorithmic point of view, though. Having an increasing or oscillating temperature makes the notion of a cooling schedule questionable, however.

Stopping criteria are often very simple. The algorithm may be stopped after some pre-defined number of steps or, equivalently, after the current temperature has dropped to some pre-defined value. Slightly less simple stopping criteria stop the algorithm if the function value of the current search point has not been decreased for some pre-defined number of steps. In practice it is not unusual to let simulated annealing run and stop the algorithm manually when the current solution is acceptable or waiting longer is infeasible for practical reasons. For theoretical run time analyses simulated annealing is sometimes considered as an infinite random process and the first point of time where an optimal value is found is considered. This random point of time $\tau := \min\{t \in \mathbb{N}_0 \mid f(x_t) = \min\{f(s) \mid s \in S\}\}$ is called the *optimization time*.

Simulated annealing was introduced by Kirkpatrick *et al.* (1993). It is interesting to note that the only novel aspect they introduced was employing a cooling schedule. In fact, simulated annealing with constant temperature $T \in \mathbb{R}^+$ was known as the *Metropolis algorithm* already since the early 1950s (Metropolis *et al.*, 1953). While the Metropolis algorithm can accurately be described as a degenerate case of simulated annealing where the cooling schedule degenerated to some constant temperature it is not at all obvious that the increased flexibility of simulated annealing is actually needed. It has turned out to be surprisingly difficult to actually prove that it is. The fact that it has been proved demonstrates that simulated annealing, while being introduced as a practical optimization tool in the context of placement problem in digital circuit design, has also attracted theoretical interest and has been subject of many studies.

In this chapter we touch some of these issues and review selected parts of the relevant literature. In the next section we discuss the crucial choice of a cooling schedule and what guidelines may be applied. We shortly discuss means of increasing simulated annealing's efficiency in practice and touch the most fundamental theoretical issue, global optimization there. In Section 6.3 we consider different concrete analyses of the optimization time. We summarize findings for fixed temperatures, in fact concerned with the Metropolis algorithm, as well as results on non-trivial cooling schedules. This section also contains the proof that making use of a cooling schedule can be crucial for the success of simulated annealing. Simulated annealing bears strong resemblance with a number of different randomized search heuristics. These are highlighted and discussed in Section 6.4.

6.2. Practical and Theoretical Issues

When simulated annealing is to be applied for optimization we can assume that the problem is already modeled in form of an objective function $f\colon S \to \mathbb{R}$ that needs to be minimized. For many search spaces S natural neighborhood concepts N exist. Nevertheless, the influence of the choice of the neighborhood N should not be underestimated. After these basic definitions are made one still needs to come up with an appropriate cooling schedule before the algorithm can actually be started.

When optimizing an objective function $f\colon S \to \mathbb{R}$ one is looking for some search point $s^* \in S$ such that $f(s^*) = \min\{f(s) \mid s \in S\}$ holds. Finding such a global optimum can be made more difficult if local optima exist. A point $s \in S$ is called a *local optimum* if for all $s' \in N(s)$ we have $f(s') \geq f(s)$. Simulated annealing is able to escape such local optima if $e^{-h(s)/T(t)}/|N|$ is not too small, where $h(s) := \min\{f(s') - f(s) \mid s' \in N(s)\}$. The probability $e^{-h(s)/T(t)}$ is the probability of accepting the most probable move out of the local optimum s and $1/|N|$ is a lower bound on moving to such an $s' \in N(s)$. Since making the move and accepting it are independent we obtain $e^{-h(s)/T(t)}/|N|$ as lower bound for the probability to escape from s. The value $h(s)$ is called the barrier height.

Kirkpatrick *et al.* (1993) introduced an exponential cooling schedule $T(t) = T_0 \cdot \alpha^t$ for some constant $\alpha \in (0;1)$ as well as a linear cooling schedule $T(t) = T_0 - \alpha \cdot t$ for some constant $\alpha \in \mathbb{R}^+$. Moreover, logarithmic cooling schedules $T(t) = T_0 \cdot \ln(2)/\ln(2+t)$ are important. It is rather obvious that even the initial temperature T_0 cannot be chosen completely independent of the objective function f. One needs the initial temperature sufficiently large such that the probability of accepting a worsening move, $e^{-\Delta/T(t)}$ is not too small. Clearly, the difference Δ depends on the objective function f and the temperature needs to be adjusted so that it matches these differences in function value. If we have a maximal barrier height $h = \max\{h(s) \mid s \in S \text{ local optimum}\}$ then at least e^{-h/T_0} needs to be reasonably large. Given an appropriate initial temperature T_0 the parameter α needs to be set appropriately. It needs to be set in a way that the temperature is decreased sufficiently slowly so that simulated annealing has sufficient time to escape from local optima. On the other hand, the temperature needs to be decreased sufficiently fast so that in the vicinity of a global optimum no large worsening leaving away from it are accepted. Since in practice the features of the objective function f are usually not completely known the cooling schedule is difficult to set. Even for completely known example functions $f\colon S \to \mathbb{R}$ it can be difficult to find optimal or even good cooling schedules. Barriers are not the only property of an objective function that can be taken into account for this (Flamm *et al.*, 2007).

If no good cooling schedule is found simulated annealing may get stuck in a local optimum. If this happens only with reasonably bounded probability this may not be a problem in practice. Consider the optimization time $\tau = \min\{t \in \mathbb{N}_0 \mid f(x_t) = \min\{f(s) \mid s \in S\}\}$ and assume that $\Pr(\tau \leq p(n)) \geq 1/q(n)$ holds for two polynomials p and q and $n := \log(|S|)$. Then we can start simulated annealing, stop it after $p(n)$ steps and start again. On average, after at most $q(n)$ starts and thus after at most $p(n) \cdot q(n)$ steps (and thus in polynomial expected time) a global optimum is found. We see that simulated annealing may be efficient in practice if combined with an appropriate restart mechanism even if the success probability $\Pr(\tau \leq t(n))$ converges to zero polynomially fast for any polynomial number of steps t. It is not even necessary to know about an appropriate number of steps p to achieve a polynomial expected optimization time. We can perform 2^i runs of simulated annealing of exactly 2^i steps for increasing $i = 1, 2, \ldots$ This way, the expected optimization time is bounded above by $O(\max\{p(n)^2, q(n)^2\})$ (Wegener, 2005).

The seemingly most important question when applying a randomized search heuristic to the task of optimization is the question of so called global convergence. Does it optimize at all? Given sufficient time will it find a global optimum? Does $\lim_{t\to\infty} \Pr(\tau \leq t) = 1$ hold? Already our discussion of restarts demonstrated that in a practical setting this question is not that important. Simulated annealing may be efficient on problems even if the probability to find a global optimum does not converge to one. The question is also not practically relevant for another reason. If global convergence is to be proved for arbitrary objective functions $f\colon S \to \mathbb{R}$ it also needs to hold for the most difficult functions. Since we cannot hope to optimize the most difficult functions in reasonable time such proofs of global convergence necessarily make statements over unreasonably long runs. Thus, global convergence is of very limited practical relevance. Nevertheless, it is theoretically interesting and not at all trivial to see whether simulated annealing achieves global convergence for arbitrary objective functions $f\colon S \to \mathbb{R}$. Hajek (1988) proves that this is indeed the case when an appropriate logarithmic cooling schedule is applied. A very accessible proof of the same fact is presented by Michiels *et al.* (2007).

6.3. Run Time Analyses

Several analyses of the run time simulated annealing achieves for different types of problems have been presented. While randomized search heuristics in general and simulated annealing in particular are not meant to replace problem-specific algorithms it still makes sense to analyze their performance for specific problems and classes of problems. An analysis for a combinatorial optimization problem or some other class of objective functions can shed light on the performance of simulated annealing in typical situations. An analysis for a specific example function can be used to highlight assets and drawbacks of simulated annealing or to demonstrate differences from other randomized search heuristics in a particularly clear way. We consider examples for both kinds of analyses in this section. For some results a non-trivial cooling schedule needs to be employed in order to achieve good performance. In other cases simulated annealing fails for any cooling schedule or a trivial cooling schedule suffices. We consider the latter case first.

6.3.1. *Results Not Depending on Non-Trivial Cooling Schedules*

Sasaki and Hajek (1988) considered the performance of simulated annealing when applied to the problem of computing a maximum matching. For an undirected graph $G = (V, E)$ a matching is a subset of the edges $M \subseteq E$ such that no two edges of the matching share a common node: $\forall e \neq e' \in M \colon e \cap e' = \emptyset$. A maximum matching is a matching M such that no other matching contains more edges. Thus, M has maximum cardinality among all matchings. It has been long known that the computation of maximum matchings is possible in polynomial time (see (Vazirani, 1994) for an efficient algorithm for the general case). When solving this problem by means of simulated annealing the most natural encoding describes the problem as selecting edges in the following way. Let $E = \{e_1, e_2, \ldots, e_n\}$ be the set of edges. The search space $S = \{0,1\}^n$ is the set of all characteristic vectors of this set E, so that $s \in S$ defines the matching $M(s) := \{e_i \mid s_i = 1\}$. For the sake of convenience we identify s and $M(s)$ in the following. The objective function can be simply defined by $f(s) := -\sum_{i=1}^{n} s_i$ if s is a matching and $f(s) = \sum_{i=1}^{n} s_i$ otherwise. This way infeasible search points can easily be transformed into feasible search points by removing single edges. Such removals are rewarded by a decrease in function value by one. It is easy to see that the set of all maximum matchings corresponds exactly to the set of all global optima. The neighborhood may be defined as the set of all Hamming neighbors $N(s) := \{s' \in S \mid H(s, s') = 1\}$ where $H(s, s') = \sum_{i=1}^{n} |s_i - s'_i|$ denotes the Hamming distance of s and s', i.e., the number of differing bits. After deciding these details one can now ask if simulated annealing is able to find a maximum matching for an arbitrary graph on average in polynomial time. Sasaki and Hajek (1988) proved that this is not the case by presenting a graph that turns out to be a particularly difficult instance for simulated annealing as well as for other randomized search heurstics (Giel and Wegener, 2003). The graph is defined over $3k$ nodes for some even $k \in \mathbb{N}$. Let $V^* = \{v_{i,1}, v_{i,2}, v_{i,3} \mid i \in \{1, 2, \ldots, k\}\}$ be the set of nodes. The nodes are arranged in three lines by $L_i = \{\{v_{j,i}, v_{j+1,i} \mid j \in \{1, 2, \ldots, k-1\}\}$ for $i \in \{1, 2, 3\}$. Moreover, the graph contains $(k/2) - 1$ complete bipartite graphs by having $E_i = \{\{v_{2i,j}, v_{2i+1,j'} \mid j, j' \in \{1, 2, 3\}\}\}$ for $i \in \{1, 2, \ldots, (k/2) - 1\}$. The complete graph is defined as $G^* = (V^*, E^*)$ with $E = \left(\bigcup_{i=1}^{(k/2)-1} E_i\right) \cup L_1 \cup L_2 \cup L_3$. In order to see why this graph is difficult for simulated annealing one needs to understand how the algorithm finds a solution.

Consider some graph $G = (V, E)$ and some matching $M \subseteq E$. A path is a sequence of edges $e_1, e_2, \ldots, e_l$ with $e_i = \{e_{i,1}, e_{i,2}\}$ such that $e_{i,2} = e_{i+1,1}$ holds for all $i \in \{1, 2, \ldots, l-1\}$ and each node appears in this path at most two times. A path is called an alternating path if $e_1, e_3, e_5, \cdots \notin M$ and $e_2, e_4, e_6, \cdots \in M$ holds. An alternating path is called an augmenting path if $e_1 \notin M$ and $e_l \notin M$. It is known that if and only if M is a maximum matching there does not exist an augmenting path in G (Cormen *et al.*, 2001). Simulated annealing can increase the size of a matching by 'following' such augmenting paths flipping edges in and out. The probability to achieve this decreases with increasing length of the augmenting path. The crucial property of the graph G^* is found at the complete bi-partite subgraphs. There an augmenting path can be increased in length in two directions and decreased in length only in one direction. Thus, with probability close to one augmenting paths tend to be increased in length over time. This implies that with probability exponentially close to one simulated annealing will not find a maximum matching even in an exponential number of steps (Sasaki and Hajek, 1988). We summarize this as a theorem. It is interesting to note that the same holds for some other randomized search heuristics, in particular for a simple evolutionary algorithm known as the (1+1) EA (Giel and Wegener, 2003).

Theorem 6.1 (Sasaki and Hajek (1988)). *On the described graph $G^* = (V^*, E^*)$ with $n = 3k$ nodes the expected optimization time of simulated annealing is $e^{\Omega(n)}$ for any cooling schedule.*

While simulated annealing is unable to find maximum matchings efficiently it is able to approximate maximum matchings efficiently (Sasaki and Hajek, 1988). The key observation is that while a matching M is still far from being optimal there are many augmenting paths. This implies that the shortest augmenting path is very limited in length. This is sufficient to prove a polynomial upper bound on the expected time needed for an improvement. Again, it is interesting to note that this holds for the (1+1) EA, too (Giel and Wegener, 2003).

Improving matchings requires as only necessary worsening move the exclusion of single edges. This does not change over time so that there is no need for a non-trivial cooling schedule. What simulated annealing can achieve on the maximum matching problem can be achieved by means of the Metropolis algorithm, i.e., with a constant temperature (Sasaki and Hajek, 1988).

A different combinatorial optimization problem, namely graph bisection, was considered by Jerrum and Sorkin (1998). In graph bisection one considers an undirected graph $G = (V, E)$. A bisection is a partition $V_1 \uplus V_2$ with $|V_1| = |V_2| = |V|/2$. One is looking for a bisection that minimizes the number of edges that are cut by this partition $\sum_{v_1 \in V_1, v_2 \in V_2} |\{v_1, v_2\} \cap E|$. Jerrum and Sorkin (1998) consider a class of random instances of the graph bisection problem $\mathcal{G}_{n,p,r}$ where the graph has $|V| = n$ nodes that are partitioned into two disjoint sets V_1, V_2 of equal size. Edges are introduced randomly by considering independently each pair of distinct nodes. If both nodes are from the same set of the partition an edge is introduced with probability p, otherwise it is introduced with probability r. For $p > r$ the optimal bisection is (V_1, V_2) with probability very close to one. Jerrum and Sorkin (1998) prove that the difference $p - r$ is not too small simulated annealing is successful on a random instance of this kind with probability very close to one. Similar to the maximum matching problem the behavior of simulated annealing does not need to change over time. Thus, good performance can be achieved with a constant temperature, i.e., the Metropolis algorithm. We cite this result without giving a full proof.

Theorem 6.2 (Jerrum and Sorkin (1998)). *Consider a random instance $\mathcal{G}_{n,p,r}$ of the graph bisection problem with $|V| = n$ nodes. Let $p - r = \Theta(1/n^{\Delta})$ hold, let $0 \leq \Delta < 1/6$ be some constant. Simulated annealing finds the optimal bisection time $O(n^2)$ with probability exponentially close to 1.*

6.3.2. *Results Depending on Non-Trivial Cooling Schedules*

Both analyses, the one for maximum matching and graph bisection, show that good performance can be achieved with a constant temperature. Having a cooling schedule appears to be an unnecessary feature. This observation gives rise to the question if there are problems where the Metropolis algorithm is clearly outperformed by simulated annealing, i.e., if a cooling down of the current temperature can be essential. Note that it is not sufficient to demonstrate that simulated annealing outperforms the Metropolis algorithm for some fixed temperature. It is necessary to show that even with an optimally adjusted constant temperature for the Metropolis algorithm simulated annealing with an appropriate cooling schedule excels. This question of practical relevance since applying the Metropolis algorithm is simpler than applying simulated annealing. For the Metropolis algorithm only an appropriate fixed temperature needs to be found. For simulated annealing one has to define a complete cooling schedule.

Sorkin (1991) proves that having a non-trivial cooling schedule can in fact be essential. To this end he defines a rather complicated fractal objective function and proves in a rather involved way that with no constant temperature efficient optimization is possible. On the other hand, simulated annealing equipped with an appropriate cooling schedule optimizes this fractal function with probability close to 1. The same result is obtained by Droste *et al.* (2001) using a very simply structured example function and a rather elementary proof.

While Sorkin (1991) as well as Droste *et al.* (2001) were able to prove that having a non-trivial cooling schedule can be crucial for the success of simulated annealing both proofs make use of artificial example functions that are specifically designed for this purpose. While such results are sufficient to answer the open question it is much nicer to have such a result for a natural problem. It is good to know that simulated annealing can be clearly superior to the Metropolis algorithm. It is much nicer to know, though, that this is can also hold in practice.

The first such result was obtained by Wegener (2005). He considers the problem of computing a minimum spanning tree in a weighted undirected graph $G = (V, E)$ with weights $w \colon E \to \mathbb{N}$. A spanning subgraph is a subgraph $C = (V, E')$ of G, $E' \subseteq E$, such that for all $v, v' \in V$ there is a path connecting v and v' in G'. A spanning subgraph C is a spanning tree if in addition $|E'| = |V| - 1$ holds. The weight of a spanning subgraph C equals the sum of the weights of its edges, $w(C) = \sum\limits_{e \in E'} w(e)$. A minimum spanning tree is a spanning tree with minimal weight. The minimum spanning tree problem is one of the classical textbook combinatorial optimization problems and it is well known that minimum spanning trees can easily be computed in polynomial time (Cormen *et al.*, 2001).

In order to solve the minimum spanning tree problem by means of simulated annealing Wegener (2005) defines search space, neighborhood and objective function for this problem in the following way. The search space is chosen analogously to the modeling of the maximum matching problem due to Sasaki and Hajek (1988). Thus, let $E = \{e_1, e_2, \ldots, e_n\}$ be the set of edges. The search space $S = \{0, 1\}^n$ equals the set of all characteristic vectors of this set E, so that $s \in S$ defines the set of edges $C(s) = \{e_i \mid s_i = 1\}$. For this search space Wegener (2005) considers the direct Hamming neighbors as neighborhood function, i.e., $N(s) = \{s' \in S \mid H(s, s') = 1\}$. The objective function f is defined as $f(s) = \sum_{i=1}^{n} s_i \cdot w(e_i)$ if $C(s)$ is a spanning subgraph and $f(s) = \infty$ otherwise. In order to avoid to the problem of finding some feasible solution s with $f(s) < \infty$ simulated annealing is not started with some $x_0 \in S$ chosen uniformly at random but deterministically with $x_0 = 1^n$, the trivial spanning subgraph. Note that the weighted graph (G, w) is assumed to be connected so that G itself is a spanning subgraph of G.

Considering this modeling of the minimum spanning tree problem Wegener (2005) defines a specific instance of this problem. Let $k \in \mathbb{N}$, the set of nodes V is defined by $V := \{v_1, v_2, \dots, v_{4k+1}\}$. On these $4k + 1$ nodes exactly $2k$ connected triangles are defined. The i-th triangle is defined by the edge set $E_i := \{\{v_{2i-1}, v_{2i}\}, \{v_{2i}, v_{2i+1}\}, \{v_{2i-1}, v_{2i+1}\}\}$ and we have edge sets for $i \in \{1, 2, \dots, 2k\}$. Since this is the complete graph we have $E := \bigcup_{i=1}^{2k} E_i$. Obviously, we have $m := |E| = 6k$. The first k of these triangles are called heavy triangles. Their weights are defined by $w(\{v_{2i-1}, v_{2i+1}\}) = m^3$, $w(\{v_{2i-1}, v_{2i}\}) = m^2$, and $w(\{v_{2i}, v_{2i+1}\}) = m^2$ for $i \in \{1, 2, \dots, k\}$. The remaining k triangles are called light triangles. Their weights are defined by $w(\{v_{2i-1}, v_{2i+1}\}) = m$, $w(\{v_{2i-1}, v_{2i}\}) = 1$, and $w(\{v_{2i}, v_{2i+1}\}) = 1$ for $i \in \{k+1, k+2, \dots, 2k\}$. This completes the definition of the minimum spanning tree instance. We denote the instance as $T^*(1, m, m^2, m^3)$ collecting the four different weights used.

We see that each triangle contains exactly two light edges and one heavy edge. The weight of the heavy and light edges differ by a factor of exactly m in both kinds of triangles, the heavy and the light ones. The weights of the corresponding edges in the heavy and light triangles differ by a factor of exactly m^2. Each spanning subgraph needs to include at least two edges from each triangle. Moreover, each subgraph that includes at least two edges from each triangle is a spanning subgraph. This implies that the minimum spanning tree contains exactly all light edges. For a collection of edges $C(s)$ we classify the triangles in the following way. If all three edges are contained in $C(s)$ the triangle is called complete. If exactly one light edge and the heavy edge belong to $C(s)$ the triangle is called bad. If exactly the two light edges are contained in $C(s)$ the triangle is called good. In all other cases the triangle is called incomplete.

Simulated annealing starts with the spanning graph $x_0 = 1^n$ and $w(1^n) = \sum_{i=1}^{n} w(e_i) < \infty$ holds. Since we have $w(s) = \infty$ for any $s \in \{0,1\}^n$ where $C(s)$ contains at least one incomplete triangle we see that during a run of simulated annealing the current search point can never contain an incomplete triangle. Thus, at all times, all triangles are either complete, good or bad. In the beginning all triangles are complete. In a complete triangle any of the three edges can be removed. Since this decreases the function value such a move is accepted with probability one. This is independent of the kind of edge that is removed, it holds for heavy edges as well as for light edges. It does also make no difference if the triangle is heavy or light. Given that in a complete triangle an edge is removed we see that with conditional probability 1/3 the heavy edge is removed and with conditional probability 2/3 a light edge is removed. The removal of a light edge turns a complete triangle into a bad triangle. Application of Chernoff bounds (Motwani and Raghavan, 1995) yields that after $\Theta(m)$ steps there is a linear number of bad triangles with probability exponentially close to one. Even more, with probability exponentially close to one there is a linear number of bad heavy triangles.

In order to find the minimum spanning tree each bad triangle needs to transformed into a good triangle. This cannot happen in a single move. It is only possible by first turning a bad triangle into a complete triangle. This increases the weight of the edge selection by the weight of the light edge that needs to be included. Thus, for a heavy triangle the probability to accept such a move equals $e^{-m^2/T(t)}$. For low temperatures $T(t) \leq m$ this probability is bounded above by e^{-m}. Thus for any fixed temperature $T \leq m$ the Metropolis algorithm needs an exponential number of steps to find a minimum spanning tree for this instance with probability exponentially close to one.

But the instance is difficult for the Metropolis algorithm not only if small temperatures $T \leq m$ are used. For large temperatures $T > m$ the light triangles become difficult to optimize. Consider an arbitrary light triangle. If the triangle is complete it is turned into a good triangle with probability $1/m$ in the next step. With probability $2/m$ it is turned into a bad triangle in the next step. Thus, with probability $1 - 3/m$ it remains complete. If the light triangle is bad it is turned into a complete triangle with probability $e^{-1/T}/m$ since the move that increases the weight by the weight of the light edge needs to be made and accepted. The light bad triangle cannot be turned into a good triangle in a single step. Thus it stays bad with probability $1 - e^{-1/T}/m$. Analogously, if the light triangle is good it cannot be turned into a bad triangle in a single step. It is turned into a complete triangle with probability $e^{-m/T}/m$ and it stays good with probability $1 - e^{-m/T}/m$. Let G_t denote the random number of good light triangles in the t-th step. We have $G_0 = 0$ and $G_t \in \{0, 1, \ldots, k\}$. We are interested in the first point of time when $G_t = k$ holds. Since in each step only a single edge can be added or removed we have $G_{t+1} \in \{G_t - 1, G_t, G_t + 1\}$ for all time steps t. According to our observations about transforming triangles we have $\Pr(G_{t+1} = G_t + 1) \leq (k - G_t)/m$ since there can be at most $k - G_t$ complete triangles. Moreover, we have $\Pr(G_{t+1} = G_t - 1) = G_t \cdot e^{-m/T}/m > G_t/(3m)$ since we consider the case of a large temperature $T > m$. Now we consider the case that G_t changes. The conditional probability that G_t is decreased is bounded by below by $G_t/(3k - 2G_t)$. Since we have $G_0 = 0$ and want to increase G_t to k we need to have a phase where G_t is increased from $(10/11)k$ to k without being decreased to $(9/11)k$. In this phase the conditional probability to decrease G_t is bounded below by $3/5$. Accordingly, the conditional probability to increase G_t is bounded above by $2/5$. Applying results on the gambler's ruin problem (Feller, 1957) we see that this does not happen with probability exponentially close to one even in an exponential number of steps. We summarize this finding in the following theorem.

Theorem 6.3 (Wegener (2005)). *Consider the described triangle instance $T^*(1, m, m^2, m^3)$ and the Metropolis algorithm with any fixed temperature. The probability that the Metropolis algorithm finds the minimum spanning tree within e^{cm} steps is $e^{-\Omega(-m)}$ for any constant $c > 0$.*

We see that this specific triangles instance of the minimum spanning tree problem cannot be efficiently solved by the Metropolis algorithm regardless of the temperature T. Now Wegener (2005) considers simulated annealing with an exponential cooling schedule with initial temperature $T_0 = m^3$ and cooling factor $\alpha = 1 - 1/(cm)$ for a sufficiently large constant $c > 0$. After $O(m \log m)$ steps the current temperature $T(t)$ has dropped to one or below. Our aim is to prove that by this time the minimum spanning tree is found with probability close to one. Clearly, in this time there is a phase where for the current temperature $m^2 \leq T(t) \leq m^2\sqrt{m}$ holds and another phase where $1 \leq T(t) \leq \sqrt{m}$ holds. The length of each of these phases is bounded below by $(c/4)m \ln m$. Note that the boundary conditions of the two phases differ by a factor of exactly m^2. This coincides with the factor the weights in the heavy and light triangles differ. This, of course, is not a coincidence.

Consider the phase with $m^2 \leq T(t) \leq m^2\sqrt{m}$. In this phase we only consider heavy triangles and do not care at all about the light triangles. The probability to accept a move that turns a heavy good triangle into a complete triangle is bounded above by $e^{-m^3/(m^2\sqrt{m})} = e^{-\sqrt{m}}$. If this happens the run is considered to be a failure. On the other hand, the probability to accept a move that turns a heavy bad triangle into a complete triangle is bounded below by $e^{-m/m^2} = e^{-1/m} = \Omega(1)$. It follows from results on the coupon collector's problem (Motwani and Raghavan, 1995) that in $(c/4)m \ln m$ steps (with c sufficiently large) each of the $k = m/6$ heavy triangles was subject to at least some positive number of moves with probability very close to 1. We see that in this phase for heavy triangles it is easy to transform a bad triangle into a good triangle but very hard to turn a good triangle into a bad one. We can conclude that at the end of this phase all heavy triangles are good triangles with probability very close to one. The probability that after this phase a heavy good triangle is changed again is bounded by $e^{-\Omega(\sqrt{m})}$. Thus, with probability very close to one at the end after $O(m \log m)$ steps all heavy triangles are good.

Now consider the phase $1 \leq T(t) \leq \sqrt{m}$. Clearly, we only have to consider the light triangles in this phase. Since the weights in the light triangles differ from the corresponding weights in the heavy triangles by the factor m^2 and since the bounds on the current temperature in this phase differ by exactly the same factor from the corresponding bounds in the other phase everything said about heavy triangles in the other phase applies to light triangles in this phase. Thus, with probability very close to one after $O(m \log m)$ steps simulated annealing with temperature $T(t) = m^3 \cdot (1 - 1/(cm))^t$ ($c > 0$ a sufficiently large constant) finds the minimum spanning tree for this instance of the minimum spanning tree problem Wegener (2005). Again we summarize this in the following theorem.

Theorem 6.4 (Wegener (2005)). *Consider the described triangle instance $T^*(1, m, m^2, m^3)$, a polynomial p, and the exponential cooling schedule $T(t) = m^3 \cdot (1 - 1/(cm))^t$ where $c > 0$ is a sufficiently large constant. The probability that simulated annealing with this cooling schedule finds the minimum spanning tree in $O(m \log m)$ steps is bounded below by $1 - 1/p(m)$.*

The reason why the Metropolis algorithm is unable to cope with this specific instance is that no appropriate constant temperature exists. For the heavy triangles it is necessary that it is not too unlikely to include a light edge that has weight m^2. Since the probability to accept such a move equals $e^{-m^2/T}$ we need to have $T = \Omega(m^2/\log m)$. For the light triangles it is necessary that it is not too likely to include a heavy edge since otherwise good triangles become complete again too easily. Since the probability to accept such a move equals $e^{-m/T}$ we need to have $T = o(m)$. Clearly, we cannot have a temperature satisfying both, $T = \Omega(m^2/\log m)$ and $T = o(m)$. Thus, the Metropolis algorithm is bound to fail on this instance of the minimum spanning tree problem. Simulated annealing, on the other hand, can start with a large temperature and first solve the part of the instance where a high temperature is needed. After that the temperature is lowered and the other part of the solution where the temperature needs to be lower can be optimized. It thus suffices to identify two different instances of the same problem, one where low temperatures fail and another one where high temperatures fail. If these two instances can be combined into one we obtain an instance where provably simulated annealing excels and the Metropolis algorithm fails regardless of the temperature. We refer to this method of separating simulated annealing and the Metropolis algorithm as the Wegener method since it is due to Wegener (2005). It can and has been applied to other combinatorial optimization problems.

Meer (2007) considers the traveling salesperson problem (TSP). In the TSP we consider a complete graph on k nodes with weights $w_{i,j} \in \mathbb{N}_0 \cup \{\infty\}$ for all $i, j \in \{1, 2, \ldots, k\}$. A tour is a permutation $\pi = (v_1, v_2, \ldots, v_k)$ over $\{1, 2, \ldots, k\}$. Such a permutation is called a tour because it corresponds to visiting all nodes exactly once in the order $v_1, v_2, \ldots, v_k$ and returning from v_k to the first v_1, again. The length of a tour π is given by the sum of the weights of the edges used, i.e., $l(\pi) = w_{v_k,v_1} + \sum_{i=1}^{k-1} w_{v_i,v_{i+1}}$. One is looking for a tour with minimal length. Meer (2007) models this problem in order to tackle it by means of simulated annealing in the following way. The search space S is the set of all permutations over $\{1, 2, \ldots, k\}$. The neighborhood N is defined for a permutation $\pi = (v_1, v_2, \ldots, v_k)$ by $N(\pi) := \{(v_1, v_2, \ldots, v_i, v_j, v_{j-1}, \ldots, v_{i+1}, v_{j+1}, v_{j+1}, \ldots, v_k) \mid i, j \in \{1, 2, \ldots, k\}, i < j\}$. Thus, two permutations π and π' are neighbors, if we can transform π into π' be removing the two edges (v_i, v_{i+1}), (v_j, v_{j+1}) and adding the two edges (v_i, v_j), (v_{i+1}, v_{j+1}) for some $i < j$. The objective function f is simply defined by $f(\pi) := l(\pi)$.

Using this modeling Meer (2007) defines a specific TSP instance applying the Wegener method. The construction is based on a small graph that will play the same role the triangles did in Wegener's construction for his instance of the minimum spanning tree problem. The small graph contains six nodes $v_1, v_2, v_3, \ldots, v_6$ and the nine edges $\{v_1, v_2\}$, $\{v_1, v_3\}$, $\{v_1, v_5\}$, $\{v_2, v_3\}$, $\{v_2, v_4\}$, $\{v_3, v_4\}$, $\{v_3, v_6\}$, $\{v_4, v_5\}$, and $\{v_5, v_6\}$. We define these small graphs in a heavy and a light way. All edges that are not present are assigned weight ∞. The edge $\{v_1, v_2\}$ will be a light edge. The edge $\{v_1, v_3\}$ is a heavy edge. All other edges are always assigned weight 1. In a heavy graph, the light edges have weight $m^3 - m^2 + 1$ and the heavy edge has weight $m^3 + 1$. In a light graph, the light edges have weight m and the heavy edge has weight $m + 1$. We connect the small graphs in a way that in each tour the small graph needs to be entered at node v_1 and left at node v_6 (or the other way round). No other node has any edge with finite weight connecting it to any other part of the graph. There are only three ways this can be done inferring finite weight. We call the permutation $(v_1, v_2, v_3, v_4, v_5, v_6)$ the middle way. It contains the light edge and thus its total weight is the weight of this edge increased by 4. The permutation $(v_1, v_3, v_2, v_4, v_5, v_6)$ is called the bad way. It contains the heavy edge and thus its total weight is the weight of this edge increased by 4. The permutation $(v_1, v_5, v_4, v_2, v_3, v_6)$ is called the good way because it only contains edges of weight 1. Its total weight is 5. We see that the good way and the bad way are not neighbors. The middle way is neighbor of both other ways. Using $i = 1$, $j = 2$ it can be turned into the bad way. Using $i = 1$, $j = 4$ it can be turned into the good way. Thus, the good way corresponds to a good triangle, the bad way corresponds to a bad triangle, and the middle way corresponds to a complete triangle. Using a linear number of heavy and light such small graphs in one instance we obtain the results here as we obtained for the specific minimum spanning tree instance. For this proof we assume that the initial tour is deterministically set in a way that it contains all middle ways in the small graphs. Using this starting point the proofs carry over almost completely (Meer, 2007).

Theorem 6.5 (Meer (2007)). *Consider the described TSP instance with $n = m/20$ vertices and the Metropolis algorithm with any fixed temperature. The probability to compute the optimal TSP tour within e^{cm} steps (where $c > 0$ is a sufficiently small constant) is $e^{-\Omega(m)}$.*

Consider some polynomial p and simulated annealing with the annealing schedule $T(t) = m^3 \cdot (1 - 1/(cm^2))^t$ where $c > 0$ is a sufficiently large constant. The probability that simulated annealing with this cooling schedule *finds the optimal TSP tour within $O(m^2 \log m)$ is bounded below by $1 - 1/p(m)$.*

Having the result that simulated annealing outperforms the Metropolis algorithm even with optimally tuned constant temperature for combinatorial optimization problems is very nice. Clearly, these results hold for very specific instances only and these instances have been designed exactly for this purpose. It is obviously desirable to obtain a more complete understanding. For the minimum spanning tree problem Wegener (2005) demonstrates how this can be done.

Recall the specific instance of the minimum spanning tree problem containing $2k$ triangles and $m = 6k$ edges. We modify the weights of the edges in the following way. In all $2k$ triangles all light edges $\{v_{2i-1}, v_{2i}\}$, $\{v_{2i}, v_{2i+1}\}$ are set to weight w for all $i \in \{1, 2, \ldots, 2k\}$. All heavy edges $\{v_{2i-1}, v_{2i+1}\}$ are set to weight $(1 + \varepsilon(m))w$. We see that all $2k$ triangles now have equal weights and now consider the Metropolis algorithm for this instance. In order to be successful accepting a move including a light edge must not be to unlikely. Thus, we need $e^{-w/T}$ to be reasonably bounded from below and thus can only consider temperatures $T = \Omega(w/\log m)$. On the other hand it must not be too likely to accept a move that includes a heavy edge. Thus, we need $e^{-(1+\varepsilon(m))w/T}$ to be reasonably bounded from above. If we have $\varepsilon(m) \geq \varepsilon$ for some constant $\varepsilon > 0$ this can be done. Setting $T := (\varepsilon/3) \cdot w/\ln m$ we have that the minimum spanning tree is found with probability very close to one in a polynomial number of steps.

If, on the other hand, $\varepsilon(m) = o(1)$ holds things are different. We already argued that the situation is hopeless if we do not have $T = \Omega(w/\log m)$. The crucial observation is that the event of accepting a move including a heavy edge is not sufficiently less likely than accepting a move including a light edge. If we consider the quotient of the two probabilities we have $e^{-w/T}/e^{-(1+\varepsilon(m))w/T} = e^{\varepsilon(m)w/T} = e^{O(\varepsilon(m)\log m)} = m^{O(\varepsilon(1))}$. Since we have $\varepsilon(m) = o(1)$ this quotient approaches one. Thus, there is only an insignificant advantage for light edges. Let B_t denote the number of bad triangles in the t-th step and let C_t denote the number of complete triangles in the t-th step. We consider $\Phi(t) := 2B_t + C_t$ as a kind of potential function. We have $\Phi(0) = n$ due to initialization with the trivial feasible solution 1^n and $\Phi(t) = 0$ if and only if the minimum spanning tree is found. Considering a single step we remember that it may turn a bad triangle into a complete one (decreasing Φ by 1), turn a complete triangle into a good one (also decreasing Φ by 1), turn a complete triangle into a bad one (inreasing Φ by 1), or turn a good triangle into a complete one (also increasing Φ by 1). We only consider steps that change Φ. For sufficiently large m the conditional probability to increase Φ is bounded below by 3/5 for $B_t \leq \sqrt{m}$. As a consequence, the probability to decrease Φ is bounded above by 2/5 in this situation. We can again apply results for the gambler's ruin problem (Feller, 1957) and see that with probability exponentially close to one the potential is not decreased to 0 even in an exponential number of steps and thus the minimum spanning tree is not found. Note that this holds for arbitrary temperatures. Thus, it does not only hold for constant temperatures but also for arbitrary cooling schedules. We see that for $\varepsilon(1) = o(1)$ neither the Metropolis algorithm nor simulated annealing are able to solve this instance of the minimum spanning tree problem efficiently even with optimal temperature and optimal cooling schedule.

Theorem 6.6 (Wegener (2005)). *Consider the described triangle instance $T^*(w, (1+\varepsilon(m))w, w, (1+\varepsilon(m))w)$, the Metropolis algorithm with any fixed temperature, and simulated annealing with any cooling schedule. For both algorithms the probability that the algorithm finds the minimum spanning tree within e^{cm} steps is $e^{-\Omega(m)}$ for any constant $c > 0$.*

Wegener (2005) accompanies this negative result by a positive result for simulated annealing. An instance of the minimum spanning tree problem is called ε-separated if for any weights w, w' of the instance either $w = w'$ or $\max\{w/w', w'/w\} \geq 1+\varepsilon$ holds. We consider instances that are ε-separated for some constant $\varepsilon > 0$ and where all weights are bounded above by 2^m, m being the number of edges. Wegener (2005) proves that for such instances simulated annealing equipped with an appropriate cooling schedule finds a minimum spanning tree within a polynomial number of steps with probability close to one. The proof strategy is very similar to the proof strategy that simulated annealing is able to optimize the specific minimum spanning tree instance that the Metropolis algorithm cannot optimize efficiently. The cooling schedule is exponential, starts with initial temperature 2^m (equal to the largest allowed value for a weight) and is decreased in each step by a factor that depends on ε. We consider phases, for each weight contained in the instance one such phase is considered. In the phase for one specific weight it is unlikely to include edges with larger weights while edges of this specific weight can still be included with a probability that is not too small. One can prove that with probability very close to one all edges with this specific weight that need to be present in the minimum spanning tree will in fact be included. Doing this for all weights of the instance the result follows. We cite the theorem without giving a complete proof.

Theorem 6.7 (Wegener (2005)). *Consider any instance of the minimum spanning tree problem with m edges. Let all weights be bounded above by 2^m. Let $\varepsilon > 0$ be some constant and let $(w > w') \Rightarrow (w \geq (1+\varepsilon)w')$ hold for all weights w, w'. There exists a cooling schedule such that simulated annealing finds a minimum spanning tree with probability $1 - O(1/m)$ within a polynomial number of steps.*

Note that together with the result on the connected triangles with weights w for the light edges and $(1+\varepsilon(m))w$ for the heavy edges this is a complete characterization of the minimum spanning tree problem for simulated annealing. If the instance is ϵ-separated for some constant $\varepsilon > 0$ simulated annealing can be efficient. As soon as we do not have such an ϵ-separation there exists an instance such that simulated annealing fails to be efficient. Moreover, it is necessary to have simulated annealing. The instance with heavy and light triangles demonstrates that using a constant temperature is insufficient and thus the Metropolis algorithm is much less effective on the minimum spanning tree problem (Wegener, 2005).

6.4. Comparison with other Search Heuristics

Simulated annealing is just one instance of a randomized search heuristic. There are countless other instances. Simulated annealing is a particularly simply structured randomized search heuristic that bears strong similarity with other equally simple randomized search heuristics. In this section we point out such similarities and consider the other similar search heuristics. Recognizing such similarities can be useful for a number of reasons. It helps to understand the role simulated annealing plays in comparison to other search heuristics. The differences in similar search heuristics give hints for what kind of problems a specific heuristic is more suitable than the other. Analytical results for a specific randomized search heuristic (like the analyses for simulated annealing reviewed in this chapter) may be transferrable to similar randomized search heuristics. Even in the cases where the attempt to transfer such analytical results fails the specific way the attempt fails may yield useful insights in the particular ways the search heuristics under consideration function. Assets and drawbacks are easier to recognize in comparison. In the long run such comparisons contribute towards a taxonomy of randomized search heuristics.

The first observation is that simulated annealing bears strong resemblance with randomized local search. In randomized local search we also have a current search point x_t and choose a new search point y from the neighborhood $y \in N(x_t)$. Such a move is accepted if y is not worse than x_t, i.e., $f(y) \leq f(x_t)$. We observe that this corresponds to simulated annealing with temperature 0. Thus, randomized local search can be considered as a special (and degenerate) case of simulated annealing. For randomized local search there is a rich and developed theory (Michiels *et al.*, 2007) including an appropriate complexity theory for local search (Papadimitriou *et al.*, 1990). Unfortunately, theoretical results about local search almost always rely heavily on the fact that worsening moves are never accepted. This is the crucial difference to simulated annealing. Thus results from one algorithm do not transfer easily to the other except for the most simple objective functions. In spite of their obvious similarities the two search heuristics are very different, in particular their performance can be very different.

A second randomized search heuristic that is very similar to simulated annealing is the Metropolis algorithm (Metropolis *et al.*, 1953). As we already pointed out the Metropolis algorithm is the special case of simulated annealing with a fixed temperature. We see that randomized local search is actually the special case of the Metropolis algorithm with temperature 0. Differently from randomized local search the Metropolis algorithm and simulated annealing perform in some sense similar. We already have seen examples where for a good performance the Metropolis algorithm is sufficient and it is not known if the performance can be provably improved by making use of simulated annealing with a non-trivial cooling schedule. Concrete examples are approximation of maximum matchings (Sasaki and Hajek, 1988) and optimal bisections for a specific class of random instances of the graph bisection problem (Jerrum and Sorkin, 1998). On the other hand, the difference between the Metropolis and simulated annealing can be enormous. This is not only demonstrated by means of artificial example functions (Sorkin, 1991; Droste *et al.*, 2001) but also for combinatorial optimization problems (Wegener, 2005; Meer, 2007). In spite of these differences analyzing the simpler Metropolis algorithm can sometimes yield useful results for simulated annealing. Remember the proof that the Metropolis algorithm is unable to efficiently find a minimum spanning tree for the instance of connected triangles with weights w and $(1+\varepsilon(m))w$ for

$\varepsilon(m) = o(1)$ (Wegener, 2005). This proof is carried out for the Metropolis algorithm but for arbitrary temperatures. This way it translates directly to a result on simulated annealing.

Another quite different class of randomized search heuristics are evolutionary algorithms. There is a multitude of different evolutionary algorithms ranging from rather simple ones to very involved algorithms combining many different mechanisms. Perhaps the most simple evolutionary algorithm known is the so-called (1+1) evolutionary algorithm ((1+1) EA). Just like simulated annealing it uses a single search point x_t and creates in each step one new search point y. Differently from simulated annealing this new search point is not chosen from a neighborhood of bounded size but generates randomly according to some probability distribution over the search space S. For the search space $S = \{0,1\}^n$ this probability distribution is defined by performing n independent random experiments, one for each bit y_i. With probability $1 - 1/n$ the bit y_i equals the corresponding bit in x_t, with the remaining probability $1/n$ it is its complement. While on average x_t and y differ in exactly one bit the Hamming distance between x_t and y may be as large as n. The second difference between simulated annealing and the (1+1) EA lies in the decision about accepting the search point y. Like it is the case in local search the move to y is accepted only if $f(y) \leq f(x)$ holds. For the sake of clarity we give a precise description of the (1+1) EA in a form that can be used for minimization of objective functions $f\colon \{0,1\}^n \to \mathbb{R}$.

Algorithm 17 (1+1) EA

---

1. Set $t := 0$. Choose $x_t \in \{0,1\}^n$ uniformly at random.
2. Repeat
3. $\quad y := x_t$. For each $i \in \{1, 2, \ldots, n\}$
   $\quad\quad$ With probability $1/n$ set $y_i := 1 - y_i$.
4. $\quad$ If $f(y) \leq f(x)$ set $x_{t+1} := y$
   $\quad$ else set $x_{t+1} := x_t$.
5. $\quad t := t + 1$
6. Until some stopping criterion is fulfilled.

While simulated annealing and the (1+1) EA share some structural similarity they differ very much in the way they escape from local optima and, consequently, in the way they conduct search. Simulated annealing restricts the choice of new search points to a neighborhood and escapes from local optima by accepting worsening moves with a probability that decreases exponentially fast with increasing difference in function values. The (1+1) EA never accepts worsening moves. It escapes from local optima by selecting new search points that can be arbitrarily far away from the current search point while the probability of selecting such a search point decreases exponentially with increasing Hamming distance. Seeing this enormous difference it does not come as a surprise that simulated annealing and the (1+1) EA can have drastically different performance.

The (1+1) EA is clearly superior to simulated annealing in situations where the current search point is a local minimum, all neighbors have much smaller function values but better search points have only small Hamming distance to the current search point. While the (1+1) EA can 'jump' to such search points easily simulated annealing has to accept a worsening that is very unlikely due to the large decrease in function value. It is not difficult to come up with example functions where this situation implies a small polynomial expected optimization time for the (1+1) EA while simulated annealing is unsuccessful with probability exponentially close to one even in an exponential number of steps (Jansen and Wegener, 2007).

On the other hand, simulated annealing is clearly superior to the (1+1) EA in situations where the current search point is a local minimum, the nearest point with larger function value has a very large Hamming distance but the differences in function values for all points between this point and the current search point are very small. Again, it is not difficult to construct example functions where this situation implies a small polynomial expected optimization time for simulated annealing while the (1+1) EA is unsuccessful with probability exponentially close to one even in an exponential number of steps (Jansen and Wegener, 2007).

While such enormous differences do not come as a surprise since the ways simulated annealing and the (1+1) EA conduct search are so different it is surprising to observe that their different ways of search can lead to surprisingly similar performance. Jansen and Wegener (2007) consider functions $f\colon \{0,1\}^n \to \mathbb{N}$ that are functions of unitation, i.e., the function value $f(x)$ does only depend on the number of 1-bits in x and not on x itself. They call such functions smooth if for all $x, x' \in \{0,1\}^n$ $|f(x) - f(x')| \leq H(x, x')$ holds where $H(x, x')$ denotes the Hamming distance of x and x'. For such smooth functions of unitation they define obstacles for the two algorithms in the following way.

Let f be such a smooth function of unitation where the all ones bit string 1^n is the unique global optimum. Let v_i be its function value for all $x \in \{0,1\}^n$ with exactly i 1-bits. Due to our assumptions we have $v_n = \max\{v_i \mid i \in \{0,1,\ldots,n\}\}$. The function has a (k,d)-obstacle for simulated annealing if $v_k > v_{k+1} > \cdots > v_{k+d-1}$ holds. The function has a (k,d)-obstacle for the (1+1) EA if $v_{k+d} \geq v_k$ and $v_i < v_k$ for all $i \in \{k+1, k+2, \ldots, k+d-1\}$.

For both algorithms a (k,d)-obstacle is an area in the search space that is located in the area of all bit strings with at least k 1-bits and less than $k+d$ 1-bits. Its location is thus characterized by k and its size or span by d. Since the unique global optimum is located at 1^n both algorithms need to get past such an obstacle in order to optimize the function f. The concrete definitions of (k,d)-obstacles differ for the two algorithms such that they reflect the different ways they perform search. While for the (1+1) EA it suffices that all function values between v_k and v_{k+d} are smaller than v_k, for simulated annealing we need the function values to be strictly decreasing.

We say that an algorithm, either simulated annealing or the (1+1) EA, overcomes such an obstacle if the following holds. For any $x_0 \in \{0,1\}^n$ with exactly k 1-bits in x_0 we consider the first point of time t when x_t contains at least $k+d$ 1-bits. We modify both algorithms such that any search point with less than k 1-bits is again replaced by the initial search point. If the expected first point of time when a bit string with at least $k+d$ 1-bits is reached is polynomially bounded we say that the algorithm is able to overcome such a (k,d)-obstacle.

Using this notion of (k,d)-obstacles for such smooth functions of unitation and this notion of overcoming them, Jansen and Wegener (2007) prove the following interesting facts.

Theorem 6.8 (Jansen and Wegener (2007)).
Simulated annealing can overcome $(k(n),d(n))$-obstacles for simulated annealing with $k(n) = (n/2) + \Omega(\sqrt{n})$ if and only if $d(n) = O(1 + \log(n-k(n))/(\log(n) - \log(n-k(n))))$ holds.

For the proof of Theorem 6.8 we consider a current search with with exactly m 1-bits, $k(n) < m < k(n) + d(n)$. With probability m/n the next search point contains $m-1$ 1-bits, with probability $(n-m)/(n \cdot e^{1/T(t)})$ the next search point contains $m+1$ bits. Overcoming the obstacle requires to increase the number of 1-bits. Thus we can set $T(t) := 0$. First we optimistically replace the probabilities to $k(n)/n$ for going one level down and $(n-k(n))/n$ for going one level up. Then we are in the situation of the gambler's ruin problem (Feller, 1957). Straightforward calculations yield that the probability not to be ruined is only polynomially large if $d(n) = O(1 + \log(n - k(n))/(\log(n) - \log(n - k(n))))$ holds. For the other direction we consider the level below the target level. Now we get $(n - k(n) - d(n) + 1)/n$ as bound on the probability for increasing the number of 1-bits. It is easy to see that this yields the same bound on $d(n)$.

Theorem 6.9 (Jansen and Wegener (2007)). *The (1+1) EA can overcome $(k(n), d(n))$-obstacles for the (1+1) EA with $k(n) = (n/2) + \Omega(\sqrt{n})$ if and only if $d(n) = O(1 + \log(n - k(n))/(\log(n) - \log(n - k(n))))$ holds.*

The proof of Theorem 6.9 one needs to calculate the probability to perform a set of n independent random experiments in the creation of y such that y contains at least d more 1-bits than k. The calculations are rather tedious but not difficult. The result demonstrates that seemingly similar search heuristics that in fact are very different can have surprisingly similar performance in a rather wide range of circumstances.

Acknowledgment

This material is based upon work supported by the Science Foundation Ireland under Grant No. 07/SK/I1205.

References

Cormen, T. H., Leiserson, C. E., Rivest, R. L. and Stein, C. (2001). *Introduction to Algorithms. 2nd Edition* (MIT Press).

Droste, S., Jansen, T. and Wegener, I. (2001). Dynamic parameter control in simple evolutionary algorithms, in W. N. Martin and W.M.Spears (eds.), *Proceedings of the 6th International Workshop on Foundations of Computer Sience (FOGA 2000)* (Morgan Kaufmann), pp. 275–294.

Feller, W. (1957). *An Introduction to Probability Theory and Its Applications. Volume I* (John Wiley).

Flamm, C., Hofacker, I. L., Stadler, B. M. R. and Stadler, P. F. (2007). Saddles and barrier in landscapes of generalized search operators, in C. R. Stephens, M. Toussaint, D. Whitley and P. F. Stadler (eds.), *Proceedings of the 9th International Workshop on Foundations of Computer Sience (FOGA 2007)* (Springer), pp. 194–212, LNCS 4436.

Giel, O. and Wegener, I. (2003). Evolutionary algorithms and the maximum matching problem, in *Proceedings of the 20th Annual Symposium on Theoretical Aspects of Computer Science (STACS 2003)* (Springer), pp. 415–426, LNCS 2607.

Hajek, B. (1988). Cooling schedules for optimal annealing, *Mathematics of Operations Research* **13**, pp. 311–329.

Jansen, T. and Wegener, I. (2007). A comparison of simulated annealing with simple evolutionary algorithms on pseudo-boolean functions of unitation, *Theoretical Computer Science* **386**, pp. 73–93.

Jerrum, M. and Sorkin, G. B. (1998). The Metropolis algorithm for graph bisection, *Discrete Applied Mathematics* **82**, pp. 155–175.

Kirkpatrick, S., Gelatt, C. D. and Vecchi, M. P. (1993). Opimization by simulated annealing, *Science* **220**, pp. 671–680.

Meer, K. (2007). Simulated annealing versus metropolis for a TSP instance, *Information Processing Letters* **104**, 6, pp. 216–219.

Metropolis, N., Rosenbluth, A., Rosenbluth, M., Teller, A. and Teller, E. (1953). Equation of state calculations by fast computing machines, *Journal of Chemical Physics* **21**, pp. 1087–1092.

Michiels, W., Aarts, E. and Korst, J. (2007). *Theoretical Aspects of Local Search* (Springer).

Motwani, R. and Raghavan, P. (1995). *Randomized Algorithms* (Cambridge University Press).

Papadimitriou, C. H., Schäffer, A. A. and Yannakakis, M. (1990). On the complexity of local search, in *Proceedings of the 22nd Annual ACM Symposium on Theory of Computing* (ACM Press), pp. 438–445.

Sasaki, G. H. and Hajek, B. (1988). The time complexity of maximum matching by simulated annealing, *Journal of the ACM* **35**, 2, pp. 387–403.

Sorkin, G. B. (1991). Efficient simulated annealing on fractal energy landscapes, *Algorithmica* **6**, pp. 367–418.

Vazirani, V. V. (1994). A theory of alternating paths and blossoms for proving correctness of the $O(\sqrt{V}E)$ general graph maximum matching algorithm, *Combinatorica* **14**, 1, pp. 71–109.

Wegener, I. (2005). Simulated annealing beats Metropolis in combinatorial optimization, in *Proceedings of the 32nd International Colloquium on Automata, Languages and Programming* (Springer), pp. 589–601, LNCS 3580.

Chapter 7

Theory of Particle Swarm Optimization

Carsten Witt

DTU Informatics
Technical University of Denmark
2800 Kgs. Lyngby
Denmark
cfw@imm.dtu.dk

Particle swarm optimization (PSO) is a relatively young bio-inspired search heuristic whose biological ideal is the behavior of bird flocks, fish schools and similar swarms. This chapter elaborates on the theoretical approaches to PSO via convergence analyses and runtime analyses. It is shown how obstacles to rigorous analyses were overcome in recent years and how techniques for the runtime analysis of bio-inspired search heuristics are cross-fertilizing the domain of PSO algorithms.

Contents

7.1. Introduction

The probably most popular subclass of randomized search heuristics is formed by the bio-inspired search heuristics. This subclass itself encompasses a huge variety of different approaches, amongst them evolutionary computation, ant colony optimization (ACO) and particle swarm optimization (PSO), to each of which this book devotes a chapter on its own.

We focus in this chapter on the probably youngest approach from the above list, namely PSO. This class of heuristics shares some similarities with ACO in that the biological ideal that has inspired the heuristics is a *swarm* in both cases. Recently, this has coined yet another umbrella term, *swarm intelligence*, under which ACO, PSO and some further, less known approaches are subsumed. The swarm that inspired PSO might be a bird flock, fish school, insect swarm, human swarm etc. Kennedy and Eberhart (1995) introduced PSO in 1995 in order to put the optimization processes observed in such swarms into an algorithmic framework. Typically, the swarm follows a common goal, yet allows the swarm members to behave to a certain extent selfishly. A main criterion of distinction between ACO and PSO might be that PSO was initially designed for the optimization in high-dimensional, *continuous* search spaces, e.g., for functions of the kind $f\colon \mathbb{R}^n \to \mathbb{R}$. Comprehensive treatments also of the history of PSO are given in the text books by Kennedy, Eberhart and Shi (2001) and Clerc (2006).

The aim of this chapter is to give an overview on the theoretical foundations of PSO that are available nowadays, roughly 15 years after the first applications. Is does not necessarily come as a surprise that these theoretical foundations lag far behind practical evidence. Not only is PSO a relatively young class of search heuristics, but also seems it harder to analyze than the famous evolutionary algorithms. Possible reasons are manifold. For example, the attempt to model the selfish and global behavior of a swarm at the same time leads to a system whose state space is complex and whose transitions are hard to control. Unlike in evolutionary computation (and, to some extent, simple ACO algorithms), we are not confronted with a relatively simple sequence of mutation-selection cycles that can be cast into a Markov chain which is, in principle, easy to describe. Additional complexity might be added by the fact that PSO lives in continuous search spaces.

Despite all the mentioned obstructions to theory, the last years have brought forward remarkable advances in the theoretical foundations of PSO. In addition to fully rigorous convergence analyses (without simplifying assumptions), the first analyses of the runtime, i.e., the efficiency, of PSO approaches are available now. This motivates us to lay down the following structure for this book chapter. In Section 7.2, we present necessary definitions of the most important variants of PSO. Section 7.3 describes convergence analyses and analyses of the dynamics of PSO variants. Benefits and limitations of different approaches will be discussed. Sections 7.5 and 7.6 deal with the few runtime analyses, available as of now for specific

PSO variants. We finish with conclusions and an outlook for future research.

7.2. Definitions

A common principle of PSO is a swarm consisting of so-called *particles*. These particles are located somewhere in the search space of the underlying optimization problem. Attracted by both a selfish and a common goal, the particles "fly" probabilistically through the search space so as to find an optimum to the objective function.

Typically, the objective function is of the kind $f\colon \mathbb{R}^n \to \mathbb{R}$. The PSO algorithm maintains m triples $(\boldsymbol{x}^{(i)}, \boldsymbol{x}^{*(i)}, \boldsymbol{v}^{(i)})$, $1 \leq i \leq m$, as particles. Each particle i consists of its current position $\boldsymbol{x}^{(i)} \in \mathbb{R}^n$, its own best-so-far position $\boldsymbol{x}^{*(i)} \in \mathbb{R}^n$ and its velocity $\boldsymbol{v}^{(i)} \in \mathbb{R}^n$. The movement for each particle is influenced by the best particle in its neighborhood. In this work, we only use the trivial neighborhood consisting of the whole swarm. This means that all particles are influenced by a single globally best particle, whose position is denoted by $\boldsymbol{x}^*$; for simplicity, we often use the symbol $\boldsymbol{x}^*$ and the term *globally best* also for the particle whose own best position determined the globally best position. The earliest PSO variant from the literature (Kennedy and Eberhart, 1995) fits this framework.

The idea to update velocities is as follows. The velocity vector is changed towards the particle's own best solution and towards the globally best solution $\boldsymbol{x}^*$. The impact of these two components is determined by so-called learning factors $c_1, c_2 \in \mathbb{R}_0^+$, which are fixed parameters to be chosen in advance (time-varying extensions are outside the scope of this book chapter). Additionally, a so-called inertia weight $\omega \in \mathbb{R}_0^+$ is used as a third parameter in order to continuously decrease the impact of previous velocity values. While the inertia weight (or, equivalently, a so-called restriction factor, see Poli, Kennedy and Blackwell, 2007a, for more details) was missing from the initial PSO variant, it is nowadays considered essential for a standard implementation of PSO (Bratton and Kennedy, 2007). In particular, choosing a small ω is meant to address a problem known as "velocity explosion" in early PSO variants. Without additional measures, velocity vectors tend to get longer and longer in the run of the PSO, possibly leading to a highly unstable behavior.

We now describe the precise update formulas. In each iteration of the PSO algorithm, the particles' velocities and positions are recomputed according to

$$\boldsymbol{v}^{(i)} := \omega \boldsymbol{v}^{(i)} + c_1 \boldsymbol{r} \otimes (\boldsymbol{x}^{*(i)} - \boldsymbol{x}^{(i)}) + c_2 \boldsymbol{r}' \otimes (\boldsymbol{x}^{*} - \boldsymbol{x}^{(i)}),$$
$$\boldsymbol{x}^{(i)} := \boldsymbol{x}^{(i)} + \boldsymbol{v}^{(i)}$$

for $1 \leq i \leq m$, where $\boldsymbol{r} \in (U[0,1])^n$ and $\boldsymbol{r}' \in (U[0,1])^n$ are vectors drawn independently and componentwise from the uniform distribution $U[0,1]$ and $\otimes$ denotes the componentwise multiplication.

Afterwards, the own best solutions are exchanged if the newly constructed solution is strictly better, i.e.,

$$\boldsymbol{x}^{*(i)} := \begin{cases} \boldsymbol{x}^{(i)} & \text{if } f(\boldsymbol{x}^{(i)}) < f(\boldsymbol{x}^{*(i)}) \\ \boldsymbol{x}^{*(i)} & \text{otherwise,} \end{cases}$$

and $\boldsymbol{x}^*$ is set to the best among the $\boldsymbol{x}^{*(i)}$. The PSO algorithm then proceeds to the next iteration unless some stopping criterion is fulfilled.

Although the above presentation might be considered as "standard" in some sense, many variants of PSO have been proposed that might excel in different applications or might be more amenable to theory. Examples are the so-called *bare-bones* PSO (Kennedy, 2003), which gets along without explicit velocities. Instead, the probabilistic behavior of the particles is driven by an operator that is reminiscent of a mutation operator in evolution strategies: a new search point is sampled around a point that corresponds to the expected next position of the particle. The sampling distribution might be uniform, Gaussian, Cauchy, with the variance controlling the degree of perturbation. A similar idea is used in the *globally convergent PSO (GCPSO)* proposed by van den Bergh (2002). Here standard update equations are used (including velocities) except for the globally best particle. In a simplified form, GCPSO samples a tentative solution $\tilde{\boldsymbol{x}}$ around the globally best solution $\boldsymbol{x}^*$ according to

$$\tilde{\boldsymbol{x}} := \boldsymbol{x}^* + \boldsymbol{p},$$

where the mutation vector $\boldsymbol{p}$ is a perturbation with each component drawn independently according to a probability distribution with zero mean. The motivation for this will become apparent when we discuss available theoretical foundations of the different approaches.

The last type of PSO algorithms that will treated in more detail in this book chapter originates from the wish to apply PSO in discrete search spaces. The so-called *Binary PSO*, first presented by Kennedy and Eberhart (1997), lives on the set $\{0,1\}^n$ of binary strings of length n. Velocities are updated in the usual way, but a trick is needed in order to translate the real-valued velocity vectors $\boldsymbol{v}^{(i)} \in \mathbb{R}^n$ into binary strings. For this purpose, one uses the sigmoid function

$$\text{sig}(\boldsymbol{v}_j^{(i)}) := \frac{1}{1 + e^{-\boldsymbol{v}_j^{(i)}}},$$

sketched in Figure 7.1, for all n components (indexed by j) of all particles. Positive velocity components bias the corresponding bit towards 1-values while negative velocities favor 0-values. Given initial velocity $\mathbf{0}$ (i.e., the all-zeros vector), each bit is completely random, hence the first created solution is uniformly distributed over $\{0, 1\}^n$.

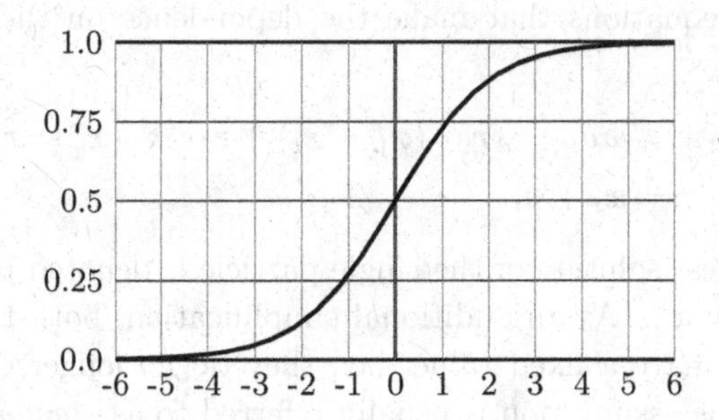

Fig. 7.1. The sigmoid function $\text{sig}(v) = \frac{1}{1+e^{-v}}$.

Of course, our presentation of PSO variants is by no means exhaustive. In practice, the success of a PSO algorithm depends heavily on a careful choice of its parameters and also fine-tuning of its components. From the perspective of a practitioner, the standard PSO might be considered as the basic algorithm which is then customized or engineered towards its application. An excellent overview on the sheer enormity of applications of PSO, not only in engineering disciplines, is given by Poli (2008). Just to name one example, that paper identifies antenna design as the still most prevalent domain where PSO is used.

7.3. Convergence Analyses and Further Theoretical Approaches

PSO is certainly an algorithmic technique whose theoretical foundations lag far behind practical evidence of its success. Notwithstanding, the first theoretical investigations of PSO can be traced back to the late 1990s, i.e., only few years after its invention. Like in evolutionary computation, these first studies put emphasis on the dynamics and convergence of simple PSO variants. Already Ozcan and Mohan (1998, 1999) conduct such an analysis. Under several simplifying assumptions, most notably the absence of stochasticity, they investigated how the sum $c_1 + c_2$ of the learning rates impacts the next position of a particle. At the time of publishing those works, PSO did typically not make use of inertia weight, i.e., the parameter ω of the standard PSO was implicitly set to 1. In a similar flavor but taking inertia into account, later works by Clerc and Kennedy (2002), van den Bergh (2002) and Trelea (2003) treat a model of PSO as a discrete-time dynamical system. Broken down to a single particle $\boldsymbol{x}$ with velocity $\boldsymbol{v}$, we obtain update equations that make the dependence on the time index t explicit:

$$\begin{aligned}\boldsymbol{v}_{t+1} &:= \omega\boldsymbol{v}_t + c_1\boldsymbol{r}\otimes(\boldsymbol{x}_{\mathrm{l}}^* - \boldsymbol{x}_t) + c_2\boldsymbol{r}'\otimes(\boldsymbol{x}_{\mathrm{g}}^* - \boldsymbol{x}_t),\\ \boldsymbol{x}_{t+1} &:= \boldsymbol{x}_t + \boldsymbol{v}_t.\end{aligned}$$

Here the own best solution of the single particle is denoted by $\boldsymbol{x}_{\mathrm{l}}^*$, and the globally best by $\boldsymbol{x}_{\mathrm{g}}^*$. As an additional simplification, both these solutions are assumed to have a fixed value, i.e., they do no longer depend on the time index. This assumption is usually referred to as *stagnation* (van den Bergh, 2002).

Rewriting the second equation as $\boldsymbol{v}_{t-1} = \boldsymbol{x}_t - \boldsymbol{x}_{t-1}$ and substituting in the first equation, adjusted to time index t, yields

$$\boldsymbol{v}_t := \omega(\boldsymbol{x}_t - \boldsymbol{x}_{t-1}) + c_1\boldsymbol{r}\otimes(\boldsymbol{x}_{\mathrm{l}}^* - \boldsymbol{x}_{t-1}) + c_2\boldsymbol{r}'\otimes(\boldsymbol{x}_{\mathrm{g}}^* - \boldsymbol{x}_{t-1}).$$

Inserting this into the second equation, the previously two equations can now be combined into a single non-homogeneous recurrence:

$$\boldsymbol{x}_{t+1} = (1+\omega)\boldsymbol{x}_t - (c_1\boldsymbol{r} + c_2\boldsymbol{r}')\otimes\boldsymbol{x}_t - \omega\boldsymbol{x}_{t-1} + c_1\boldsymbol{r}\otimes\boldsymbol{x}_{\mathrm{l}}^* + c_2\boldsymbol{r}'\otimes\boldsymbol{x}_{\mathrm{g}}^*,$$

which represents the dynamical system underlying the above-mentioned works. Neglecting stochasticity and replacing $c_1\boldsymbol{r}$ and $c_2\boldsymbol{r}'$ by their expectations $c_1/2$ and $c_2/2$, we obtain a deterministic system whose dynamics only depend on the two learning rates and the inertia weight ω. Then typically a stability analysis is performed. For example, Trelea (2003) derives three different regions in the parameter space according to the resulting behavior of the dynamical system: stable convergence to a fixed point, harmonic oscillations, or zigzagging around the fixed point.

Although the dynamical-systems model allows meaningful insights into the behavior and stability of PSO w.r.t. the choice of its parameters, it suffers from a severe drawback. Namely, the negligence of stochasticity deprives the search heuristic of its driving force. Depending on the strength of these forces, a completely different behavior from the deterministic system might be observed.

This is addressed in the first complete convergence analysis of the standard PSO. Jiang, Luo and Yang (2007) study this algorithm in the presence of stochasticity. Their convergence analysis first proves conditions for the expectations of the particles' positions to converge towards a fixed point. Afterwards, the same is done for the variances of the random positions. Combining these insights, Jiang, Luo and Yang (2007) show that the whole PSO system converges – under a reasonable *stochastic* convergence mode, namely convergence in mean square – for certain parameter settings. In detail, their theorem reads:

Theorem 7.1 (Jiang *et al.* (2007)). *Given* $\omega, c_1, c_2 \geq 0$, *if* $0 \leq \omega < 1$, $c_1+c_2 > 0$ *and* $0 < -(c_1+c_2)\omega^2+(\frac{c_1^2}{6}+\frac{c_2^2}{6}+\frac{c_1c_2}{2})\omega+c_1+c_2-\frac{c_1^2}{3}-\frac{c_2^2}{3}-\frac{c_1c_2}{2} \leq c_2^2(1+\omega)/6$ *are all satisfied together, the standard PSO system determined by parameter tuple* $\{\omega, c_1, c_2\}$ *will converge in mean square to* $\boldsymbol{x}^*$.

Despite the complicated-looked formulation, two necessary conditions for the theorem to hold might be called intuitive. First, $\omega < 1$ is needed for the inertia weight. This means that the impact of previous values of the velocity vector steadily vanishes over time. Second, the theorem contains a lower bound on c_2. This can be interpreted as a lower bound on the attractiveness of the globally best solution, which determines the common goal of the whole swarm. Also the notion of convergence in mean square reflects this common goal. Formally, the statement that "the PSO system converges" means that $\lim_{t\to\infty} E(|\boldsymbol{x}_t - \boldsymbol{x}_g^*|^2) = 0$ holds for all instantiations of $\boldsymbol{x}_t$ with the current positions of a particle.

The work by Jiang, Luo and Yang (2007) is one of the few analyses of the standard PSO that take randomness into account. This analysis is heavily based on the fact that particles behave independently if stagnation occurs, i.e., $\boldsymbol{x}^*$ remains fixed. Similarly, this observation enabled Poli and Broomhead (2007) to provide a first fixed-point analysis of the above dynamical system in the presence of stochasticity. Using exact formulas for the sampling distribution and numerical integration, they give insights on the behavior of the standard PSO also for parameter settings where the theorem by Jiang, Luo and Yang (2007) does not apply.

All convergence analyses that we have discussed so far make an intrinsic problem of PSO more or less explicit. Namely, PSO might suffer from stagnation, i.e., the globally best position does not improve, even though it has not yet found an optimum. This becomes most apparent in the theorem by Jiang, Luo and Yang (2007), where the PSO system is proved to converge to the globally best position $\boldsymbol{x}^*$. For an unfortunate (or pathological) swarm configuration, $\boldsymbol{x}^*$ may be arbitrarily far from optimality, which means that PSO – particle swarm "optimization" – is not an optimizer. Such pathological configurations are addressed by van den Bergh (2002). One readily checks that stagnation occurs in the standard PSO when $\boldsymbol{x}^* = \boldsymbol{x}^{(1)} = \boldsymbol{x}^{*(1)} = \cdots = \boldsymbol{x}^{(m)} = \boldsymbol{x}^{*(m)}$ and $\boldsymbol{v}^{(1)} = \cdots = \boldsymbol{v}^{(m)} = \vec{0}$. Despite looking like a really pathological configuration, such a stagnation becomes visible in experiments for swarms of small size even on very simple functions (Angeline, 1998).

Van den Bergh (2002) replies to the stagnation phenomenon by defining the GCPSO mentioned in Section 7.2. The update formula for the globally best particle (in a simplified* form $\tilde{\boldsymbol{x}} := \boldsymbol{x}^* + \boldsymbol{p}$) allows for a change even if all velocities have reached zero. The concrete implementation of the random perturbation $\boldsymbol{p}$ proposed by van den Bergh (2002) seems to be inspired by the randomly weighted vectors pointing towards the locally and globally best solutions. Actually, a time-dependent choice of $\boldsymbol{p}$ is proposed, more precisely a uniform choice from a hypercube with side lengths $2\rho(t)$.

$$\boldsymbol{p}_t \in (U[-\rho(t), \rho(t)])^n.$$

Many other ways of drawing $\boldsymbol{p}_t$ may be imagined, e.g., according to the typical distributions of mutation operators in evolutionary computation or the above-mentioned bare-bones PSO (Kennedy, 2003). The "hypercube" distribution has the advantage of being easy to implement. However, also due to the simplicity of the distribution, GCPSO might be very inefficient at making progress towards (local or global) optima if $\rho(t)$ is set inappropriately. Van den Bergh (2002) cures this by allowing $\rho(t)$ to be adapted in the course of optimization in a similar way as in evolutionary computation.

The framework to adapt $\rho(t)$ is as follows. The run of the GCPSO is divided into observation phases consisting of ℓ iterations. In each phase, the number s of successes (i.e., replacements of the globally best position)

*The original formulation of GCPSO actually maintains also a residual velocity for the globally best particle, however, this is not used for the convergence analyses by van den Bergh (2002) since it vanishes in worst-case configurations where the swarm collapses to a single position.

and the number of failures $f = \ell - s$ is observed. If s is greater than a certain threshold then $\rho(t)$ is doubled, reflecting the observation that it was easy to make progress and that larger progress might be achieved by larger steps. Conversely, if the number of failures f exceeds a threshold, $\rho(t)$ is halved so as to sample from a smaller hypercube and to prevent the algorithm from overshooting regions with better fitness.

After presenting all these modifications of the standard PSO leading to the GCPSO, van den Bergh (2002) studies convergence of the new algorithm to optima, without loss of generality being *minima*, of the objective function. Here a crucial distinction has to be made between local optima and global optima. Van den Bergh (2002) first sets up two conditions to be met in order for a search algorithm to be convergent to local minima. Informally, the first condition demands that the search algorithm is elitist, i.e., maintains a memory of the so far best seen solution. The second condition requires a positive probability of decreasing the distance – for a reasonable distance measure – to a local optimum by at least a constant amount in a step of the search algorithm. Van den Bergh (2002) proves that the GCPSO is able to meet both conditions, i.e., is a local optimizer, while the standard PSO does not converge in this sense.

Finally, van den Bergh (2002) studies convergence to global optima of an objective function. To this end, a stronger variant of the above-mentioned second condition is needed. It turns out that neither the standard PSO nor the GCPSO are global optimizers. The main reason for this is that $\boldsymbol{p}_t$ has a bounded support. More precisely, if only inferior solutions are contained in the hypercube with side lengths $2\rho(t)$ around the globally best solutions, then superior search points outside the hypercube cannot be reached. This is not healed by the adaptation of the side length since the adaptation scheme would result in halving $\rho(t)$. Instead, a suitable restart scheme is required in order to explore different optima of multimodal functions. Van den Bergh (2002) proposes the so-called multi-start PSO (MPSO) that restarts itself with a random initialization of the particles' positions once it has converged to a local optimum. This PSO variant satisfies the proposed conditions for global convergence, mostly as a result of the repeated random initialization. However, the variant is interesting more from a theoretical than from a practical point of view. In particular, a theoretically infinite number of restarts is required, and it is unclear how the restarts should be initiated. In any case, unlike the adaptation scheme used in the GCPSO, the MPSO does not include any components to allow *efficient* optimization on well-behaved functions. A more careful study of the efficiency of PSO variants will be the common subject of the rest of this book chapter.

7.4. Runtime Analysis

Indubitably, convergence analyses of algorithms provide a solid theoretical ground to explain that the algorithm actually works, i.e., that it eventually will find a solution of a certain quality. A proof of convergence is closely related to the question about the effectiveness of the approach: Does it do the things it is supposed to do? If the answer is no, we should try to correct the algorithm in order to make it at least effective.

From this perspective, a convergence analysis of an algorithm, i.e., an analysis of its effectiveness, is an essential prerequisite for the study of its efficiency: How fast is the approach at reaching its goal? In the theory of algorithms (see, e.g., Cormen, Leiserson, Rivest and Stein, 2001), this corresponds to the second step in the following classical procedure: First prove correctness, then analyze the runtime, which is a natural measure of efficiency.

As already indicated, a main distinguishing feature between randomized search algorithms and classical algorithms is their design. A problem-specific algorithm has almost always been conceived with its analysis in mind, whereas randomized search heuristics have typically been inspired by a model from nature, e.g., swarm dynamics. In most cases, one did not think of a theoretical analysis at the time when the heuristics were designed. However, research from the last two decades has shown that many randomized search heuristics can be analyzed both w.r.t. their effectiveness (convergence) and, in particular, w.r.t. their efficiency, i.e., runtime. For example, the runtime analysis of evolutionary computation has evolved at rapid pace since the early 1990s and found its place in theoretical computer science. Also the runtime analysis of ant colony optimization has made remarkable progress in recent years. Chapters of this book on their own have been devoted to related results. The first runtime analyses of PSO have been available since 2008. Before we cover two works from this area in greater depth, we elaborate on the framework for a runtime analysis.

The typical domain of application for randomized search heuristics, in particular for the bio-inspired ones, is the so-called *black-box* scenario. This means that the function for which an optimum is sought is completely unknown to the algorithm and information can be gathered solely by evaluating it on certain inputs. In particular, no information on the gradient and other characteristics is available, which excludes classical search methods such as the Newton method from our considerations.

The black-box assumption is met, e.g., in complex systems where the quality of a solution can only be determined as the outcome of a simulation of the system for a given parameter setting. This also motivates the typical cost measure, namely, the number of evaluations of the black-box function until the goal of optimization has been reached. So the basic question is: given an objective function $f\colon D^n \to \mathbb{R}$ mapping points from an n-dimensional space to an objective value, how many f-evaluations take place until the objective value has reached a certain quality? We formally capture this by the following definition.

Definition 7.2. The *runtime* $T_{A,f}$ of an algorithm A optimizing f is the number of times that f is evaluated on a search point until an optimum is found.

If randomized search heuristics are studied, then $T_{A,f}$ is typically a random variable. As usual in the analysis of randomized algorithms, we therefore often focus on the expectation $E(T_{A,f})$, the so-called expected runtime of A on f. However, the expected runtime is not necessarily a reliable measure of the heuristic's efficiency. Even if the expected runtime is exponentially big, the probability of a small polynomial runtime might still be sufficient to allow efficient optimization by using multistart variants of the heuristic. Hence, we are also interested in the distribution of $T_{A,f}$, e.g., we try to estimate $\Pr(T_{A,f}) \leq D$ for a given time bound D.

Runtime analyses are usually conducted in an asymptotic framework where the number of dimensions n is the parameter that determines the size (or difficulty) of the problem. Similarly to an asymptotic analysis in computer science, where n is the size of the input (e.g., the number of cities in a TSP instance), we would therefore like to represent the required or sufficient runtime as a function of n. As the last sentence suggests, often no exact formula is available but at least upper and lower bounds in O-notation.

One of the first steps towards a runtime analysis of PSO was made by Poli and Langdon (2007) and, in particular, by Poli, Langdon, Clerc and Stephens (2007b). Their work is based on the insight that PSO and many other randomized search heuristics can be modeled as Markov chains. A description of the current state of the system is, in principle, enough in order to predict the next state accurately. This allows for the extraction of the figures we are interested in, e.g., the expected runtime. However, despite there being a powerful theory of Markov chains, an exact approach is typically not feasible in practice. The state space is too complex for an exact formula of the expected runtime to be derived. As mentioned above, the typical resort is a coarse-graining, i.e., we give up accuracy in favor of approximate analyses.

Poli and Langdon (2007) and Poli, Langdon, Clerc and Stephens (2007b) introduce such a coarse-grained Markov chain by collapsing several states with similar fitness values to a single one. This approach, which was gleaned from finite-element methods in physics, eases the computation of the transition matrix of the approximate system and allows predictions of the runtime of PSO approaches (in particular, the above-mentioned bare-bones PSO) that coincide with experimental validations in a remarkable way. However, the analyses do not allow for asymptotic results. If the dimensionality of the search spaces grows, it is unclear to how to control the error. In this framework, asymptotic results excel. Using estimations whose error can be bounded, one obtains theorems that hold for all problem dimensions. Up to now, not many analyses of this kind exist for PSO.

7.5. Runtime Analysis of the Binary PSO

To the best of the author's knowledge, the first work proving rigorous theorems on the runtime of a PSO variant originates from 2008. Sudholt and Witt (2008, 2010) study the Binary PSO mentioned in Section 7.2, i.e., a variant for the optimization of pseudo-boolean functions $f\colon \{0,1\}^n \to \mathbb{R}$. One motivation of the authors to consider a discrete search space is the expertise in the runtime analysis of different randomized search heuristics like evolutionary algorithms and ant colony optimization. Most runtime analysis obtained in these areas refer to discrete search spaces.

In order to present the main results by Sudholt and Witt (2010), we give in Algorithm 18 a precise definition for the Binary PSO with a swarm size of m and learning factors c_1, c_2. By lower indices we address the n components of the three parts of a particle $(\boldsymbol{x}^{(i)}, \boldsymbol{x}^{*(i)}, \boldsymbol{v}^{(i)})$, $1 \le i \le m$. The algorithm starts with an initialization step (Step 1), where all velocities are set to all-zeros vectors and all solutions, including own best and globally best, are undefined, represented by the symbol $\perp$. The subsequent loop (Steps 2–5) chooses the random vectors $r_1 \in (U[0, c_1])^n$ and $r_2 \in (U[0, c_2])^n$ with each component drawn independently and uniformly from the given interval. These values are then used as weights for the cognitive and the

Algorithm 18 Binary PSO

(1) Initialize velocities with 0^n and all solutions with $\perp$.
(2) Choose $r_1 \in (U[0, c_1])^n$ and $r_2 \in (U[0, c_2])^n$.
(3) For $j := 1$ to m do
  For $i := 1$ to n do
    Set $\boldsymbol{x}_i^{(j)} := 1$ with probability $\text{sig}(\boldsymbol{v}_i^{(j)})$, otherwise set $\boldsymbol{x}_i^{(j)} := 0$.
  If $f(\boldsymbol{x}^{(j)}) > f(\boldsymbol{x}^{*(j)})$ or $\boldsymbol{x}^{*(j)} = \perp$ then $\boldsymbol{x}^{*(j)} := \boldsymbol{x}^{(j)}$.
  If $f(\boldsymbol{x}^{*(j)}) > f(\boldsymbol{x}^*)$ or $\boldsymbol{x}^* = \perp$ then $\boldsymbol{x}^* := \boldsymbol{x}^{*(j)}$.
(4) For $j := 1$ to m do
  Set $\boldsymbol{v}^{(j)} := \boldsymbol{v}^{(j)} + r_1 \otimes (\boldsymbol{x}^{*(j)} - \boldsymbol{x}^{(j)}) + r_2 \otimes (\boldsymbol{x}^* - \boldsymbol{x}^{(j)})$.
  For $i := 1$ to n do
    Set $\boldsymbol{v}_i^{(j)} := \max\{\boldsymbol{v}_i^{(j)}, -v_{\max}\}$.
    Set $\boldsymbol{v}_i^{(j)} := \min\{\boldsymbol{v}_i^{(j)}, v_{\max}\}$.
(5) Goto 2.

social component, respectively. Note that for the default choices $c_1 = c_2 = 2$ the expected weight for each component is 1.

In Step 3, the velocity is probabilistically translated into a new position for the particle, i.e., a new solution using the sigmoid function mentioned above. Afterwards, the own best and globally best are exchanged if the newly constructed solution is better. Note that the selection is strict, i.e., a best solution is only exchanged in case the new solution has strictly larger fitness. The initial solutions constructed for the particles, however, are accepted in any case. Recall that these solutions are uniformly distributed over $\{0, 1\}^n$ since the initial velocity vector is $\mathbf{0}$.

In Step 4, the Binary PSO updates the velocity vectors using the well-known update formula with the inertia ω implicitly set to 1. The Binary PSO is a relatively old variant defined before inertia weight was introduced into the standard PSO. However, the Binary PSO uses a different trick to prevent the velocity vectors from suffering the "explosion" phenomenon addressed in Section 7.2. Namely, every velocity vector is bounded componentwise by minimum and maximum values, usually according to a symmetrical interval $[-v_{\max}, v_{\max}]$. This reflects the choice of a maximum absolute velocity as studied by Shi and Eberhart (1998). For practical purposes, often velocities in the interval $[-4, 4]$ are proposed. In order to conduct an asymptotic analysis, Sudholt and Witt (2010) allow the maximum velocity to grow with the problem dimension n and confine the components to logarithmic values by letting $v_{\max} := \ln(n-1)$. This choice will be explained later. Finally, after having bounded velocities, the Binary PSO jumps back to Step 2 and starts a new so-called generation. Each generation contains m evaluations of the fitness function.

Sudholt and Witt (2010) deal with different parametrizations of the Binary PSO, differing in the swarm size m and the learning factors c_1 and c_2. They investigate the heuristic on classes of functions and specific simple functions like the example problem $\text{ONEMAX}(x_1, \ldots, x_n) = x_1 + \cdots + x_n$ and obtain both general lower bounds as well as specific upper bounds on the runtime. The first lower bound is based on the following insight on the choice of $v_{\max}$. If all velocities are restricted to constant values, then the Binary PSO is too close to random search and the algorithm fails badly, even given exponential time and a large number of global optima.

Theorem 7.3 (Sudholt and Witt (2010)). *Consider the Binary PSO with arbitrary values for m, c_1, and c_2, where $v_{\max}$ is redefined to a constant value. Then there is a constant $\alpha = \alpha(v_{\max})$ such that the following holds. If f contains at most $2^{\alpha n}$ global optima, the probability that the Binary PSO finds a global optimum on f within $2^{\alpha n}$ constructed solutions is $2^{-\alpha n}$.*

The proof idea for the preceding theorem is as follows. If $v_{\max}$ is a constant value, i.e., does not depend on n, then this is the case also for $\text{sig}(v_{\max})$ and $\text{sig}(-v_{\max})$. In particular, the probability of setting a bit to a given value is bounded from above by $1-\Omega(1)$. The probability of sampling a certain binary string is then bounded from above by $(1-\Omega(1))^n = 2^{-\Omega(n)}$, and this does not happen within an exponential number $2^{\alpha n}$ steps with a probability of $1 - 2^{-\Omega(n)}$ if α is a small constant. Taking a union bound (Lemma 1.1) over $2^{\alpha n}$ different optima still leads to an exponentially small success probability of $2^{-\Omega(n)}$ if α is small enough.

Hence, in order to obtain polynomial optimization times, we must allow $v_{\max}$ to depend on n. As already mentioned, Sudholt and Witt (2010) propose the choice $v_{\max} = \ln(n-1)$. As now $\text{sig}(-v_{\max}) = 1/n$ and $\text{sig}(v_{\max}) = 1 - 1/n$, the probability of setting a bit to 1 is always in the interval $[1/n, 1-1/n]$. This is inspired by standard mutation operators in evolutionary computation, where incorrectly set bits have a probability of $1/n$ of being corrected. The probability of sampling a certain solution can be as high as $(1-1/n)^n = \Omega(1)$.

An important step towards obtaining good bounds on the runtime for the Binary PSO is to understand the dynamics of the probabilistic model underlying PSO, that is in particular the behavior of the velocity vector. Consider a single bit that is set to 1 both in the own best and in the globally best. Then, as long as these solutions are not exchanged, its velocity value v is guided towards the upper bound $v_{\max}$. An important observation is that the velocity is only increased in case the bit is set to 0 in the next constructed solution. The probability that this happens is given by

$$1 - \text{sig}(v) = 1 - \frac{1}{1+e^{-v}} = \frac{1}{1+e^{v}},$$

and this probability decreases rapidly with growing v. Hence, the closer the velocity is to the bound $v_{\max}$, the harder it is to get closer. A symmetrical argument holds for velocities that are guided towards $-v_{\max}$.

As long as the v-values are not too close to the velocity bounds $-v_{\max}$ and $v_{\max}$, the search of the Binary PSO is too random for it to find single optima with high probability. This idea is made precise by the following, general lower bound, which holds for all practical choices of the learning factors c_1 and c_2 and a polynomial swarm size m. The maximum change of a velocity per generation will be abbreviated by $c := c_1 + c_2$ hereinafter. Recall that a generation of the Binary PSO corresponds to m fitness evaluations.

Theorem 7.4 (Sudholt and Witt (2010)). *Let f be a function with a unique global optimum, and let $m = \text{poly}(n)$ and $c = O(1)$. Then the expected number of generations of the Binary PSO on f is $\Omega(n/\log n)$.*

Proof. Since the search heuristic treats zero- and one-bits symmetrically, we can w. l. o. g. assume that the global optimum is the all-ones string. Let $t := \alpha n/\ln n$ for a small constant $\alpha > 0$, which is chosen later. We show that the probability of not creating the optimum within t generations is $1 - o(1)$, which implies the claim of the theorem.

We consider an arbitrary bit in an arbitrary particle. The event of creating a one at this bit is called *success*. Let a bit be called *weak* if its success probability has been at most $p := 1 - \frac{e\ln(\mu n)}{n}$ up to and including the current generation. Let a set of bits be called weak if it contains only weak bits. We will show that with probability $1 - 2^{-\Omega(n)}$, after t generations of the 1-PSO, each particle still contains a weak subset of bits of size at least n/e. The probability of setting all bits of such a weak subset to 1 simultaneously is bounded from above by $p^{n/e} \le 1/(\mu n)$ for each particle. Note that this event is necessary to create the all-ones string in a particle. Thus, the probability of finding the optimum within t generations creating

μ new solutions each is still less than $t\mu/(\mu n) = O(1/\log n) = o(1)$, which will prove the theorem.

We still have to show that with probability $1 - 2^{-\Omega(n)}$, after t generations, there is a weak subset of size at least n/e in each particle. One generation can increase the velocity by at most c. Note that $p = 1 - O((\ln n)/n)$ since $\mu = \text{poly}(n)$. To reach success probability at least p, the current velocity must be between $s^{-1}(p) - c = \ln(p/(1-p)) - c$ and $s^{-1}(p)$ at least once, where s^{-1} denotes the inverse sigmoid function. Pessimistically assuming the first value as current velocity, the probability of not increasing it in a single step is at least

$$\frac{1}{1+e^{-\ln(p/(1-p))+c}} = 1 - \frac{e^c(1-p)}{p+e^c(1-p)} \geq 1 - 2e^c(1-p)$$

if n is large enough for $p \geq 1/2$ to hold. The last expression equals $1 - 2e^{c+1}\ln(\mu n)/n$ by definition of p. Hence, along with $c = O(1)$ and again $\mu = \text{poly}(n)$, the probability of not increasing the velocity within t generations is at least

$$\left(1 - \frac{2e^{c+1}\ln(\mu n)}{n}\right)^t = \left(1 - \frac{O(\ln n)}{n}\right)^{\alpha n/\ln n} \geq 2e^{-1}$$

if α is chosen small enough. This means that each bit in each particle independently has a probability of at least $2e^{-1}$ of being weak at generation t. Using Chernoff bounds (Corollary 1.10), the probability of not having a weak set of size at least n/e in a specific particle is at most $e^{-\Omega(n)}$. As $\mu = \text{poly}(n)$, the probability that there exists a particle without weak subset at generation t is still $\mu e^{-\Omega(n)} = e^{-\Omega(n)}$. □

The lower bound from Theorem 7.4 relied on the fact that a velocity that is guided towards $v_{\max}$ does not reach this value in short time and so the Binary PSO cannot find a single target efficiently. On the other hand, if we consider a longer period of time, velocities may reach the bounds $-v_{\max}$ or $v_{\max}$ they are guided to. In this case, we say that the velocity has been "frozen" as the only chance to alter the velocity again is to have an improvement of the own best or the globally best. The random time until a bit is frozen is called *freezing time*. Using a detailed analysis, Sudholt and Witt (2010) bound the expected number of generations until k bits freeze towards velocity bounds by $O(n\ln(2k))$. Hence, all bits are frozen within $O(n\log n)$ generations.

The analysis of the expected freezing time allows the derivation of upper bounds for the Binary PSO in a similar way as for evolutionary algorithms.

The idea is to concentrate on the globally best particle. Due to the strict selection in the Binary PSO, this is only exchanged in case a better solution is discovered. This means that after some time either the globally best has improved or all velocities in the globally best are frozen. In the latter case, the probability of creating a 1 for every bit is now either $\mathrm{sig}(-v_{\max}) = 1/n$ or $\mathrm{sig}(v_{\max}) = 1-1/n$. The distribution of constructed solutions now equals the distribution of offspring for the above-mentioned (1+1) EA, with the globally best $\boldsymbol{x}^*$ as the current search point. This similarity between PSO and evolutionary algorithms can be used to transfer a well-known method for the runtime analysis from evolutionary algorithms to PSO, namely the fitness-level method. We present this method, also called the method of f-based partitions (see Wegener, 2002), in a restricted formulation. Note that this method has recently been transferred to ACO as well (see, e.g., Gutjahr and Sebastiani, 2008; Neumann, Sudholt and Witt, 2009).

Let $f_1 < f_2 < \cdots < f_\ell$ be an enumeration of all values of the underlying fitness function f, and let A_i, $1 \le i \le \ell$, contain all solutions with fitness f_i. We also say that A_i is the i-th fitness level. Note that the last fitness level A_ℓ contains only optimal solutions. Now, let s_i, $1 \le i \le \ell - 1$, be a lower bound on the probability of the (1+1) EA to create an offspring in $A_{i+1} \cup \cdots \cup A_\ell$, provided the current population belongs to A_i. The expected waiting time until such an offspring is created is at most $1/s_i$ (cf. Theorem 1.6) and then the i-th fitness level is left for good. As every fitness level has to be left at most once, the expected number of generations for the (1+1) EA to optimize f is bounded from above by

$$\sum_{i=1}^{\ell-1} \frac{1}{s_i}. \tag{7.1}$$

A similar bound holds for the Binary PSO.

Theorem 7.5 (Sudholt and Witt (2010)). *Let A_i form the i-th fitness level of f and let s_i be the minimum probability for the (1+1) EA to leave A_i towards $A_{i+1} \cup \cdots \cup A_\ell$. If $c = \Theta(1)$, the expected number of generations for the Binary PSO to optimize f is bounded from above by*

$$O(\ell n \log n) + \sum_{i=1}^{\ell-1} \frac{1}{s_i}.$$

The right-hand sum is the upper bound obtained for the (1+1) EA from (7.1). The left-hand term goes back to the following idea. On each fitness level, we pessimitically wait for the velocity vector of the globally best particle to freeze according to the globally best solution $\boldsymbol{x}^*$. As mentioned above, new solutions for this particle are then constructed with the same distribution as for the (1+1) EA and we can subsequently apply the argumentation of the fitness-level method to bound the number of generations until the considered fitness level is left. Since the expected number of generations until n bits freeze towards velocity bounds is $O(n \log n)$ and ℓ fitness levels are considered, the additional term $O(\ell n \log n)$ follows.

An an application, Sudholt and Witt (2010) concretize the bound for unimodal functions. We say that a function f is unimodal if it has exactly one local optimum with respect to Hamming distance. Hence, if the global best $\boldsymbol{x}^*$ is not the unique optimum, there is always at least one Hamming neighbor (a solution with Hamming distance 1 to $\boldsymbol{x}^*$) with larger fitness. The probability for the (1+1) EA to create a specific Hamming neighbor as offspring equals $(1/n)(1-1/n)^{n-1} \geq 1/(en)$. We conclude $s_i \geq 1/(en)$ for every non-optimal fitness level. Theorem 7.5 yields the following bound.

Corollary 7.6. *Let f be a unimodal function with m different function values. Then the expected number of generations for the Binary PSO to optimize f is bounded by*

$$O\left(mn \log n + \sum_{i=1}^{m-1} en\right) = O(mn \log n + m \cdot en) = O(mn \log n).$$

We discuss the limits of the approach. A straightforward application of Theorem 7.5 w.r.t. the simple function OneMax leads to $\ell = n+1$ and, therefore, to the relatively large upper bound of $O(n^2 \log n)$ on the number of generations. This is a consequence of the pessimistic assumptions the fitness-level relies upon, in particular the assumption that the Binary PSO spends $\Theta(n \log n)$ generations on each fitness level until all bits have been frozen to velocity bounds according to the globally best. Sudholt and Witt (2010) show in a more detailed analysis that this is far too pessimistic.

The in-depth analysis concentrates on the so-called 1-PSO using just one particle. This heuristic can be considered as a counterpart to the well-known (1+1) EA, which gets along with only one individual and is nevertheless surprisingly efficient on many functions. Also the 1-PSO is provably efficient on many functions, e.g., Corollary 7.6 applies to it. Similarly to the (1+1) EA, the 1-PSO keeps track of exactly one solution with best-so-far value. This simplified algorithm is considerably easier to study than the general case since globally best and own best solution have collapsed to a single search point and there is only a single velocity vector.

Considering the 1-PSO, Sudholt and Witt (2010) improve the optimization time bound on OneMax to $O(n \log n)$, which is the same asymptotic upper bound as for the (1+1) EA and a simple ant colony optimization algorithm called 1-ANT. This analysis is crucial to gain an understanding of the probabilistic behavior of velocities and to show that not all of them need to stay at $\pm v_{\max}$ in the course of optimization. Hence, in terms of velocity values vs. mutations, the stochastic processes behind Binary PSO and the (1+1) EA differ significantly but lead to the same upper bounds on the runtime. It is also conjectured that the bound $O(n \log n)$ for the 1-PSO on OneMax is asymptotically tight. The currently best known lower bound is the $\Omega(n/\log n)$ result from Theorem 7.4.

7.6. Runtime Analysis of the Guaranteed Convergent PSO

The work by Sudholt and Witt (2010) studies one of the few discrete variants of PSO, which lives in finite search spaces and optimizes each function in expected finite time $O(n^n)$. Typically, however, PSO is applied to real-parameter optimization, which is reflected by the standard search space $\mathbb{R}^n$. In this domain, not only the methods for the analysis of randomized search heuristics differ significantly from the discrete domain. Even worse, one usually has to sacrifice the goal of exact optimization that is implicit in Definition 7.2. The reason is that $\mathbb{R}^n$ is an infinite domain. Unless the objective function is extremely simple, we can only expect black-box algorithms to obtain approximations of optima in this setting. A well-known example is the Sphere function, a continuous "cousin" of OneMax, defined by

$$\text{Sphere}(\boldsymbol{x}) := \|\boldsymbol{x}\|^2 = x_1^2 + \cdots + x_n^2$$

with its global optimum in $\mathbf{0}$. Since the black-box algorithm does not know that it is optimizing a Sphere function, it will almost surely take infinitely long to sample the optimum if starting from a different search point. Therefore we usually wait for it to sample a solution y of distance ϵ from the optimum, i.e., $\|y\| \leq \epsilon$. Note that the distance can be measured in the search space rather than the objective-value space due to the one-to-one correspondence between these measures. A fundamental question studied in the analysis of black-box algorithms for the Sphere function is therefore: express the "runtime", i.e., the number of Sphere-evaluations, until a solution in an ϵ-neighborhood of $\mathbf{0}$ has been sampled, as a function of n and ϵ. Simple evolutionary algorithms for continuous optimization such as the so-called (1+1) evolution strategy ((1+1) ES) work in this scenario since they are randomized black-box algorithms. The runtime of the (1+1) ES on SPHERE with high probability can be bounded by $O(n \log(1/\epsilon))$, given an appropriate initialization of the algorithm (Jägersküpper, 2007a).

Posing the same question for PSO algorithms can lead to unsatisfactory answers since many PSO variants suffer from the stagnation phenomenon described in Section 7.3. As said before, the PSO algorithm is not an optimizer in this case. This also holds for many variants which introduce more stochasticity (e.g., the bare-bones PSO) but still stagnate when the whole swarm has collapsed to a single, possibly non-optimal point. Fortunately, this does not hold for the GCPSO, as an analysis showing convergence at least to local optima is available, see Section 7.3. This motivated Witt (2009) to take GCPSO as the starting point for a first runtime analyses of PSO variants in continuous search spaces. In the following, we summarize the results that were obtained in this study.

Witt (2009) investigates the GCPSO in the case $m = 1$, i.e., he studies a swarm of size 1 since only the globally best solution is subject to mutation while the remaining particles might still suffer from stagnation. As mentioned above, the effects of residual velocities are ignored since these do not contribute to the convergence of the heuristic. For simplicity, the side length of the hypercube used in the mutation is denoted by 2ℓ, i.e., the mutation vector satisfies $\boldsymbol{p} \in (U[-\ell, \ell])^n$. Hence, given an arbitrary initialization of $\boldsymbol{x}^*$ and ℓ, the GCPSO with $m = 1$ and without residual velocity breaks down to Algorithm 19.

Algorithm 19 Framework for GCPSO_1

(1) $\tilde{\boldsymbol{x}} := \boldsymbol{x}^* + \boldsymbol{p}$, where $\boldsymbol{p} \in (U[-\ell, \ell])^n$
(2) If $f(\tilde{\boldsymbol{x}}) < f(\boldsymbol{x}^*)$ set $\boldsymbol{x}^* := \tilde{\boldsymbol{x}}$
(3) Update ℓ.
(4) Go to 1.

Recall that the update strategy for ℓ proposed by van den Bergh and Engelbrecht (2002) is to double it if the number of successes (i.e., replacements of $\boldsymbol{x}^*$) in a predefined observation phase is above a certain threshold and to halve it if it is below another threshold. As an example, the update strategy might be:

After an observation phase consisting of n steps has elapsed, double ℓ if the total number of successes was at least $n/5$ in the phase and halve it otherwise. Then start a new phase.

The proposed strategy is called 1/5-*rule* in the theory of evolutionary computation (see Jägerskupper, 2007a, and references therein). Regarding the 1/5, it is only important to note that $1/5 \in [\Omega(1), 1/2 - \Omega(1)]$; the motivation of the exact value is beyond the scope of this book chapter. In the following, GCPSO will be instantiated with 1/5-rule, $m = 1$ and without residual velocity for the particle. Let the obtained heuristic be called GCPSO_1. This is reminiscent of the (1+1) ES using 1/5-rule (see also Jägerskupper, 2007a), yet the "cube" mutation seems to make a significant difference. The (1+1) ES draws the mutation vector according to $\boldsymbol{p} \in (N(0, \ell))^n$, with each component independently following a Normal distribution with zero mean and standard deviation ℓ. Then $\boldsymbol{p}$ is omnidirectional (or *isotropic*), which means that, conditioned on a fixed length $\|\boldsymbol{p}\| = \ell^*$, it has the same density in all directions.

Isotropy is obviously not the case if $\boldsymbol{p} \in (U[-\ell, \ell])^n$. For example, only the 2^n vectors in $\{-\ell, \ell\}^n$ represent points at distance $\ell\sqrt{n}$ having non-zero density. In this sense, the cube mutation favors the "corners" of the hypercube. GCPSO is therefore not invariant w.r.t. rotations of the coordinate system, nor, incidentally, is the standard PSO (Angeline, 1998; Hansen, Ros, Mauny, Schoenauer and Auger, 2008). From a black-box algorithm, we would not expect its search strategy to be biased towards certain directions.

Surprisingly, it turns out that GCPSO_1 has the same asymptotic runtime behavior as the (1+1) ES on SPHERE irrespectively of rotations of the coordinate system. In order to halve the distance from the globally best particle to the optimum, a typical number of $\Theta(n)$ steps is needed. More generally, the following theorem is obtained.

Theorem 7.7 (Witt (2009)). *Consider the GCPSO_1 on* SPHERE. *Given that it holds $\ell = \Theta(\|\boldsymbol{x}^*\|/n)$ for the initial solution $\boldsymbol{x}^*$, the runtime until the distance to the optimum is no more than $\epsilon\|\boldsymbol{x}^*\|$ is $O(n \log(1/\epsilon))$ with probability at least $1 - 2^{-\Omega(n)}$ provided that $2^{-n^{O(1)}} \le \epsilon \le 1$.*

Again a complete proof would be beyond the scope of this book chapter. However, the underlying proof idea is interesting and relatively easy to describe. Witt (2009) derives a general scenario to inspect the probability of the GCPSO_1 making progress on the SPHERE function. First he exploits that the GCPSO$_1$ is invariant w.r.t. shifts and scalings (but not rotations) of the coordinate system, i.e., if we replace the SPHERE function by the function $a\|\boldsymbol{x} - \boldsymbol{y}\|^2$ for $a \in \mathbb{R}$ and $\boldsymbol{y} \in \mathbb{R}^n$ and perform a corresponding transformation for the particle and ℓ, the behavior of GCPSO does not change.

Hence, for a step of the runtime analysis, we can w. l. o. g. assume that the globally best position is the origin, i.e., $\boldsymbol{x}^* = \boldsymbol{0}$, and the optimum is a distance 1, i.e.,

$$\boldsymbol{opt} \in \{\boldsymbol{x} \in \mathbb{R}^n \mid \|\boldsymbol{x}\| = 1\}.$$

Then the new position of GCPSO is a random sample $\tilde{\boldsymbol{x}} = (r_1, \ldots, r_n)$ with $r_i \in U[-\ell, \ell]$ independently for $1 \leq i \leq n$. An exchange of $\boldsymbol{x}^*$, called success, is equivalent to

$$\mathrm{dist}(\tilde{\boldsymbol{x}}, \boldsymbol{opt}) < 1 \Leftrightarrow (r_1 - o_1)^2 + \cdots + (r_n - o_n)^2 < 1^2 = 1,$$

where $(o_1, \ldots, o_n) = \boldsymbol{opt}$ and $\mathrm{dist}(\boldsymbol{a}, \boldsymbol{b}) = \|a - b\|$ denotes the Euclidean distance. This inequality describes the volume of the intersection of the hypercube corresponding to the outcome of the mutation and the hypersphere corresponding to the success region, i.e., the region of better search points. Since $o_1^2 + \cdots + o_n^2 = \|\boldsymbol{opt}\|^2 = 1$, the event of a success simplifies to

$$\underbrace{r_1^2 + \cdots + r_n^2}_{=:L} < \underbrace{2(r_1 o_1 + \cdots + r_n o_n)}_{=:R(\boldsymbol{opt})}. \tag{7.2}$$

The last inequality makes the event of a success equivalent to having two random variables in a specific order. The left-hand side L is composed of the sum of n identically distributed random variables with bounded support, which implies by the Central Limit Theorem that L follows a sharp-concentration result around its expectation $n\ell^2/3$. This allows us to conceptually replace L by a deterministic value in the analysis. The random variable $R(\boldsymbol{opt})$ governing the right-hand side is more complicated to study. The exact location of the optimum goes in, which accounts for the fact that the mutation operator is not isotropic. Interestingly, it is possible to study the distribution of $R(\boldsymbol{opt})$ in an asymptotic exact way without knowing the exact location of $\boldsymbol{opt}$. Witt (2009) shows the following results.

Lemma 7.8. *Let $\|\boldsymbol{opt}\|^2 = 1$ and let $\alpha(n) > 0$ be an arbitrary positive number, which may depend on n. Then*

(1) $\Pr(R(\boldsymbol{opt})) \geq \alpha(n) \cdot \ell) \leq e^{-\alpha(n)^2/2}$,
(2) $\Pr(R(\boldsymbol{opt})) \geq \alpha(n) \cdot \ell) \geq 1/2 - O(\alpha(n))$.

The first statement of Lemma 7.8 shows that we cannot hope for progress that is asymptotically bigger than ℓ, which is made precise using Hoeffding's inequality. On the other hand, the second statement shows us that $R(\boldsymbol{opt})$ takes a value in the order $\Omega(\ell)$ with a constant probability; just choose $\alpha(n)$ as a constant that is small enough to compensate the constant in the O-term. The proof of this statement is more involved, using geometrical arguments.

Recalling that $L = \Theta(n\ell^2)$ is valid with high probability and contrasting this with Lemma 7.8, we see that the decisive inequality (7.2) can only hold with sufficient probability if we choose $\ell = c/n$ for a constant c. Taking into account the second part of the lemma, there is also a choice of c that allows for a progress in the order of $\Omega(1/n)$. Switching back to an arbitrary $\boldsymbol{x}^*$ at arbitrary distance $\|\boldsymbol{x}^*\|$, this translates to $\ell = \Theta(\|\boldsymbol{x}^*\|/n)$ as an initial condition for optimal progress. Additionally, properties of the 1/5-rule that were proven by Jägersküpper (2007a) before make explicit that this rule controls ℓ in the right way. As GCPSO approaches the optimum and $\|\boldsymbol{x}^*\|$ takes smaller and smaller values, the asymptotic identity $\ell = \Theta(\|\boldsymbol{x}^*\|/n)$ is still maintained. This is the key to the proof of Theorem 7.7, which does not only reveal insights into the time to halve the distance to the optimum but also into the time until ϵ-approximations are attained, for a wide range of ϵ-values.

It is remarkable that the runtime $O(n\log(1/\epsilon))$ seems to be the best possible that can be obtained in this setting. Jägersküpper (2007b) has shown that a broad class of black-box algorithms with isotropic mutations needs an expected number of $\Omega(n)$ steps to halve the distance to the optimum. Of course, GCPSO does not search fully isotropically, yet matches the lower bounds. Intuitively, the upper bound on the runtime from Theorem 7.7 is in the same order as for the (1+1) ES since the mutation operator of GCPSO is still "close" to being isotropic for the appropriate choice of ℓ. Namely, ℓ should be chosen so small that basically the portion of the hypercube corresponding to a success is almost half the hypercube itself. If ℓ is too big, the curse of dimensionality with high probability prevents the mutation from hitting the "right" region of the hypercube. The same intuition holds in the analysis of the (1+1) ES with the isotropic mutation operator (Jägersküpper, 2007a). However, the geometric objects appearing in the latter analysis, namely intersections of hyperspherical volumes, are much more difficult to handle than the intersection of a hypersphere and a hypercube that is implicit in Witt's (2009) analysis of the GCPSO_1. In this respect, GCPSO_1 combines asymptotically optimal progress with a considerably simplified analysis. This result might come as a surprise also to those who are familiar with the theory of evolution strategies.

7.7. Outlook and Future Research

The theory of PSO algorithms has made remarkable progress in recent years, however, especially runtime analyses are in their early infancy. This book chapter has elaborated on recent trends in the theory of PSO. The very first theoretical treatments relate to the standard PSO in the domain $\mathbb{R}^n$ and describe its convergence to fixed points. These convergence analyses were by no means trivial. It took several years until the first convergence analysis coping with the full amount of stochasticity intrinsic in PSO was finally published in 2007.

Experiences from the theory of evolutionary computation and ACO have shown that convergence analyses of a heuristic often represent precursors to their runtime analysis. In this chapter, we have argued why the theory of randomized search heuristics should follow this path from convergence to runtime analyses. At the time of writing, the results regarding the runtime of PSO variants are still very limited. The Binary PSO studied in Section 7.5 is a relatively uncommon PSO variant. Similar objections hold for the GCPSO studied in Section 7.6. Apart from that, all results available so far have their drawbacks in terms of generality, tightness of upper bounds, lack of lower bounds at all, etc.

Nevertheless, there is reasonable hope that especially the runtime results presented in this book chapter pave the ground for future important and insightful analyses. Having the vision of a unified theory of randomized search heuristics, a small piece of the puzzle was laid by the analyses of single-individual heuristics such as the simple ACO algorithm 1-ANT and the (1+1) EA on the one hand and the simple 1-PSO on the other hand. All of these heuristics build probabilistic models for good solutions in the course of optimization. Moreover, in these simple cases, this distribution is a product of independent distributions for the single bits. This scenario was studied with similar techniques both for the 1-ANT and Binary PSO algorithms. In particular, the method of fitness-based partitions known from the analysis of evolutionary algorithms could be carried over in both cases. From a more general perspective, 1-ANT and the Binary PSO are nothing else than probabilistic model-building algorithms, also known as estimation-of-distribution algorithms (EDAs). They follow a common principle and can be analyzed as specific instantiations of EDAs with different update mechanism. The methods for the analysis of simple ACO and PSO algorithms might cross-fertilize the theory of EDAs and contribute to a classification of useful update mechanisms in such algorithms.

To some extent, parallels between PSO and other search heuristics could also be drawn in the analysis of the $GCPSO_1$ on the SPHERE function. Existing results on the runtime of a simple evolutionary algorithm were extremely useful in order to analyze the parametrization of the hypercube mutation operator. Unfortunately, $GCPSO_1$ is quite different from standard PSO algorithms. Given a non-trivial swarm, the impact of its size and the interaction of particles on the runtime is widely misunderstood so far. At the moment, it is not very obvious either how existing methods for the analysis of randomized search heuristics allow for the runtime analysis of standard PSO algorithms. Therefore, a first result on the time for the standard PSO in continuous search spaces to converge to its swarm leader, assuming a non-trivial parametrization of the learning factors, is a challenging open problem.

References

Angeline, P. J. (1998). Evolutionary optimization versus particle swarm optimization: Philosophy and performance differences, in *Proc. of Evolutionary Programming VII, LNCS*, Vol. 1447 (Springer), pp. 601–610.

Bratton, D. and Kennedy, J. (2007). Defining a standard for particle swarm optimization, in *Proc. of Swarm Intelligence Symposium (SIS 2007)* (IEEE Press), pp. 120–127.

Clerc, M. (2006). *Particle Swarm Optimization* (ISTE).

Clerc, M. and Kennedy, J. (2002). The particle swarm – explosion, stability, and convergence in a multidimensional complex space, *IEEE Transactions on Evolutionary Computation* **6**, 1, pp. 58–73.

Cormen, T. H., Leiserson, C. E., Rivest, R. L. and Stein, C. (2001). *Introduction to Algorithms, Second Edition* (The MIT Press and McGraw-Hill Book Company).

Gutjahr, W. J. and Sebastiani, G. (2008). Runtime analysis of ant colony optimization with best-so-far reinforcement, *Methodology and Computing in Applied Probability* **10**, pp. 409–433.

Hansen, N., Ros, R., Mauny, N., Schoenauer, M. and Auger, A. (2008). PSO facing non-separable and ill-conditioned problems, Research Report 6447, INRIA, France.

Jägersküpper, J. (2007a). Analysis of a simple evolutionary algorithm for minimization in Euclidean spaces, *Theoretical Computer Science* **379**, 3, pp. 329–347.

Jägersküpper, J. (2007b). Lower bounds for randomized direct search with isotropic sampling, *Operations Research Letters* **36**, pp. 327–332.

Jiang, M., Luo, Y. P. and Yang, S. Y. (2007). Stochastic convergence analysis and parameter selection of the standard particle swarm optimization algorithm, *Information Processing Letters* **102**, 1, pp. 8–16.

Kennedy, J. (2003). Bare bones particle swarms, in *Proc. of the IEEE Swarm Intelligence Symposium*, pp. 80–87.

Kennedy, J. and Eberhart, R. C. (1995). Particle swarm optimization, in *Proc. of the IEEE International Conference on Neural Networks* (IEEE Press), pp. 1942–1948.

Kennedy, J. and Eberhart, R. C. (1997). A discrete binary version of the particle swarm algorithm, in *Proc. of the World Multiconference on Systemics, Cybernetics and Informatics (WMSCI)*, pp. 4104–4109.

Kennedy, J., Eberhart, R. C. and Shi, Y. (2001). *Swarm Intelligence* (Morgan Kaufmann).

Neumann, F., Sudholt, D. and Witt, C. (2009). Analysis of different MMAS ACO algorithms on unimodal functions and plateaus, *Swarm Intelligence* **3**, 1, pp. 35–68.

Ozcan, E. and Mohan, C. K. (1998). Analysis of a simple particle swarm optimization system, *Intelligent Engineering Systems Through Artificial Neural Networks* , 1, pp. 253–258.

Ozcan, E. and Mohan, C. K. (1999). Particle swarm optimization: Surfing the waves, in *Proc. of the Congress on Evolutionary Computation (CEC '99)* (IEEE Press), pp. 1939–1944.

Poli, R. (2008). Analysis of the publications on the applications of particle swarm optimisation, *Journal of Artificial Evolution and Applications* **2008**, 10 pages.

Poli, R. and Broomhead, D. (2007). Exact analysis of the sampling distribution for the canonical particle swarm optimiser and its convergence during stagnation, in *Proc. of the Genetic and Evolutionary Computation Conference (GECCO '07)* (ACM Press), pp. 134–141.

Poli, R., Kennedy, J. and Blackwell, T. (2007a). Particle swarm optimization, *Swarm Intelligence* **1**, 1, pp. 33–57.

Poli, R. and Langdon, W. B. (2007). Markov chain models of bare-bones particle swarm optimizers, in *Proc. of the Genetic and Evolutionary Computation Conference (GECCO '07)* (ACM Press), pp. 142–149.

Poli, R., Langdon, W. B., Clerc, M. and Stephens, C. R. (2007b). Continuous optimisation theory made easy? Finite-element models of evolutionary strategies, genetic algorithms and particle swarm optimizers, in *Proc. of Foundations of Genetic Algorithms 9 (FOGA '07)*, pp. 165–193.

Shi, Y. and Eberhart, R. C. (1998). Parameter selection in particle swarm optimization, in *Proc. of the Seventh Annual Conference on Evolutionary Programming*, pp. 591–600.

Sudholt, D. and Witt, C. (2008). Runtime analysis of binary PSO, in *Proc. of the Genetic and Evolutionary Computation Conference (GECCO '08)* (ACM Press), pp. 135–142.

Sudholt, D. and Witt, C. (2010). Runtime analysis of a binary particle swarm optimizer, *Theoretical Computer Science* **411**, 21, pp. 2084–2100.

Trelea, I. C. (2003). The particle swarm optimization algorithm: convergence analysis and parameter selection, *Information Processing Letters* **85**, 6, pp. 317–325.

van den Bergh, F. (2002). *An Analysis of Particle Swarm Optimizers*, Ph.D. thesis, Department of Computer Science, University of Pretoria, South Africa.

van den Bergh, F. and Engelbrecht, A. P. (2002). A new locally convergent particle swarm optimiser, in *Proc. of the IEEE International Conference on Systems, Man and Cybernetics.*

Wegener, I. (2002). Methods for the analysis of evolutionary algorithms on pseudo-Boolean functions, in R. Sarker, X. Yao and M. Mohammadian (eds.), *Evolutionary Optimization* (Kluwer), pp. 349–369.

Witt, C. (2009). Why standard particle swarm optimisers elude a theoretical runtime analysis, in *Proc. of Foundations of Genetic Algorithms 10 (FOGA '09)* (ACM Press), pp. 13–20.

Chapter 8

Ant Colony Optimization: Recent Developments in Theoretical Analysis

Walter J. Gutjahr

Dept. of Statistics and Decision Support Systems
University of Vienna
Universitaetsstrasse 5/9, A-1010 Wien, Austria
walter.gutjahr@univie.ac.at

This chapter complements available surveys on the theoretical analysis of Ant Colony Optimization (ACO) by (i) providing proof ideas for some selected results, and (ii) placing special emphasis on new developments during the last few years. The chapter starts with a quick formal introduction into ACO and a classification scheme for ACO algorithms. Then, results on convergence of ACO as well as on the runtime behavior of some ACO variants on certain test problems are recapitulated. In particular, upper and lower bound results for expected optimization times are discussed, and asymptotic approximation techniques for the analysis of the ACO process are sketched. A short outline of some open problems concludes the chapter.

Contents

8.1. Introduction

Nature-inspired randomized search algorithms can be roughly classified into methods drawing from the paradigm of genetic evolution, i.e., the evolution of species, and methods relying on individual evolution, the last being based on the biological mechanism of *learning*. In an attempt to give a finer classification, one observes that the two most important types of learning also find their counterparts in current techniques of randomized search: Whereas imitation learning is a key feature of Particle Swarm Optimization (PSO), reinforcement learning is the fundamental principle behind *Ant Colony Optimization* (ACO).

The ACO paradigm was developed by Dorigo *et al.* (1991, 1996). The first applications of ACO concern the famous Traveling Salesperson Problem (TSP). Suppose that n points in a plane have to be visited in a closed tour, each point exactly once. We can imagine an *ant* moving randomly from one of these points to some other point, then to a third, etc., always avoiding still visited points. After the visit of the last point, the ant returns to its start position. This defines a tour, which can also be described by a sequence of arcs (i, j), where i, j are point indices. Now, assume that if the tour happens to be comparably short, the ant increases its probability to traverse arcs (i, j) that lie on that tour in the future, and it decreases the probability of traversing other arcs. In other terms, successful moves are *reinforced.* Computationally, this is done by an increment of real numbers assigned to the arcs, the so-called *pheromone values*, along the arcs of the tour. Then, the ant starts its walk from anew, applying transition probabilities that are proportional to the current pheromone values. By this mechanism, the average quality of a tour improves over time, which can be used to obtain approximate solutions to the TSP.

After first successes on the TSP, it turned out that the range of application of the ACO paradigm was much broader: The paradigm can be extended to the entire area of combinatorial optimization (CO) and even beyond this field, which means that ACO algorithms may be considered as *metaheuristics*, cf. Dorigo and Caro (1999). A survey on diverse variants of ACO algorithms, their applications and properties is given in Dorigo and Stuetzle (2004).

The present chapter reviews some results that have been obtained in the *theoretical* performance analysis of ACO algorithms; cf. also the previous surveys (Dorigo and Blum, 2005; Gutjahr, 2007). Section 2 recapitulates the basics of ACO in formal terms. In Section 3, the issue of convergence is treated. Sections 4 and 5 review runtime results for ACO algorithms with best-so-far and iteration-best reinforcement, respectively. In Section 6, we deal with some specific ACO implementations, and Section 7 gives some suggestions for possible future research.

8.2. What is Ant Colony Optimization?

In this section, we introduce ACO in formal terms. Let us start with a concept that allows it to generalize the idea of ACO (as outlined in the Introduction) to problems different from the TSP. Throughout the chapter, we shall restrict ourselves to the CO case. The general form of a CO problem is

$$\max f(x) \;\; s.t. \;\; x \in S, \tag{8.1}$$

where S (often called the search space) denotes a finite set with some combinatorial structure. E.g., S can be the set $\{0,1\}^n$ of all binary vectors $x = (x_1, \ldots, x_n)$ of length n, or it can be the set of all permutations of $1, \ldots, n$.

8.2.1. *The Construction Graph Concept*

For treating a given CO problem by ACO, we have to define a graph on which the fictitious computational unit called "ant" performs its random walks, *and* to encode each solution $x \in S$ as a path in this graph (not necessarily in a unique way), such that the trajectories of the ants can be decoded to solutions in S. This graph, which obviously has to depend on the problem instance, is called the *construction graph* (CG). Two different suggestions for formally defining the CG have been given in the ACO literature: a *pheromone-on-arcs* version (Gutjahr, 2000) and a *pheromone-on-nodes* version (Dorigo and Blum, 2005). Since the pheromone-on-arcs version is more intuitive in the context of the TSP, we choose this version here.

Definition 1. Let an instance (S, f) of a CO problem (8.1) be given. Furthermore, suppose that $\mathcal{C} = (\mathcal{V}, \mathcal{A})$ is a directed graph with node set $\mathcal{V}$ and arc set $\mathcal{A}$. Let a unique node in $\mathcal{V}$ be marked as the so-called *start node*, and let $\mathcal{W}$ be a set of directed paths w in $\mathcal{C}$, called *feasible paths*, that satisfy the following conditions: (i) w starts at the start node of $\mathcal{C}$; (ii) w contains each node of $\mathcal{C}$ at most once; (iii) w is maximal in $\mathcal{W}$, i.e., it cannot be prolonged (by appending edges and nodes) to a longer feasible path in $\mathcal{W}$.

Moreover, let Φ be a function mapping the set $\mathcal{W}$ of feasible paths onto the search space S of the given problem instance: To each feasible path $w \in \mathcal{W}$, there corresponds a feasible solution $x = \Phi(w) \in S$, and to each feasible solution $x \in S$, there corresponds at least one feasible path in $\mathcal{W}$ such that $\Phi(w) = x$. Then, the graph $\mathcal{C}$, endowed with the function Φ, is called a *construction graph* for the problem instance (S, f).

Example 1. The natural CG for a TSP with n nodes (customers) $v_1, \ldots, v_n$ is a complete graph on node set $\mathcal{V} = \{v_1, \ldots, v_n\}$. We can choose v_1 as the start node. The decoding function Φ takes a feasible path w and assigns to it a feasible solution (i.e., a closed tour) of the TSP by adding an additional move from the last node of w back to the start node, such that the tour becomes closed. $\bowtie$

Example 2. For the optimization of pseudo-boolean functions, i.e., in the case where $S = \{0, 1\}^n$, the following CG has been introduced in Gutjahr (2006) under the name *chain* graph: Arrange nodes $0, \ldots, n$ from left to right. From node $i - 1$, draw an upper arc as well as a lower arc to node i $(i = 1, \ldots, n)$. In order to avoid multiple arcs, for $i = 1, \ldots, n$, divide the upper arc from $i - 1$ to i by a dummy node i^+ into arc $(i - 1, i^+)$ and arc (i^+, i). Analogously, divide the lower arc by the introduction of a dummy node i^- into arc $(i - 1, i^-)$ and arc (i^-, i). The chain graph is illustrated in Figure 8.1. Node 0 is the start node. Whenever the ant chooses node i^+, this is decoded as an assignment $x_i = 1$, and whenever the ant chooses node i^-, this is decoded as an assignment $x_i = 0$. As soon as the ant has reached the rightmost node, a complete solution $x \in \{0, 1\}^n$ has been constructed. Pheromone values on arcs (i^+, i) and (i^-, i) are irrelevant since these arcs must be traversed anyway. Thus, only the pheromone values $\tau^+(i)$ on arcs $(i-1, i^+)$ and the pheromone values $\tau^-(i)$ on arcs $(i-1, i^-)$, respectively, count. $\bowtie$

By convention, the fitness function value assigned to a feasible path w is defined as that of the corresponding feasible solution $x = \Phi(w)$. Based on this convention, we can identify feasible paths and solutions and denote them by the same symbol x.

Typically, there are several possible encodings of a combinatorial optimization problem in the form of a construction graph. E.g., in Gutjahr (2006), two alternatives to the chain graph for the optimization of pseudo-boolean functions have been analyzed. The investigation shows that the

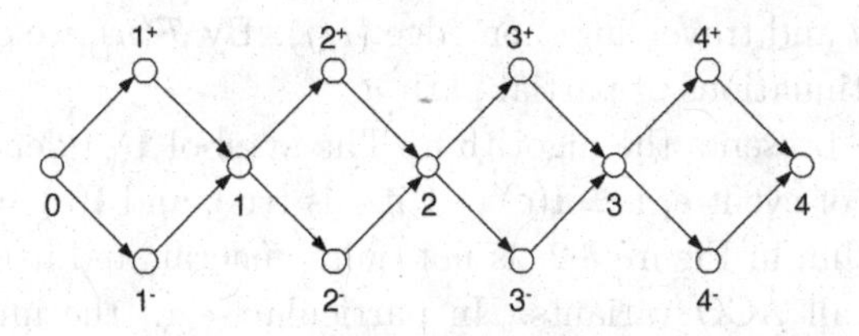

Fig. 8.1. Chain construction graph.

efficiency of an ACO algorithm on a given optimization problem can vary dramatically, depending on the chosen construction graph.

8.2.2. *A Basic ACO Algorithm*

A few additional notions are required to present a basic algorithmic scheme of ACO. By a *partial path* in $\mathcal{C}$, we understand a path u satisfying properties (i) – (ii) in Definition 1, but not necessarily property (iii). A *continuation* of a partial path u with node i as its last node is an edge $(i, j) \in \mathcal{A}$. A continuation (i, j) of u is called *feasible* if there exists a feasible path $x \in \mathcal{W}$

Procedure ACO
Initialize pheromone values τ_{ij} on the arcs $(i, j) \in \mathcal{A}$;
for iteration $m = 1, 2, \ldots$ **do**
  for ant $\sigma = 1, \ldots, s$ **do**
    set i, the current position of the ant, equal to the start node of $\mathcal{C}$;
    set u, the current partial path of the ant, equal to the empty list;
    while ($\mathcal{F}(u) \neq \emptyset$) **do**
      select successor node j with probability
        $p_{ij} = \mathbf{I}\{(i, j) \in \mathcal{F}(u)\} \cdot \tau_{ij} \; / \; \sum_{(i,r) \in \mathcal{F}(u)} \tau_{ir}$;
      append arc (i, j) to u and set $i = j$;
    end while
    set $x^\sigma = u$;
  end for
  update the pheromone values τ_{ij} based on the current solutions x^σ;
end for

Fig. 8.2. Pseudocode of the basic ACO procedure.

starting with u and traversing then edge (i, j). By $\mathcal{F}(u)$, we denote the set of feasible continuations of partial path u.

Figure 8.2. presents the algorithm. The symbol $\mathbf{I}(e)$ denotes the indicator function of event e, i.e., $\mathbf{I}(e) = 1$ if e is true, and $\mathbf{I}(e) = 0$ otherwise.

The algorithm in Figure 8.2. is not quite general, and it is not claimed that it covers all ACO variants. In particular, e.g., the important ACS variant (Dorigo and Gambardella, 1997) is not covered. As it can be seen, the algorithm allows it to use several ants instead of only one, such that in each iteration, several solutions are constructed. To apply $s > 1$ ants is the normal case in the practice of ACO. Nevertheless, for some ACO variants, already the investigation of the case $s = 1$ yields interesting insights.

Readers familiar with ACO will miss a typical component of implemented ACO versions in the formulation above, the use of problem-specific *heuristic values*, usually denoted by the symbol η_{ij}. These values are multiplied by the τ_{ij} in the equation for the computation of the transition probabilities p_{ij}. The heuristic component often considerably improves the performance of ACO, but in available theoretical investigations, it usually play only a minor role for two reasons: First, the specific contribution of ACO lies in the reinforcement learning component working with the τ-values rather than with the η-values. Secondly, performance results including the last would heavily depend on the chosen problem-specific heuristic and would thus be hard to generalize.

Pheromone initialization is usually done in a uniform way: each arc obtains the same initial value of τ_{ij}. The step "pheromone update" will be addressed in the next subsection.

8.2.3. *A Classification Scheme for ACO*

We use the scheme *#ants / reinf / reward / pherbound* to classify ACO variants with a basic structure corresponding to that in Figure 8.2.

(I) The parameter *#ants* denotes the number of ants.

(II) The parameter *reinf* indicates which paths are reinforced. It can take the following values:

ac (all of current iteration)
bf (best-so-far)
bf* (best-so far with strict improvements)
bf_0 (best-so-far on exchange only)

bf_0^* (best-so-far with strict improvements on exchange only)
ib(r) (iteration-best with r ranked ants)

(1) *reinf* = ac: Each path traversed by some ant in the current iteration is reinforced (usually by different rewards). This is done as follows: For each $(i, j) \in \mathcal{A}$, set

$$\tau_{ij} = (1-\rho) \cdot \tau_{ij} + \frac{\rho}{s} \cdot \sum_{\sigma=1}^{s} \mathbf{I}\{(i,j) \in x^{\sigma}\} \cdot \text{reward}(x^{\sigma}),$$

where $\rho \in [0, 1]$ is the so-called *evaporation rate.* Multiplication of the pheromone values by $1 - \rho$ in each iteration is also used in the other pheromone update schemes; this step is called *evaporation.*

(2) *reinf* = bf, bf*, bf_0 or bf_0^*: Only the best path found up to now in any of the previous iterations (including the current one) by any ant is reinforced. Let $\hat{x}$ be this best path. The four sub-variants differ by the question whether or not $\hat{x}$ is also updated if an equally good solution x is found, and by the question whether or not the pheromone update is done in each iteration or only in iterations where $\hat{x}$ has been exchanged. Let $\bar{x}$ be the best path found by some of the ants in the *current* iteration, i.e., the path with $f(\bar{x}) \geq f(x^{\sigma})\ \forall \sigma$. (If there is more than one best path, take the first found.) The symbol $\succ$ below is to be replaced by the symbol $\geq$ in the cases *reinf* = bf and *reinf* = bf_0, and by the symbol $>$ in the cases *reinf* = bf* and *reinf* = bf_0^*.

(a) For *reinf* = bf and *reinf* = bf*, the update rule is:
if $(f(\bar{x}) \succ f(\hat{x}))$ set $\hat{x} = \bar{x}$ **end if**
for each $(i, j) \in \mathcal{A}$, set $\tau_{ij} = (1-\rho) \cdot \tau_{ij} + \rho \cdot \mathbf{I}\{(i,j) \in \hat{x}\} \cdot \text{reward}(\hat{x})$;

(b) For *reinf* = bf_0 and *reinf* = bf_0^*, the update rule is:
if $(f(\bar{x}) \succ f(\hat{x}))$
set $\hat{x} = \bar{x}$;
for each $(i, j) \in \mathcal{A}$, set $\tau_{ij} = (1-\rho) \cdot \tau_{ij} + \rho \cdot \mathbf{I}\{(i,j) \in \hat{x}\} \cdot \text{reward}(\hat{x})$
end if

(3) *reinf* = ib(r): Let $x^{[1]}, x^{[2]}, \ldots, x^{[r]}$ denote the r best paths found in the current iteration. Then the following pheromone update rule is applied: For each $(i, j) \in \mathcal{A}$, set

$$\tau_{ij} = (1-\rho) \cdot \tau_{ij} + \rho \cdot \sum_{k=1}^{r} (r-k+1) \cdot \mathbf{I}\{(i,j) \in x^{[k]}\} \cdot \text{reward}(x^{[k]}).$$

Reinforcement rules can also be combined. E.g., bf* + ib(1) means that both the best-so-far and the iteration-best solution obtain pheromone increments (for arcs lying on both solutions, the contributions are added).

(III) The parameter *reward* describes the amount of reward that is added to the pheromone value of a reinforced arc. This can be co (constant or conserving) or fp (fitness-proportional).

(1) *reward* = co: In this situation, for all paths x, either $\text{reward}(x) = C$ with some constant C, or $\text{reward}(x) = C/\text{length}(x)$, where $\text{length}(x)$ is the number of arcs on path x. In the last case, the parameter C can be chosen in such a way that the overall amount of pheromone on all arcs remains constant over the iterations, leading to a "pheromone-conserving" update scheme. Some typical construction graphs as those presented in Examples 1 and 2 in Subsection 2.1 have the property that all feasible paths possess the same length, such that the constant and the conserving reward strategy coincide. This fact suggests a subsumption of the two strategies under the same group.

(2) *reward* = fp: In this case, $\text{reward}(x) = C \cdot f(x)$ with some constant C for all paths x. A natural extension consists in providing a reward of size $C \cdot g(f(x))$ with some nondecreasing function g; we subsume also this situation under the case *reward* = fp.

(IV) The parameter *pherbound* indicates whether or not upper or lower bounds $\tau_{\max}$ and $\tau_{\min}$ on the pheromone values are used. The parameter can take the values nb (no bound), ub (upper bound), lb (lower bound), or ulb (upper and lower bound). In the case where an upper pheromone bound $\tau_{\max}$ is applied, a pheromone update rule of the form $\tau_{ij} = \text{expression}$ is modified to the rule $\tau_{ij} = \min(\text{expression}, \tau_{\max})$. Analogously, in the case where a lower pheromone bound $\tau_{\min}$ is applied, the rule is modified to $\tau_{ij} = \max(\text{expression}, \tau_{\min})$, and if both bounds are applied, the rule becomes $\tau_{ij} = \max(\min(\text{expression}, \tau_{\max}), \tau_{\min})$.

Let us describe some ACO algorithms from literature by this scheme. The classical ACO algorithm, *Ant System* (Dorigo *et al.*, 1991, 1996), is the variant s/ac/fp/nb. *MAX-MIN Ant System* (short: MMAS), developed in Stuetzle and Hoos (1997, 2000), encompasses the variants s/bf*/fp/ulb, s/ib(1)/fp/ulb, and s/bf*+ib(1)/fp/ulb. *Rank-based Ant System*, developed in Bullnheimer *et al.* (1999), comprises s/ib(r)/fp/nb and s/bf*+ib(r)/fp/nb with $r \geq 2$.

In theoretical research, a simplified ACO algorithm called GBAS has been designed in order to study convergency properties. The GBAS variant investigated in Gutjahr (2000) is $s/\text{bf}_0/\text{co}/\text{nb}$. Later in Gutjahr (2002), two GBAS variants were analyzed which can be classified as $s/\text{bf}^*/\text{co}/\text{nb}$ and $s/\text{bf}^*/\text{co}/\text{lb}$, respectively. (The second of these variants is already of MMAS type.) For runtime analysis purposes, Neumann and Witt (2006) introduced a variant called 1-ANT, which can be described as $1/\text{bf}_0/\text{co}/\text{ulb}$, whereas in Gutjahr (2005, 2008a) and in Gutjahr and Sebastiani (2008), the runtime analysis of the MMAS-variants $1/\text{bf}^*/\text{co}/\text{ulb}$ and $1/\text{bf}^*/\text{fp}/\text{ulb}$ has been started.

8.3. Convergence

For several of the ACO variants described in the previous section, it can happen that with probability larger than zero, the optimal solution to a given problem instance is never constructed by any ant in any of the iterations of the algorithm. Of course, even for metaheuristics the main purpose of which is rather to find "good" solutions than to compute the best solution, this is an undesirable behavior, and it would be interesting to identify conditions under which an ACO variant is ensured to converge to an optimal solution. One may distinguish two forms of convergence of an ACO variant to optimality, which will be indicated in Definition 2 below.

In the following, $\hat{x} = \hat{x}(m)$ denotes the best solution found until iteration m according to the procedure for *reinf* = bf or for *reinf* = bf* in Subsection 2.3. (The detailed meaning of the convergence definitions will depend on the last choice.) The symbol S^* denotes the set of optimal solutions to the problem (8.1). Furthermore, we consider the stochastic process (X_m) $(m = 1, 2, \ldots)$ where $X_m = (\tau(m), \hat{x}(m-1))$ is the state of the ACO process during iteration m with $\tau(m) = (\tau_{ij}(m))_{(i,j)\in\mathcal{A}}$ denoting the vector of pheromone values in iteration m. It is easy to see that for the ACO variants described in the previous section, (X_m) is a Markov process, provided that the meaning of $\hat{x}(m)$ is adapted to the variant under consideration.

Definition 2. For an ACO variant on an instance of a CO problem, (a) *best-so-far convergence* holds if there exists an $x^* \in S^*$ such that with probability one, $\hat{x}(m) \to x^*$ as $m \to \infty$, (b) *model convergence* holds if there exists (i) a pheromone vector τ^* allowing only the generation of optimal solutions $x \in S^*$, and (ii) a special solution $x^* \in S^*$, such that with probability one, $X_m \to (\tau^*, x^*)$ as $m \to \infty$.

Because of the finiteness of S, we may also say that best-so-far convergence holds if with probability one, there is an integer $m_0 \geq 1$ such that $\hat{x}(m) = x^* \in S^* \ \forall m \geq m_0$. Best-so-far convergence is a relatively weak property since it is even exhibited by *random search*. (Random search is obtained, e.g., as the special case of an ACO algorithm of type s/bf/co/nb with $\rho = 0$.) Model convergence requires that best-so-far convergence holds *and* that the current probabilistic model for generating candidate solutions converges to a model the support of which is the set S^* of optimal solutions. From a practical point of view, the advantage of a model-convergent ACO variant lies in the property that such a variant has not only a proven *exploration* capacity (as also random search has), but also a proven *exploitation* capacity: If the model is ensured to gradually approach an optimal model, the randomized search will focus more and more around the (global) optimizers of the problem instance.

Several convergence results of different types have been shown for diverse ACO variants (Gutjahr, 2000; Stuetzle and Dorigo, 2002; Gutjahr, 2002, 2003; Sebastiani and Torrisi, 2005). To outline the proof ideas applied for showing the stronger property of model convergence, we present an example of such a result from Gutjahr (2002).

Theorem 1. Consider the ACO variant s/bf*/co/nb, where the choice of the evaporation rate $\rho = \rho_m$ depends on the iteration m $(m = 1, 2, \ldots)$. Take the reward function reward$(x) = 1/\text{length}(x)$ and initial pheromone values of size $1/|\mathcal{A}|$ on each arc $(i, j) \in \mathcal{A}$, such that the sum of all pheromone values remains identical to unity in each iteration. Furthermore, assume that the sequence $(\rho_1, \rho_2, \ldots)$ satisfies, for some $m_0 > 1$,

$$\rho_m \leq 1 - \frac{\log m}{\log(m+1)} \quad (m \geq m_0) \quad \text{and} \quad \sum_{m=1}^{\infty} \rho_m = \infty. \tag{8.2}$$

Then for each instance of an arbitrary CO problem, model convergence holds.

Proof. First, we verify that on the indicated conditions, best-so-far convergence holds. We do this by showing that for each solution $x \in S$ (and hence also for each optimal solution), there exists with probability one an iteration m and an ant σ such that x is traversed in iteration m by ant σ. This property (which is immediate for variants with *pherbound* = lb or ulb, but nontrivial in the present context) is sufficient for ensuring best-so-far convergence, since $\hat{x}$ stores the best found solution and is only updated in the case of a strict improvement.

Let $\nu = |\mathcal{V}|$ denote the number of nodes in the construction graph, let $L(x) = \text{length}(x)$, and let $\tau(m)$ with components $\tau_{ij}(m)$ be the vector of pheromone values during iteration m. Furthermore, denote by $T_m(x)$ the event that path x is traversed in iteration m by some ant, and by $T_m^c(x)$ the complementary event. Choose some fixed $x \in S$. The probability that there is no iteration in which x is traversed by some ant is

$$\Pr\left(\bigcap_{m=1}^{\infty} T_m^c(x)\right) = \Pr(T_1^c(x)) \cdot \prod_{m=2}^{\infty} \Pr\left(T_m^c(x) \mid \bigcap_{k=1}^{m-1} T_k^c(x)\right). \tag{8.3}$$

Now we derive a lower bound for the pheromone value on a fixed arc $(i,j) \in \mathcal{A}$ at the end of iteration m. In the worst case, (i,j) has never been reinforced in any of the iterations $1, \ldots, m$ and only lost pheromone by evaporation. Using the first condition in (8.2), this gives

$$\tau_{ij}(m) \geq \left(\prod_{k=1}^{m-1} (1-\rho_k)\right) \tau_{ij}(1) \geq \left(\prod_{k=1}^{m_0-1} (1-\rho_k)\right) \left(\prod_{k=m_0}^{m-1} \frac{\log k}{\log(k+1)}\right) \frac{1}{|\mathcal{A}|}$$

$$= \left(\prod_{k=1}^{m_0-1} (1-\rho_k)\right) \frac{\log m_0}{\log m} \frac{1}{|\mathcal{A}|} = \frac{K}{\log m}$$

with a constant K not depending on m. From the last inequality, we get a lower bound on the probability $p_{ij}(m)$ that in iteration $m \geq m_0$, an ant located in node i traverses arc (i,j), using $\tau_{ij}(m) \leq 1$ and the fact that i has at most $\nu - 1$ feasible successor nodes: $p_{ij}(m) \geq K/(\nu \log m)$ $(m \geq m_0)$. As a consequence, the probability that a fixed ant σ traverses x in a fixed iteration $m \geq m_0$ is $\prod_{(i,j)\in x} p_{ij}(m) \geq [K/(\nu \log m)]^{L(x)}$. It is easy to see that this bound holds also conditionally on arbitrary events in iteration $1, \ldots, m-1$. This gives us the following upper bound for the r.h.s. of (8.3):

$$\prod_{m=m_0}^{\infty} \left[1 - \left(\frac{K}{\nu \log m}\right)^{L(x)}\right].$$

The product in the last equation is zero exactly if $\sum_{m=m_0}^{\infty} (K/(r \log m)^{L(x)}) = \infty$, which, however, is true since $\sum_m (\log m)^{-L}$ diverges for each positive integer L. Therefore, the r.h.s. and thus also the l.h.s. of (8.3) is zero, which shows best-so-far convergence: With probability one, there is an iteration m where $\hat{x}(m)$ becomes equal to some $x^* \in S^*$. In iterations $t \geq m$, $\hat{x}(t) = x^*$ by the chosen update rule.

It remains to show that as $t \to \infty$, the vector $\tau(t)$ tends to some limiting vector τ^* that supports only the generation of optimal solutions. The vector

τ^* defined by $\tau^*_{ij} = \mathbf{I}\{(i,j) \in x^*\}/L(x^*)$ only supports the generation of solution x^*. We show $\tau_{ij}(t) \to \tau^*_{ij}$ $(t \to \infty)$ for all $(i,j) \in \mathcal{A}$. Let m be the index of the iteration where x^* is traversed for the first time. In all iterations $t > m$, only path x^* is reinforced. It is easy to verify by induction that as a consequence, for an arc $(i,j) \in x^*$ and some $r \geq 1$,

$$\tau_{ij}(m+r) = \left[\prod_{t=m}^{m+r-1} (1-\rho_t)\right] \tau_{ij}(m) + \frac{1}{L(x^*)} \sum_{k=0}^{r-1} \rho_{m+k} \prod_{\ell=k+1}^{r-1} (1-\rho_{m+\ell}). \tag{8.4}$$

Because of the second condition in (8.2), the series $\sum_t \rho_t$ diverges, and hence $\prod_{t=1}^{\infty}(1-\rho_t) = 0$. Therefore, the limes of the first term on the r.h.s. of (8.4) as $r \to \infty$ is zero. Since the second term on the r.h.s. of (8.4) does not depend on (i,j), we conclude that $\limsup_{r\to\infty} \tau_{ij}(m+r)$ has the same value on all arcs $(i,j) \in x^*$, and the same holds for $\liminf_{r\to\infty} \tau_{ij}(m+r)$.

Next, consider an arc $(i,j) \notin x^*$. In iterations $t > m$, such an arc is never reinforced; it is immediately seen that this implies $\tau_{ij}(m+r) \to 0$ $(r \to \infty)$ for $(i,j) \notin x^*$, which is the first part of the convergence property still to be shown. Finally, also the *sum* of the pheromone values on arcs $\notin x^*$ tends to zero, with the consequence that the sum of the pheromone values on arcs $\in x^*$ must tend to unity. Because of the independence of the lim sup and lim inf values on the specifically chosen arc $(i,j) \in x^*$, we can conclude that $\limsup_{r\to\infty} \tau_{ij}(m+r) = \liminf_{r\to\infty} \tau_{ij}(m+r) = \lim_{r\to\infty} \tau_{ij}(m+r) = 1/L(x^*)$ for $(i,j) \in x^*$, which completes the proof. □

An example for an evaporation rate scheme satisfying the conditions of Theorem 1 is $\rho_m = c/(m \log(m+1))$ with $0 < c < 1$.

The proof of Theorem 2 decomposes the sequence of iterations into a first, exploration-oriented part where an optimal solution has not yet been found, and a second, exploitation-oriented part in which an already found optimal solution is gradually reinforced. It may seem that the second part is irrelevant and that each variant that is convergent in the best-so-far sense can be made model-convergent simply by letting it switch from exploration to exploitation at the moment when an optimal solution is hit for the first time. However, this idea fails on the fact that the algorithm cannot recognize whether or not it has found an optimal solution. Therefore, in order to achieve model convergence, an ACO variant must perform exploitation (at least to some degree) already *before* finding the optimum, which, in turn, induces the risk of hampering exploration and thus preventing best-so-far convergence. Theorem 1 shows that by a suitable parameter scheme, an

ACO variant can keep the right balance between exploration and exploitation.

In Sebastiani and Torrisi (2005), the result above has been generalized and extended to the practically important case of ACO variants of type s/bf*/fp/nb.

8.4. Runtime Behavior of Variants with Best-So-Far Reinforcement

Let us now turn to the analysis of the *expected optimization time*, where the *optimization time* (also known as *first hitting time*) is defined as the number of fitness function evaluations required until for the first time, an optimal solution $x^* \in S^*$ is constructed. Since in each iteration, s solutions are evaluated, the optimization time is equal to sm^*, where m^* denotes the index of the first iteration where an optimal solution is found. We are mainly interested in the asymptotic dependence of the optimization time on the problem instance size n.

It is obvious that the expected value of the optimization time can only be finite if the variant under consideration converges (at least in the sense of best-so-far convergence) to an optimal solution with probability one. In this section, we investigate ACO variants with best-so-far reinforcement and pheromone bounds. These variants facilitate the runtime analysis already insofar as best-so-far convergence is always guaranteed for them, since the pheromone bounds ensure a certain minimum amount of random-search-type behavior, such that each solution will be visited infinitely often during the process with probability one. As we shall see, however, these variants lend themselves very well for a runtime analysis also for the additional reason that during process, the currently achieved solution quality can never decrease, which allows to apply a modification of the classical analytical method of fitness-based partitions.

Contrary to the situation in convergence analysis, in runtime analysis, it cannot be expected that meaningful results valid for all CO problems (e.g., showing consistently better performance of an algorithm over another algorithm) can be derived. The reason lies in No-Free-Lunch theorems (see Wolpert and Macready (1997) or Chapter 9 of this book), which basically state that in the average over all CO problems, no search algorithm is better than random search. (Of course, this needs not to hold for special problem classes of practical interest. For a more detailed discussion, see Gutjahr (2008b).) Therefore, the runtime behavior of an ACO variant has to be investigated for special functions or at least for more restricted classes of functions.

8.4.1. *Upper Bounds for Expected Optimization Times*

Let us start with the presentation of a general approach that can be applied to derive upper runtime bounds for variants with best-so-far reinforcement, the *method of fitness-based partitions.* This method is well established in the analysis of evolutionary algorithms (see, e.g., Droste *et al.* (2002)). Its application to ACO requires a technical modification since in its original form, the method works on discrete finite state spaces, whereas by the pheromone vectors, a state of an ACO process contains a component from an infinite continuous space. In Gutjahr and Sebastiani (2008) it has been shown that the method can be transferred to the analysis of ACO algorithms in a general and mathematically rigorous way. Also to other swarm intelligence metaheuristics as PSO, the modified method is applicable.

The approach partitions the set S of feasible solutions into classes defined by the fitness function values. Certain ranges of the fitness function ("levels") correspond to certain subsets ("level sets") of the search space S. In the simplest form, the level set A_k ($k = 1, \ldots, D$) is defined as the set of all solutions x with $f(x) = f_k$, where $f_1 < f_2 < \ldots < f_D$ are the possible different fitness values, arranged in ascending order.

Consider variants of the types s/bf*/co/ulb or s/bf*/fp/ulb and the optimization of pseudo-boolean functions using the *chain* construction graph explained in Example 2. Each of the pheromone values $\tau^+(i)$ and $\tau^-(i)$, as defined in Example 2, is initialized to the value 1/2. One immediately verifies that if $0 < \tau_{\min} < 1 - \tau_{\min} = \tau_{\max} < 1$, then during all iterations, we always have $\tau^+(i) + \tau^-(i) = 1$ for $i = 1, \ldots, n$. As a consequence, the probability that an ant chooses the upward arc $(i-1, i^+)$ and thus sets $x_i = 1$ is simply $\tau^+(i)$, and the probability that an ant chooses the downward arc $(i-1, i^-)$ and thus sets $x_i = 0$ is $\tau^-(i) = 1 - \tau^+(i)$. Extending a term coined in Neumann *et al.* (2009), we call the pheromone values *frozen in* x, if $\tau^+(i) = \tau_{\max}$ for each i with $x_i = 1$ and $\tau^+(i) = \tau_{\min}$ for each i with $x_i = 0$. It is easy to see that provided that the current best-so-far solution $\hat{x}$ does not change, the pheromone values finally freeze in $\hat{x}$ by the reinforcement mechanism.

The following result has been proven in Gutjahr and Sebastiani (2008). There, the statement is presented in a more general context and in strict formal terms. We confine ourselves here to a special case and give a less formal presentation in the vein of the recapitulation in Neumann *et al.* (2009).

Theorem 2. Assume an ACO variant of the type above. Let $t^*(k)$ denote a deterministic upper bound on the number of iterations until all pheromone values are frozen in $\hat{x}$, given that in the beginning, the current best-so-far solution $\hat{x}$ is in A_k and that is not exchanged in the meantime. Let q_k be a lower bound on the probability that during one iteration in which the pheromone values are frozen in some $\hat{x} \in A_k$, a strictly better solution than $\hat{x}$ is found. Then the expected optimization time is smaller or equal to

$$\sum_{k=1}^{D-1} t^*(k) + \sum_{k=1}^{D-1} \frac{1}{q_k}. \tag{8.5}$$

Proof idea. Let us assume that in the first iteration, a best-so-far solution $\hat{x} \in A_k$ is constructed. Now consider the next $t^*(k)$ iterations. During this time, either the current best-so-far solution is changed to some solution in a higher level set A_j $(j > k)$, or the pheromone values freeze in $\hat{x}$. In the last case, after freezing, the subsequent iterations perform independent trials to construct a better solution than $\hat{x}$, based on the unchanged pheromone values frozen in $\hat{x}$. Since the success probability is larger or equal to q_k, the expected number of trials until success is smaller or equal to $1/q_k$. Therefore, $t^*(k) + 1/q_k$ is an upper bound for the expected time until the current best-so-far solution leaves level set A_k. Because each A_k can only be left in the direction of a level set A_j with $j > k$, each A_k can only be visited at most once. Thus, by summation over all $k < D$, we obtain an upper bound on the expected optimization time (note that $A_D = S^*$). □

Type 1/bf*/co/ulb. As an example, let us compute $t^*(k)$ for the special case of the variant 1/bf*/co/ulb with $\tau_{\min} = 1/n$, $\tau_{\max} = 1 - 1/n$, and constant reward $C = 1$. As in Neumann *et al.* (2009), we abbreviate this variant by MMAS* in the sequel. Assume that $\hat{x}_i = 0$ for a certain bit i, such that this bit is never reinforced as long as $\hat{x}$ remains unchanged. Then $\tau^+(i)$, which is initially smaller or equal to $1 - 1/n$, would drop to a value smaller or equal to $(1-\rho)^t(1-1/n)$ after t iterations, unless if the lower pheromone bound $1/n$ becomes active before. One easily convinces oneself that the pheromone increment for reinforced bits happens in a symmetric manner than the pheromone loss for not reinforced bits. From that, one immediately obtains the freezing time bound

$$t^*(k) = -\log(n-1)/\log(1-\rho) \leq (\log n)/\rho, \tag{8.6}$$

which is in this case independent of k. (The handy upper bound on the r.h.s. has been suggested in Neumann *et al.* (2009).)

The method of fitness-based partitions can immediately be used for obtaining upper optimization time bounds for algorithms of reinforcement type bf* on well-known test functions from the evolutionary algorithms literature, such as OneMax or LeadingOnes. Results for these two functions have been first derived in Gutjahr (2008a) and in Gutjahr and Sebastiani (2008). We shortly outline the situation for OneMax. Here, $A_k = f^{-1}(k)$ is the set of all binary vectors $x \in \{0,1\}^n$ containing k 1-bits $(k = 0, \ldots, n)$. One possibility to increase the fitness function value of a solution $x \in A_k$ consists in flipping one single 0-bit and leaving all other bits identical. The flipped bit can be selected in $n-k$ possible ways. Therefore, $q_k \geq (n-k) \cdot (1/n) \cdot (1-1/n)^{n-1} \geq (n-k)\,(en)^{-1}$ with $e = \exp(1)$. Thus, in this case, insertion of (8.6) into (8.5) (starting already with $k = 0$) yields the upper expected optimization time bound

$$(n \log n)/\rho + enH_n, \tag{8.7}$$

where $H_n = \sum_{k=1}^{n} i^{-1} = O(\log n)$ denotes the nth harmonic number. For $\rho = O(1)$ $(n \to \infty)$, this bound is of order $O(n \log n)$. The technique can be applied to the type 1/bf*/fp/ulb instead of 1/bf*/co/ulb as well: Leaving everything else unchanged, we obtain an upper bound of order $O(n(\log n)^2)$.

For LeadingOnes, the technique above yields a bound of order $O((n \log n)/\rho + n^2)$ for MMAS*. By an alternative technique, the article (Neumann *et al.*, 2009) improves this bound to two other valid upper bounds, $O(n/\rho + n^2)$ and $O(\frac{n/\rho}{\log(1/\rho)} + n^2 \cdot (1/\rho)^{\epsilon})$ for arbitrary fixed $\epsilon > 0$.

In the indicated paper, Neumann et al. are also able to prove a more general result by means of the method of fitness-based partitions. To formulate it, we start with a definition: A pseudo-boolean function f on S is called *unimodal* if to each non-optimal $x \in S$, there exists a Hamming neighbor x' with $f(x') > f(x)$.

Theorem 3. The expected optimization time of MMAS* on a unimodal function attaining D different fitness values is bounded from above by $O((n + (\log n)/\rho)D)$.

Proof. For $k < D$, the level set A_k contains only non-optimal solutions. Therefore, by unimodality, to each $x \in A_k$, a Hamming neighbor $x' \in A_j$ of x with $j > k$ can be found. The solutions x and x' differ by one single bit i. For pheromone values frozen in x, the probability that bit i is flipped, but the other bits are not, is $(1/n) \cdot (1-1/n)^{n-1} \geq (en)^{-1}$. As a consequence, $\sum_{k=1}^{D}(1/q_k) \leq enD$, whence the assertion follows. □

An interesting issue is the behavior of ACO algorithms on plateau functions, the simplest of which is the NEEDLE function with one optimal solution and all other solutions equal in fitness. Obviously, this function is not of practical interest, but a combination of it with ONEMAX, the function NEEDLE-ONEMAX$(x) = \left(\prod_{i=1}^{k} x_i\right)\left(\sum_{i=k+1}^{n} x_i + 1\right)$ introduced in Gutjahr and Sebastiani (2008), has features that often occur in applied problems: The bit combinations on some positions have to be found by trial-and-error, whereas on other positions, the fitness function gives "hints" for a solution improvement. It is shown in Gutjahr and Sebastiani (2008) that the expected optimization time of MMAS* on NEEDLE-ONEMAX with $k = \log_2 n$ and $\rho = n^{-3}$ has an upper bound of order $O(n^4 \log n)$. If ρ is not decreased with n, the expected optimization time is superpolynomial, which indicates that low evaporation rates may be necessary for some problems.

Type 1/bf/co/ulb. Next, we turn to the variant 1/bf/co/ulb with $\tau_{\min} = 1/n$, $\tau_{\max} = 1 - 1/n$, and constant reward $C = 1$ and write it shortly as MMAS in the sequel. (Note that MMAS is now used in a more special sense than in Section 2.) In Neumann *et al.* (2009), this variant is investigated in detail and compared to MMAS*. Here, other proof techniques than the method of fitness-based partitions are required. To give an example of a result, let us mention the following counterpart to Theorem 3, shown in Neumann *et al.* (2009): The expected optimization time of MMAS on a unimodal function attaining D different fitness values is bounded by $O((n^2 \log n \,/\, \rho)D)$.

For the function NEEDLE-ONEMAX, the results in Neumann *et al.* (2009) indicate a more favorable runtime behavior of MMAS compared to MMAS*, which is not surprising since on a plateau of solutions with equal fitness, the MMAS variant (accepting also equally good solutions) can still perform search by a random walk, whereas MMAS* cannot leave the current search point until a strict improvement is found. For MMAS, it already suffices to decrease ρ as 1/polylog(n), where polylog(n) denotes a polynomial in the logarithm of n, to obtain a polynomial optimization time, which is too slow for MMAS*.

Type 1/bf$_0$/co/ulb. The 1-ANT algorithm, differing from MMAS by the property that reinforcement is only done in iterations where the best-so-far solution has been exchanged, has been analyzed in Neumann and Witt

(2006), the first published work on runtime complexity analysis of ACO, and by Doerr *et al.* (2007). The results in the mentioned publications are not easy to compare with those on MMAS and MMAS*, since a different pheromone update rule has been chosen. In Gutjahr (2007) and (independently) in Doerr and Johannsen (2007), it has been shown that by a simple transformation, the update rules can be made comparable: Let $\bar{\rho}$ and $\bar{\tau}_{ij}$ denote the evaporation rate and the pheromone as introduced in Neumann and Witt (2006), respectively. Then

$$\rho = (2n\bar{\rho})/(1 - \bar{\rho} + 2n\bar{\rho}), \quad \tau_{ij} = 2n\bar{\tau}_{ij}, \tag{8.8}$$

leads us back to the framework of Section 2 used in Gutjahr and Sebastiani (2008) and in Neumann *et al.* (2009).

In Neumann and Witt (2006), it is proven that with $\bar{\rho} = \Omega(n^{-1+\epsilon})$ for some $\epsilon > 0$, the optimization time of 1-ANT on OneMax is $O(n^2)$ with probability $1 - 2^{-\Omega(n^{\epsilon/2})}$. The behavior for a scheme of $\bar{\rho}$ values that is decreasing faster in n will be discussed in the next subsection. Doerr *et al.* (2007) derive an upper bound of order $O(n^2 \cdot (6e)^{1/(n\bar{\rho})})$ for the expected optimization time of 1-ANT on LeadingOnes. The proof is rather difficult and uses the consideration of "free-rider bits" (cf. Droste *et al.* (2002)), which are right-hand-side bits in LeadingOnes that do not influence the fitness of the *current* solution (but can become essential after local changes by bit flips). Similarly, Doerr *et al.* (2007) bound the expected optimization time of 1-ANT on the BinVal function by $O(n^2 \cdot (4e^2)^{1/(n\bar{\rho})})$.

8.4.2. *Lower Bounds for Expected Optimization Times*

A general lower bound for the expected optimization time of ACO algorithms of the types 1/bf/co/ulb or 1/bf*/co/ulb on the chain construction graph has been derived in Neumann *et al.* (2009):

Theorem 4. Let $f : \{0,1\}^n \to \mathbb{R}$ have a unique global optimum. Then, if ρ is chosen as $\rho = 1/\text{poly}(n)$, the expected optimization time of MMAS and of MMAS* on f is $\Omega((\log n)/\rho - \log n)$.

Proof. Assume w.l.o.g. that $x^* = 1^n$ is the unique optimal solution. Define the success probability of bit i as the probability $\tau^+(i)$ of creating a one at this position. After t iterations, all pheromone values must be still larger or equal to $(1/2) \cdot (1-\rho)^t$ and therefore, by symmetry, smaller or equal to $1 - (1/2) \cdot (1-\rho)^t$. Choose $t = (\frac{1}{\rho} - 1) \cdot \frac{1}{2} \cdot \log \frac{n}{4}$ which is of order $\Theta((\log n)/\rho - \log n)$ and hence of polynomial order in n because of

$\rho = 1/\text{poly}(n)$. Since $\log(1-\rho) \geq -\rho/(1-\rho)$ for all $\rho \in]0,1[$, we obtain

$$\log\left((1-\rho)^t\right) = \frac{1-\rho}{\rho} \cdot \frac{1}{2} \cdot \log\frac{n}{4} \cdot \log(1-\rho) \geq -\frac{1}{2} \cdot \log\frac{n}{4},$$

such that the upper bound of the success probability becomes smaller or equal to $1 - (1/2) \cdot \exp(-(1/2) \cdot \log(n/4)) = 1 - n^{-1/2}$. The probability that x^* is generated in one of the first t steps is then smaller or equal to $t \cdot (1 - n^{-1/2})^n \leq t \cdot \exp(-\sqrt{n}) = \exp(-\Omega(\sqrt{n}))$ and tends therefore to zero exponentially fast as $n \to \infty$. Therefore, the expected optimization time cannot be smaller than $\Omega(t)$. □

For constant ρ, this gives a lower bound of order $\Omega(\log n)$. We see that on OneMax, there is still a considerable gap to the upper bound $O(n \log n)$ for MMAS*. Closing this gap by matching upper and lower bounds is an open problem.

Lower bounds for 1-ANT on specific functions are provided in Neumann and Witt (2006) for OneMax and in Doerr *et al.* (2007) for LeadingOnes and BinVal. These bounds reveal an interesting phenomenon: When decreasing $\bar{\rho}$ in dependence of n with a certain speed, one obtains a sharp phase transition separating a regime of polynomial expected optimization time (for $\bar{\rho}$ decreasing not too fast) from a regime of exponential expected optimization time occuring for a fast decrement of $\bar{\rho}$. In the case of OneMax, the phase boundary is given by the speed $\bar{\rho} = \Theta(n^{-1})$: Neumann and Witt show that for $\bar{\rho} = O(n^{-1-\epsilon})$, the optimization time of 1-ANT on OneMax is $2^{\Omega(n^{\epsilon/3})}$ with probability $1 - 2^{-\Omega(n^{\epsilon/3})}$, whereas for $\bar{\rho} = \Omega(n^{-1+\epsilon})$, as mentioned above, an $O(n^2)$ upper optimization bound holds with overwhelming probability. Note that by (8.8), the decrement speed $\bar{\rho} = \Theta(1/n)$ corresponds to $\rho = \frac{2}{3}(1+o(n))$, which means that already for small *constant* ρ, the algorithm 1-ANT is very inefficient on OneMax.

In Doerr *et al.* (2007), similar results are shown for LeadingOnes and BinVal. To give a flavor of the applied proof technique, we repeat the result for LeadingOnes and outline the proof idea:

Theorem 5. With probability $2^{-\Omega(\min\{1/(n\bar{\rho}),n\})}$, the optimization time of 1-ANT on LeadingOnes is $2^{\Omega(\min\{1/(n\bar{\rho}),n\})}$.

Proof idea. Set $k = 1/(8n\bar{\rho})$ and define the success probability of a bit as in the proof of Theorem 4. Call an iteration m *accepting* if pheromone values are changed in iteration m. Now, consider the first accepting iteration $m = m_f$ in which either the fitness f_c of the current solution reaches a value larger or equal to $n/2$, or k accepting iterations (including iteration

m_f) have occurred. In each iteration, the success probability can grow by at most $\rho \leq 2n\bar{\rho}$ (cf. (8.8)), hence k accepting iterations can increase the initial success probability by at most $2n\bar{\rho}k$, such that in iterations $1, \ldots, m_f$, the success probabilities are bounded above by $1/2 + 2n\bar{\rho}k = 3/4$ and therefore, by symmetry, bounded below by $1/4$.

At the beginning of iteration m_f, all success probabilities are between $1/4$ and $3/4$, and the current solution has a fitness smaller than $n/2$. Since all bits in the right half of x have to be set to one in order to reach the optimum, the probability that the optimal solution is found in iteration m_f is smaller or equal to $(3/4)^{n/2}$. Thus, with a probability $1 - 2^{-\Omega(n)}$, the optimal solution is not found in iterations $1, \ldots, m_f$.

Now we show that after iteration m_f, the current fitness has already become that large that a further accepting iteration within a "reasonable" time has become very improbable. At the end of iteration m_f, either $f_c \geq n/2$, or k accepting iterations have occurred. In the last case, $E(f_c) > k/4$, because each accepting iteration leads with a probability of at least $1/4$ to a fitness increase, whence one derives $\Pr(f_c \geq k/8) = 1 - 2^{-\Omega(1/(n\bar{\rho}))}$ by Chernoff bounds. Thus, with $\mu = \min(n/2, k/8)$, it holds that $f_c \geq \mu$ with a probability larger or equal to $1 - 2^{-\Omega(1/(n\bar{\rho}))}$. Assume that $f_c \geq \mu$. Then the probability that none of the iterations $m_f + 1, m_f + 2, \ldots, m_f + t$ is accepting is larger or equal to $1 - t\,2^{-\mu C}$ with some constant $C > 0$. Choosing $t = 2^{\mu c}$ with $0 < c < C$ and considering that $\mu = \Omega(\min\{n, 1/(n\bar{\rho})\})$ provides the statement of the theorem. □

One observes that here, the phase transition to exponential runtime behavior happens if $\bar{\rho}$ is decreased faster than with order $\Theta(1/(n \log n))$. (For a slower decrement, the failure probability in Theorem 5 is too large to give a meaningful runtime bound.)

Summarizing upper and lower bound results, we see that the 1-ANT algorithm is not competitive with MMAS or MMAS* on the considered simple test functions.

8.4.3. *Hybridization of ACO with Local Search*

In practice, ACO algorithms are often combined with local search. For pseudo-boolean functions, local search consists in the procedure of repeatedly going to a better Hamming neighbor (a solution differing from the current solution only in one bit) until there does not exist a neighbor with this property anymore, such that a so-called local optimum has been reached. If this is done in a single final post-optimization step, it can only improve the obtained solution (or leave it unchanged), at the price of a (small) increment of the runtime. It is tempting, however, to apply local search even in each iteration of the process, similarly as in memetic algorithms. In Neumann *et al.* (2008), this variant is analyzed, and it is shown that depending on the function to be optimized, incorporating local search can be advantageous or disadvantageous in terms of optimization time. The difference can be dramatic, even deciding between polynomial and exponential runtime.

The authors compare the MMAS* algorithm (see Subsection 4.1) to the following modification MMAS-LS*: After the construction of a current solution x^1 (step "set $x^\sigma = u$" in Algorithm 1 in Section 2; note that $\sigma = s = 1$), local search is applied to x^1, which yields a solution z. The fitness of z is compared to that of x^1. If $f(z) > f(x^1)$, one replaces x^1 by z. The remainder of the algorithm is identical to MMAS*.

Two example classes of pseudo-boolean functions are analyzed. In both classes, a *short path* defined as $\mathcal{P} = \{1^i 0^{n-i} \,|\, 0 \leq i \leq n\}$ is used. The classes are parametrized by a parameter $\gamma > 1$; each value of γ produces a concrete fitness function. The two classes mainly differ by the definition of the respective set S^* of global optima. In class 1, the set of global optima is defined as $S^* = S_{3/4} = \{x \,|\, \sum_{i=1}^n x_i \geq (3/4) \cdot n$ and $d_H(x, \mathcal{P}) \geq n/(\gamma \log n)\}$, where d_H denotes the Hamming distance. The chosen fitness function is

$$f_1(x) = \begin{cases} \sum_{i=1}^n (1 - x_i), & x \notin \mathcal{P} \cup S^*, \\ n + i, & x = 1^i 0^{n-i} \in \mathcal{P}, \\ 3n, & x \in S^*. \end{cases}$$

Observe that the fitness function f_1 gives hints to reach 0^n, the starting point of the *short path* $\mathcal{P}$, and to further proceed from there (along $\mathcal{P}$) to the local optimum 1^n. This local (but not global) optimum, however, constitutes a trap that is hard to overcome. It can be expected that MMAS-LS* will detect the trap much faster than MMAS* and will thus save time.

In class 2, the set S^* of global optima only consists of the point $0^2 1^{n-2}$. The set $S_{3/4}$ becomes now a trap and will be denoted by $\mathcal{T}$. The chosen fitness function is

$$f_2(x) = \begin{cases} \sum_{i=1}^n (1 - x_i), & x \notin \mathcal{P} \cup \mathcal{T} \cup S^*, \\ n + i, & x = 1^i 0^{n-i} \in \mathcal{P}, \\ 3n, & x \in \mathcal{T}, \\ 4n, & x \in S^*. \end{cases}$$

In this case, the unique global optimum lies very near to the local optimum 1^n (at Hamming distance two), but cannot be reached from 1^n by local search. Under these circumstances, one may expect that MMAS* will have an advantage over MMAS-LS*.

Indeed, the article (Neumann *et al.*, 2008) shows the following results on the optimization time, which hold independently of the chosen $\gamma > 1$. Here, the optimization time is measured by the number of iterations instead of the number of function evaluations; this different measurement, however, does not impair a runtime gap of polynomial vs. super-polynomial behavior.

Theorem 6.

(i) For $\rho = 1/\text{poly}(n)$, the optimization time of MMAS* on f_1 is $2^{\Omega(n^{2/9})}$ with probability $1 - 2^{-\Omega(n^{2/9})}$.

(ii) For $1/\text{poly}(n) \le \rho \le 1/16$, the optimization time of MMAS-LS* on f_1 is $O(1/\rho)$ with probability $1 - 2^{-\Omega(n)}$.

(iii) For $\rho = 1/\text{poly}(n)$, the optimization time of MMAS* on f_2 is $O((n \log n)/\rho + n^3)$ with probability $1 - 2^{-\Omega(n^{2/9})}$.

(iv) For $1/\text{poly}(n) \le \rho \le 1/16$, the optimization time of MMAS-LS* on f_2 is $2^{\Omega(n)}$ with probability $1 - 2^{-\Omega(n)}$.

We see that MMAS-LS* is (considerably) more efficient than MMAS* on f_1, but (considerably) less efficient than MMAS* on f_2.

8.5. Runtime Behavior of Variants with Iteration-Based Reinforcement

Compared to the ACO variants with *reinf* = bf, bf*, bf_0 or bf_0^* addressed in the previous section, variants with *reinf* = ac (all solutions of the current iteration are reinforced) or *reinf* = ib(r) (the r best solutions of the current iteration are reinforced) are considerably harder to analyze. The reason is that the last-mentioned schemes lack the monotonicity property of best-so-far schemes where the quality of the reinforced solutions never decreases. E.g., in the oldest ACO algorithm, *Ant System* (AS), which is s/ac/fp/nb in our classification, it can happen that in an iteration m, only comparably poor solutions are constructed, yet they will nevertheless be reinforced at the end of this iteration. This obstructs the application of the method of fitness-based partitions, but also of other techniques applied in the proofs of results cited in Section 4.

Because of the complexity of the stochastic process induced by the last-mentioned variants, it makes sense to start an analysis by studying asymptotic approximations to this process. In view of Theorem 1 showing that a decrement of the evaporation rate ρ over time can facilitate convergence to the optimal solution, the most meaningful approximation seems to be one where ρ tends to zero. Since decreasing ρ also decreases the amount of changes during one iteration, it can be expected that in order to obtain a limiting process in the mathematical sense, the number of iterations to be performed per time unit has to be increased simultaneously to the reduction of ρ. It is clear that from a computational point of view, the time required for one iteration has a lower bound ϵ larger than zero, but even if ϵ does not reach the value zero, the asymptotic approximation of the process for small ϵ can give us valuable information concerning the true process.

An approximation of this kind has been carried out in Gutjahr (2008a) for AS (preliminary results have been presented in Gutjahr (2005, 2006)). The resulting asymptotic process is a continuous deterministic process. (An approximation by a *discrete* deterministic process, the so-called *ACO model*, has already been performed in Merkle and Middendorf (2002).) In the sequel, we shortly outline the technique; for more details, the reader is referred to Gutjahr (2007, 2008a). We start with a definition.

Definition 3. The *passage fitness* of arc $(i, j) \in \mathcal{A}$ under a given pheromone vector $\tau = (\tau_{ij})$ is the random variable $\mathbf{I}\{(i, j) \in X\} \cdot f(X)$, where X is the random walk of a fixed ant under pheromone values τ_{ij}. The *expected passage fitness* of (i, j) under τ is the mathematical expectation of the passage fitness, i.e., the value $F_{ij}(\tau) = E(\mathbf{I}\{(i, j) \in X\} \cdot f(X))$.

The considered asymptotic case is the following: We let $\rho \to 0$ and simultaneously $M \to \infty$ in such a way that $\rho \cdot M = 1$, where M is the number of iterations per time unit. The last assumption produces a specific scaling of the time axis. In this scaling, an iteration takes $1/M = \rho$ time units. Re-scaled time will be denoted by the symbol t in the sequel. The described type of re-scaling for obtaining asymptotic results on "slow learning" is well-established in learning theory (see Norman, 1972).

Definition 4. For AS on a given instance, the ACDP (Associated Continuous Deterministic Process) is given as the solution $\bar{\tau}(t)$ of the system

$$d\bar{\tau}_{ij}/dt = F_{ij}(\bar{\tau}) - \bar{\tau}_{ij}, \qquad ((i, j) \in \mathcal{A}), \tag{8.9}$$

of ordinary differential equations (ODEs), where $\bar{\tau} = (\bar{\tau}_{ij})$ is the pheromone vector at time t, and $F_{ij}(\bar{\tau})$ is the expected passage fitness of arc (i, j) under $\bar{\tau}$.

Intuitively, it is easy to see that the ACDP approximates the AS process: Assume that in the interval $[t, t+dt]$ of the re-scaled time axis, K iterations take place (K denoting a large constant), such that $dt = K/M = K\rho$. By the Law of Large Numbers, this increases the pheromone value on (i, j) by $F_{ij}(\bar{\tau}) \cdot K \cdot \rho$ via reinforcements. In the same time, the pheromone value on (i, j) is reduced by $\bar{\tau}_{ij} \cdot K \cdot \rho$ via evaporation. We obtain $d\bar{\tau}_{ij} = [F_{ij}(\bar{\tau}) - \bar{\tau}_{ij}] \cdot dt$.

To make the asymptotic approximation above mathematically valid by means of a rigorous convergence argument, one requires some technical conditions (mainly Lipschitz conditions) which have to be verified for special fitness functions f under consideration. In Gutjahr (2008a), this is done for the OneMax function, making sure that the re-scaled trajectories of the AS process on OneMax converge in the considered asymptotics to those of the corresponding ACDP $\bar{\tau}(t)$. The result uses a theorem proven in Norman (1972) for more general cases of slow learning approximations.

Moreover, for OneMax, upper and lower bounds for the expected fitness of an ant's walk resulting from pheromone vector $\bar{\tau}(t)$ can be determined. Defining the *relative fitness* of a solution x by $(f(x) - f_{\min})/(f_{\max} - f_{\min})$ with $f_{\max} = \max_{x \in S} f(x)$ and $f_{\min} = \min_{x \in S} f(x)$, the following result can be shown: The required amount of (re-scaled) time t until the expected relative fitness of the ant's walk in the ACDP $\bar{\tau}(t)$ reaches a value of $1 - \epsilon$ is of order $\Theta(n \log((1 - \epsilon)/\epsilon))$. Provided that we keep ρ independent of n, this leads to a time of order $\Theta(n \log n)$ until the ant chooses the optimal solution with a probability of at least $1 - \epsilon_0$, where $\epsilon_0 \ll 1$, since for OneMax, the difference $f_{\max} - f_{\min}$ is of order $O(n)$. We see that this time complexity corresponds to that of MMAS* on OneMax. However, two differences should be kept in mind: First, the result refers to the (approximating) ACDP and not to AS itself; in this sense, the result is weaker than that for MMAS*. Secondly, however, the result does not only show that the optimal solution will found within an (expected) time of order $\Theta(n \log n)$, but also that the pheromone values focus on the optimal solution within a time of the same order, i.e., *model convergence* in the sense of Definition 2 takes place within this time. In this sense, the assertion of the result is stronger than that for MMAS*.

Borkar and Das (2009) use a related ODE technique to approximate the trajectories of a combination called MAF-ACO of an ACO algorithm with a traditional learning recursion. The ACO part is not covered by the scheme in Section 2, but rather applies a more direct "local" interpretation of the ant metaphor by assuming that each ant chooses a path, traverses it in a time proportional to its length, returns then to the start node and begins a new walk; the fitness function does not enter here in an explicit way, but implicitly by the fact that long paths (paths with low fitness) obtain less reinforcements than short paths, such that the considered ant system is able to solve shortest path problems. The authors start with considering the case where the construction graph consists of d parallel paths that are disjoint with the exception of the start node and the end node. The pheromone update mechanism of the considered ant algorithm is given by

$$\tau_i(t+1) = (1-\rho)\,\tau_i(t) + \rho\, Q\, R_i(t) \quad (i = 1, \ldots, d;\ t = 0, 1, \ldots),$$

where $\tau_i(t)$ is the pheromone value of an edge of path i at time t, Q is a constant, and $R_i(t)$ is the number of ants that have just finished traversing path i at the end of time step t. In addition to this pheromone update scheme, the following learning rule is applied:

$$X_i(t+1) = X_i(t) + a(t)\, X_i(t)\, \tau_i(t+1) \quad (i = 1, \ldots, d;\ t = 0, 1, \ldots),$$

where $\sum_{t \geq 0} a(t) = \infty$ and $\sum_{t \geq 0} (a(t))^2 < \infty$. The probabilities of selecting paths $1, \ldots, d$ are chosen proportionally to the values $X_i(t)$.

A theorem from Beneviste *et al.* (1990) is used to show that with a probability that can be driven arbitrarily close to one, the trajectory of the pheromone vector stays within a "tube" with radius δ around the solution of a suitably defined associated ODE. By letting $\delta \to 0$, the approximation becomes more and more accurate. Technically, this is a similar approach to that chosen in Gutjahr (2008a), but the investigated ACO variant is quite different. After analyzing the case of parallel paths, the authors extend their approach to the shortest path problem on a multi-stage graph. Finally, they provide simulation results for the multi-stage and for the TSP case.

8.6. ACO Variants for Specific Problems

In some sense, the MAF-ACO algorithm of Borkar and Das (2009) can be seen as an ACO variant for a specific optimization problem, the shortest path problem. Also in some other works, the basic concept of ACO has been adapted to special problems under consideration. This does not mean that these variants have been turned into problem-tailored heuristics, but rather that they use the flexibility inherent in the algorithmic design of a metaheuristic to cope well with certain particular problems.

The first theoretical runtime analysis of an ACO adaptation to a problem of shortest path type seems to be the article of Attiratanasunthron and Fakcharoenphol (2008). This work provides an ACO algorithm for the

Single-Destination Shortest Path Problem (SDSPP) which consists in finding the shortest path from each node v in a weighted directed acyclic graph $G = (\mathcal{V}, \mathcal{A})$ to a fixed destination node $v_d \in \mathcal{V}$. The authors start with the special case of the chain graph, which they re-define as a multigraph with two arcs between each pair of consecutive nodes $(i-1, i)$. In this way, the ONEMAX problem is obtained. After that, the general case is treated.

In our classification scheme, the proposed algorithm can be described as of type n/bf$_0$/co/ulb, i.e., in some aspects, it generalizes 1-ANT to the case of $n = |\mathcal{V}|$ ants. However, there are two essential deviations from the standard algorithmic procedure: First, the n ants are not positioned at a common start node at the beginning of their walks, but each ant is initially positioned at a separate node of $\mathcal{V}$. (This initialization has also been frequently used in experimental research.) Secondly, after an ant has completed its walk, it does not reinforce pheromone on all arcs of the traversed walk, but only on the first arc of this walk.

The authors show that the expected optimization time of the described variant on ONEMAX has an upper bound of order $O(n^2 \log n \,/\, \rho)$. For the general case of the SDSPP on a weighted directed acyclic graph, they prove an upper bound of order $O(n \log n \cdot |\mathcal{A}| \,/\, \rho)$. In this general case, pheromone bounds $\tau_{\min} = 1/n^2$ and $\tau_{\max} = 1 - 1/n^2$ are used, contrary to the ONEMAX case, where the usual bounds $1/n$ and $1 - 1/n$ are applied.

In a recent paper (Horoba and Sudholt, 2009), the authors extend the results of Attiratanasunthron and Fakcharoenphol (2008) into several directions: (i) They improve the optimization time bounds of Attiratanasunthron and Fakcharoenphol (2008) by replacing the 1-ANT-based algorithm with an algorithm of MMAS type, i.e., by switching from *reinf* = bf$_0$ to *reinf* = bf. (ii) The restriction to acyclic graphs is removed. (iii) In addition to the SDSPP, also the APSPP (All-Pairs Shortest Path Problem) is treated. (iv) Interaction between ants is examined. The proposed ant system for the APSPP with ant interaction is shown to have an expected optimization time of order $O(n \log n + \log(\ell) \log(\Delta\ell)/\rho)$, where Δ is the maximum degree and ℓ is the maximum number of edges on a shortest path.

For the *Minimum Spanning Tree* (MST) problem, the article of Neumann and Witt (2008) studies an ACO variant where the set of possible pheromone values is restricted to two constants. The used reinforcement rule is of type *reinf* = bf$_0$. Two types of construction graphs are chosen. In the first choice, the input graph itself is taken as the construction graph, which leads to a procedure similar to Broder's classical algorithm for the MST problem. In the second choice, the combinatorial problem is decomposed generically into its components and invalid selections during the construction process are excluded, which leads to similarities with Kruskal's algorithm. A certain parametrization of this algorithm turns out as competitive with Kruskal's problem-specific algorithm in a theoretical runtime analysis.

Purkayastha and Baras analyze a quite different application of the ant algorithm paradigm, namely *Ant Routing Algorithms* (Purkayastha and Baras, 2007). These are algorithms used in communication networks with the purpose to find suitable schemes for packet routing. To each node i in the network, a routing table $\mathcal{R}_i$ is maintained which contains the probabilities for routing an incoming packet at node i bound for destination k via the neighbor node j of node i. The routing probabilities are derived from quantities that form the counterpart to pheromone values, but are now computed as the reciprocals of variables $X^i_{j,k}$. These variables are updated by a formula analogous to the pheromone update rules in Section 2, with the reward replaced by the observed delay of the packet arrival. The paper (Purkayastha and Baras, 2007) focuses the analysis on the situation of a graph with a certain number of parallel paths between a source and a sink node, and the stochastic trajectories are approximated by means of an ODE approximation for an asymptotic case where the learning rate ρ tends to zero, using general results from (Beneviste *et al.*, 1990). The equilibrium behavior of the ant routing algorithm is studied in some detail.

8.7. Conclusions

As this chapter shows, the theory of ant colony optimization has made distinct progress during the last years. Convergence results have been deepened and some analytical runtime complexity results have been obtained, mainly for comparably simple test functions, but already for some problems of practical interest as well. It is very probable that the next years will see a considerable extension of insights into the theoretical foundations of ACO.

There is still a long list of open problems in this field; let us only shortly outline some of them: (i) Whereas the behavior of some ACO variants on unimodal functions and on plateaus is already understood rather well, deceptive functions and problems with multiple local optima have been investigated far less frequently. However, just these types of functions provide the main motivation for using metaheuristic techniques. (ii) Compared to the analysis of ACO variants with best-so-far reinforcement, the analysis of variants with iteration-best reinforcement seems to be underdeveloped at present and to move at a slower pace. This holds for convergence results as well as for runtime results. (iii) Most research has focused on the case of pseudo-boolean functions. Nevertheless, the classical applications of ACO do not deal with subset problems, but with problems of permutation type (e.g., TSPs, VRPs, job shop scheduling problems etc.) instead. (iv) As in other fields of evolutionary computation, also in the ACO domain, the theoretical analysis of NP-hard problems represents a fascinating challenge. Already to show *analytically* that on some NP-hard problem, an ACO variant outperforms *random search* with respect to (average case or worst case) expected optimization time would be a progress.

References

Attiratanasunthron, N. and Fakcharoenphol, N. (2008). A running time analysis for an ant colony optimization algorithm for shortest path on direct acyclic graphs, *Information Processing Letters* **105**, pp. 88–92.

Beneviste, A., Priouret, P. and Metivier, M. (1990). Adaptive algorithms and stochastic approximations, *Applications of Mathematics* **22**.

Borkar, V. and Das, D. (2009). A novel ACO algorithm for optimization via reinforcement and initial bias, *Swarm Intelligence* **3**, pp. 3–34.

Bullnheimer, B., Hartl, R. and Strauss, C. (1999). A new rank–based version of the ant system: A computational study, *Central European Journal for Operations Research and Economics* **7**, pp. 25–38.

Doerr, B. and Johannsen, D. (2007). Refined runtime analysis of a basic ant colony optimization algorithm, in *Proc. of the IEEE Congress on Evolutionary Computation (CEC) 2007*, pp. 501–507.

Doerr, B., Neumann, F., Sudholt, D. and Witt, C. (2007). On the runtime analysis of the 1-ANT ACO algorithm, in *Proc. GECCO 2007*, pp. 33–40.

Dorigo, M. and Blum, C. (2005). Ant colony optimization theory: a survey, *Theoretical Computer Science* **344**, pp. 243–278.

Dorigo, M. and Caro, G. D. (1999). The ant colony optimization metaheuristic, in *New Ideas in Optimization, D. Corne, M. Dorigo, F. Glover (eds.)*, pp. 11–32.

Dorigo, M. and Gambardella, L. (1997). Ant colony system: A cooperative learning approach to the traveling salesman problem, *IEEE Transactions on Evolutionary Computation* **1**, pp. 53–56.

Dorigo, M., Maniezzo, V. and Colorni, A. (1991). Positive feedback as a search strategy, Technical Report TR-POLI-91-016, Dipartimento di Elletronica, Politecnico di Milano, Milan, Italy.

Dorigo, M., Maniezzo, V. and Colorni, A. (1996). Ant system: optimization by a colony of cooperating agents, *IEEE Trans. on Systems, Man, and Cybernetics* **26**, pp. 1–13.

Dorigo, M. and Stuetzle, T. (2004). *Ant Colony Optimization* (MIT Press, Cambridge, MA).

Droste, S., Jansen, T. and Wegener, I. (2002). On the analysis of the (1+1) evolutionary algorithm, *Theoretical Computer Science* **276**, pp. 51–81.

Gutjahr, W. (2000). A graph–based ant system and its convergence, *Future Generation Computer Systems* **16**, pp. 873–888.

Gutjahr, W. (2002). ACO algorithms with guaranteed convergence to the optimal solution, *Information Processing Letters* **82**, pp. 145–153.

Gutjahr, W. (2003). A generalized convergence result for the graph–based ant system, *Probability in the Engineering and Informational Sciences* **17**, pp. 545–569.

Gutjahr, W. (2005). Theory of ant colony optimization: status and perspectives, in *MIC '05 (6th Metaheuristics International Conference), Proceedings CD-ROM*.

Gutjahr, W. (2006). On the finite-time dynamics of ant colony optimization, *Methodology and Computing in Applied Probability* **8**, pp. 105–133.

Gutjahr, W. (2007). Mathematical runtime analysis of ACO algorithms: survey on an emerging issue, *Swarm Intelligence* **1**, pp. 59–79.

Gutjahr, W. (2008a). First steps to the runtime complexity analysis of ant colony optimization, *Computers & Operations Research* **35**, pp. 2711–2727.

Gutjahr, W. (2008b). Stochastic search in metaheuristics, To appear in: Handbook of Metaheuristics, 2nd edition, Springer.

Gutjahr, W. and Sebastiani, G. (2008). Runtime analysis of ant colony optimization with best-so-far reinforcement, *Methodology and Computing in Applied Probability* **10**, pp. 409–433.

Horoba, C. and Sudholt, D. (2009). Running time analysis of ACO systems for shortest path problems, in *Proc. SLS 2009, Springer LNCS 5217*, pp. 76–91.

Merkle, D. and Middendorf, M. (2002). Modeling the dynamics of ant colony optimization, *Evolutionary Computation* **10**, pp. 235–262.

Neumann, F., Sudholt, D. and Witt, C. (2008). Rigorous analyses for the combination of ant colony optimization and local search, in *Proc. ANTS 2008, Springer LNCS 5217*, pp. 132–143.

Neumann, F., Sudholt, D. and Witt, C. (2009). Analysis of different MMAS ACO algorithms on unimodal functions and plateaus, *Swarm Intelligence* **3**, pp. 35–68.

Neumann, F. and Witt, C. (2006). Runtime analysis of a simple ant colony optimization algorithm, in *Proc. ISAAC '06, Springer LNCS 4288*, pp. 618–624, also in: Algorithmica 54, pp. 243–255 (2009).

Neumann, F. and Witt, C. (2008). Ant colony optimization and the minimum spanning tree problem, in *Proc. LION 2007, Springer LNCS 5313*, pp. 153–166.

Norman, F. (1972). *Markov Processes and Learning Models* (Academic Press, New York and London).

Purkayastha, P. and Baras, J. (2007). Convergence results for ant routing algorithms via stochastic approximation and optimization, in *Proc. 46th IEEE Conf. on Decision and Control*, pp. 340–345.

Sebastiani, G. and Torrisi, G. (2005). An extended ant colony algorithm and its convergence analysis, *Methodology and Computing in Applied Probability* **7**, pp. 249–263.

Stuetzle, T. and Dorigo, M. (2002). A short convergence proof for a class of ACO algorithms, *IEEE Trans. Evol. Comput.* **6**, pp. 358–365.

Stuetzle, T. and Hoos, H. (1997). Max-min ant system and local search for the travelling salesman problem, in *Proc. ICEC '97 (Int. Conf. on Evolutionary Computation)*, pp. 309–314.

Stuetzle, T. and Hoos, H. (2000). Max-min ant system, *Future Generation Computer Systems* **16**, pp. 889–914.

Wolpert, D. and Macready, W. (1997). No free lunch theorems for optimization, *IEEE Trans. on Evolutionary Computation* **1**, pp. 67–82.

Chapter 9

A "No Free Lunch" Tutorial: Sharpened and Focused No Free Lunch

Darrell Whitley
Colorado State University, USA
whitley@cs.colostate.edu

Jonathan Rowe
Birmingham University, UK
J.E.Rowe@cs.bham.co.uk

This tutorial presents the basic results that follow from various No Free Lunch theorems. The presentation is designed to be intuitive and accessible to a general reader. The Sharpened No Free Lunch theorem applies to black box optimization and states that no arbitrarily selected algorithm is better than another when the algorithms are compared over sets of functions that are closed under permutation. Focused No Free Lunch theorems look at comparisons of specific search algorithms; in this case algorithms can display exactly the same performance even when the set of functions used in the comparison are not closed under permutation. Also, Focused No Free Lunch results can sometimes occur even when the optimization is not black box.

Contents

9.1. Background

Wolpert and Macready presented the first "No Free Lunch Theorems" for search (Wolpert and Macready, 1995) (Wolpert and Macready, 1997). The original No Free Lunch theorems can be summarized as follows:

> *For all possible metrics, no search algorithm is better than another when their performance is compared over all possible discrete functions.*

To set the stage for discussing No Free Lunch theorems let f be the objective function. Let $\mathcal{X}$ represent the domain of the f and note that $\mathcal{X}$ also represents the search space. Let the set $\mathcal{Y}$ define the codomain and range of the search space. Without loss of generality, we can assume we wish to minimize the objective function f. Thus the objective of search is to find $x_i \in \mathcal{X}$ such that $f(x_i)$ is minimal. It is often convenient to use a more concise notation where $f(x_i)$ is denoted by $f(i)$.

A key assumption made when discussing No Free Lunch proofs is that resampling is ignored when comparing the behavior of search algorithms. If a search algorithm samples point $i \in \mathcal{X}$ and evaluates the objective function $f(i)$ then point i is never sampled again during a particular search. Another way of thinking about this is to imagine that a search algorithm keeps a *history* of points that have already been evaluated as well as the corresponding evaluation; in this case, there is no need to reevaluate a point. But this restriction is also critical to No Free Lunch theory: if an algorithm never resamples a point, then there can only be a finite number of search *behaviors in a discrete finite search space. But if sampling is allowed, then there are infinitely many* behaviors that an algorithm could display. By restricting search behaviors to a (large) finite number, we can more easily enumerate and describe those behaviors.

Wolpert and Macready model a search algorithm as a procedure that searches for m steps. This also does not restrict any of the No Free Lunch results. Furthermore, Wolpert and Macready also show that the No Free Lunch results hold for both stochastic and deterministic search algorithms.

No Free Lunch can also be restated in other variations.

The aggregate behavior of any two search algorithms is equivalent when compared over all possible discrete functions.

At the root of this observations is another more fundamental result. Consider any algorithm A_i applied to function f_j. Let $\text{APPLY}(A_i, f_j, m)$ represent a "meta-level" algorithm that outputs the order in which A_i visits m elements in the codomain of f_j after m steps. We will assume m is a constant when comparing algorithms.

For every pair of algorithms A_k and A_i and for any function f_j, there exists another function f_l such that

$$\text{APPLY}(A_i, f_j, m) \equiv \text{APPLY}(A_k, f_l, m).$$

This perspective starts to shed some light on some of the counterintuitive implications of No Free Lunch theorems. No Free Lunch asserts that no search algorithm is better than another when their performance is compared over all possible discrete functions. This also implies that no "intelligent" search algorithm is better than a search algorithm that uses blind random enumeration when the algorithms are compared over all possible discrete functions.

Consider a *Best-First version of steepest descent local search which restarts when a local optimum is encountered. Also consider a* Worst-First steepest ascent local search, also with restarts. We incorporate re-starts so that these algorithms continue searching for an arbitrary number of steps as defined by m. Then, for every function f_j there exists a function f_l such that:

$$\text{APPLY}(\textit{Best-First}, f_j, m) \equiv \text{APPLY}(\textit{Worst-First}, f_l, m).$$

Most researchers would accept that Best-First local search is a reasonable search algorithm and that it probably is useful on many real world problems. In other words, there is a subset of problems where Best-First local search is likely to be a useful search method. But there is a corresponding set of functions where Worst-First local search is equally effective. What do these functions look like? They probably are not random, but rather "structured" in some sense.

Why is Best-First search generally viewed as a reasonable search algorithm and Worst-First as an unreasonable search algorithm?

One answer is that we usually expect "local similarity" in the objective function; another ways of saying this is that we expect small steps in the search space will result in small changes in the output of the objective function. But in the space of all possible discrete functions these kinds of intuitions about the structure of the search space are generally invalid. If one picks a random function out of the space of all possible discrete functions, then in expectation the structure of that function is unpredictable and random.

This observation leads to some of the most common objections to No Free Lunch theorems.

1) The space of *all* possible discrete functions is infinitely large and is mostly composed of random functions. Almost all of the functions in the space of all possible discrete functions are uncompressible. If a function is uncompressible then the space required to write down a description of the function has the same complexity as the space required to enumerate the function. Uncompressible functions do not represent the objective functions of real problems that are of practical interest. Real-world objective functions are always compressible.

2) No Free Lunch only applies to black box optimization. In black box optimization the search algorithm is given no information about the objective function f. The objective function is indeed a black box, where domain values are passed into the "box" and codomain values come out of the "box." Thus, if a researcher is only interested in scheduling problems which are not black box optimization problems, then the researcher can assume No Free Lunch theorems do not hold over this restricted set of problems.

These objections are true over the most general forms of No Free Lunch, but not necessarily true for the "Sharpened No Free Lunch" results (Schumacher *et al.*, 2001) or for "Focused No Free Lunch" results (Whitley and Rowe, 2008). The Sharpened No Free Lunch result states that No Free Lunch theorem holds over sets of functions that are closed under permutation. Therefore, it is now widely understood that the "Sharpened No Free Lunch" theorem holds over finite sets, and some of these sets are compressible. However, it does not appear to be widely understood that "Focused No Free Lunch" results can hold over very small sets of functions. The Sharpened No Free Lunch theory still applies to arbitrary algorithms. The Focused No Free Lunch theory looks at comparisons of specific algorithms. Thus, while the Sharpened No Free Lunch theorem applies to sets of functions that are *closed under permutation, Focused No Free Lunch can holds over sets that are not* closed under permutation. Other intuitions about the limitations of No Free Lunch theory are no longer true when focused comparisons are made of specific algorithms. In some situations Focused No Free Lunch results can hold even when the optimization is no longer black box optimization.

It is useful to think of No Free Lunch results as a kind of "law of conservation" for search where algorithms compete in a zero-sum game. Recall the meta-level algorithm $\text{APPLY}(A_i, f_j, m)$ and recall that given an algorithm A_i and an algorithm A_k and an objective function f_j there must also exist a function f_l such that

$$\text{APPLY}(A_i, f_j, m) \equiv \text{APPLY}(A_k, f_l, m).$$

Thus if we have a set of functions where algorithm A_i does particularly well, there must exist another set of functions where A_j has exactly the same behaviors as A_i. We can label the first set of functions S_1 and the second set of functions S_2. If A_i is better on set S_1 and algorithm A_j is better on set S_2 then can we say that A_i is better than A_j? In this sense, one algorithm does not "win" over the other algorithm.

No Free Lunch theorems impose a stricter requirement as well. For No Free Lunch to hold two algorithms must display the *same set of behaviors on the* same set of functions. These sets can be constructed such that one algorithm cannot win over another.

Sharpened No Free Lunch proves that NFL holds overs all sets that are closed under permutation. Wolpert and MacReady's general No Free Lunch theorem holds over all possible discrete functions because the set of all possible discrete functions is also closed under permutation. Both the general No Free Lunch result and the Sharpened No Free Lunch result assumes that we do not know in advance which algorithms we are going to compare. But if we know which algorithms we are comparing, then there may exists a set S_x where algorithm A_i and A_j have the same behaviors, and where S_x is smaller than any set which is closed under permutation. In this case, we will say that a Focused No Free Lunchs holds, and that S_x is a focused set with respect to algorithms A_i and A_j.

Do the No Free Lunch results tell us anything useful or constructive? Arguably, the greatest value of No Free Lunch is that these theorems makes it impossible to declare that "algorithm A_i is better than algorithm A_j" without also considering the question "On what class of problems is algorithm A_j better than algorithm A_i?"

The state of affairs is such that hundreds (or thousands) of search algorithms have been proposed. Most are never used by anyone except the research group that created them, and most are never evaluated except on a limited set of benchmarks. Algorithm designers virtually never prove that their search algorithm is better than another for a particular target class of problems. In this context, No Free Lunch theorems raise a caution flag for search algorithm designers who uses benchmarks to demonstrate the superiority of their algorithm. Are search algorithm designers really sure their benchmarks are faithful representatives of real applications of practical interest? And can potential users of the algorithm really be sure that the performance seen on select benchmarks will translate into similar performance on their applications?

9.2. Sharpened No Free Lunch and Permutation Closure

Several researchers have suggested a connection between the permutation closure of functions and No Free Lunch (Whitley *et al.*, 1997; Rana and Whitley, 1998; Whitley, 2000; Schumacher, 2000; Schumacher *et al.*, 2001; Droste *et al.*, 2002) This idea was also inherent in observations made by Radcliffe and Surry who pointed out that No Free Lunch also holds over all possible representations of a function (Radcliffe and Surry, 1995). A change in representation can also be seen as a transformation of the objective function; the change in representation can be viewed as a reindexing of the original function. But the "new" function must be in the permutation closure of the original function. Representations will be reviewed later in this tutorial.

We need to be careful to distinguish between algorithms and their behaviors. Behaviors are defined so that there are only a finite number of behaviors when searching a finite objective function. We might define two types of behaviors. When an algorithm searches an objective function, 1) in what order does it sample points in the domain? The next question is 2) in what order does it visit the codomain values of the function?

Consider the following small example of a search problem. Let the set $\mathcal{X} = \{x_1, x_2, x_3\}$ represent the domain of the objective function. Let the set $\mathcal{Y} = \{y_1, y_2, y_3\}$ represent the codomain of the objective function. We can next ask how many bijective functions we can construct using these 3 codomain values. Obviously, only 3! or 6 bijective functions can be constructed from these values. Additionally, only 3! *behaviors* are possible for any search algorithm, assuming that an algorithm does not re-sample points. First we will look at an algorithm's behavior as a permutation over the set of domain values. We next enumerate all possible permutations over the codomain values.

```
  POSSIBLE BEHAVIORS                          POSSIBLE
  IN THE DOMAIN                               FUNCTIONS

B1:  < x_1, x_2, x_3 >                    F1:  < y_1, y_2, y_3 >
B2:  < x_1, x_3, x_2 >                    F2:  < y_1, y_3, y_2 >
B3:  < x_2, x_1, x_3 >                    F3:  < y_2, y_1, y_3 >
B4:  < x_2, x_3, x_1 >                    F4:  < y_2, y_3, y_1 >
B5:  < x_3, x_1, x_2 >                    F5:  < y_3, y_1, y_2 >
B6:  < x_3, x_2, x_1 >                    F6:  < y_3, y_2, y_1 >
```

The set of functions that can be generated by reordering the codomain values is a *permutation closure.* If a function f is a bijection over N codomain values, then there are $N!$ permutations representing different reorderings of the codomain values. Thus we can construct $N!$ functions from N codomain values. If the function f is not a bijection and some points in the domain map onto the same codomain values, we can still might construct $N!$ functions, but some of these will represent the same function. Only unique functions are members of the permutation closure.

We next can also look at the behavior of a search algorithm in the codomain. We will use the metafunction "APPLY" and will assume m is the size of the search space.

$$\text{APPLY}(A, f, m)$$

APPLY generates a permutation of codomain values; this will be referred to as the *trace* produced by algorithm A when executed on function f. If a search algorithm explores the entire search space, then each trace contains every value in the codomain of the function. We will use the more concise notation $tr_A(f)$ to represent the *trace* where

$$tr_A(f) = \text{APPLY}(A, f, m)$$

Clearly, a trace is also just a permutation of codomain values. If A is deterministic, then it always produces the same trace.

We have already used permutations to represent the set of functions in the permutation closure, where the ith element of the permutation corresponds to the codomain value when $f(i)$ is evaluated.

But the set of all possible traces when all possible algorithm (behaviors) are consider also yield the permutation closure when $m = N$. Thus the set of all possible algorithm behaviors as expressed in terms of traces when searching a single function f also generates the permutation closure.

And if we apply all the possible search behavior in the domain space to all of the functions in the permutation closure, the resulting set of traces again corresponds to the permutation closure.

One might object to such a deterministic treatment of algorithm behaviors. Real search algorithms adapt their behaviors based on the set of domain values sampled so far, and on the codomain values that are observed. We refer to the set of domain points and codomain values sampled up until some time t as the search *history*. However, when evaluating an algorithm over all functions within a permutation closure, the behavior of a search algorithm after m steps cannot be used to predict which codomain value the algorithm will encounter next. Just for the purpose of illustration, assume the objective function is a bijection. Then under black box optimization, if algorithms are compared on a set that is closed under permutation, all of the unseen values in the codomain are equally likely to occur at the next time step with equal frequency.

Assume all of the codomain values of f_j are placed in a grab-bag. (The function can be a bijection, or repeated codomain values can be placed into the grab bag according to the number of time they appear in the function.) By randomly drawing values out of the grab-bag we can construct any function in the permutation closure. The grab-bag is equivalent to the permutation closure because when all codomain values are in the grab-bag, all functions in the permutation closure can be generated by sampling from the grab-bag.

Now assume an "adversary" can select codomain values out of the grab bag. If algorithm A_j samples codomain value v_i as the qth component of the trace produced by $\text{APPLY}(A_i, f_j, m)$ then when algorithm A_j picks which point to sample next in the search space then the adversary can also assign codomain values y_i to the qth component of the trace produced by $\text{APPLY}(A_k, f_l, m)$. As the trace is defined, f_l is incrementally constructed by the adversary. Joe Culberson's first suggested a variant of this adversarial interpretation of No Free Lunch (Culberson, 1998).

Another thing that is made clear by this adversarial argument is that the search method is really a black box optimization method and there can be no constraints on the structure of the functions being optimized. For specific classes of problems, such as MAXSAT or NK-Landscapes or Traveling Salesman Problems, the general No Free Lunch and the Sharpened No Free Lunch theorems do not hold because one cannot take codomain values and then randomly assign them to domain points in an arbitrary fashion and thereby create a new instance of MAXSAT or NK-Landscape. (It is still unclear at this point if Focused No Free Lunch results might hold in these kinds of domains.)

Schumacher *et al.* (2001) present what is now called the Sharpened No Free Lunch theorem by formally relating NFL to the *permutation closure* of a set of functions. Let $\mathcal{X}$ and $\mathcal{Y}$ denote finite sets and let $f : \mathcal{X} \rightarrow \mathcal{Y}$ be a function where $f(x_i) = y_i$. Let σ be a permutation such that $\sigma : \mathcal{X} \rightarrow \mathcal{X}$. We can permute functions as follows:

$$\sigma f(x) = f(\sigma^{-1}(x))$$

Since $f(x_i) = y_i$, the permutation $\sigma f(x)$ can also be viewed as a permutation over the values that make up the codomain (the output values) of the objective function.

We can now define the permutation closure $P(F)$ of a set of functions F.

$$P(F) = \{\sigma f : f \in F \text{ and } \sigma \text{ is a permutation}\}$$

This provides the foundation for the following result.

Theorem 9.1 (The Sharpened No Free Lunch theorem).
When comparing arbitrarily chosen search algorithms the No Free Lunch theorem holds if an only if that set of functions is closed under permutation.

Proofs are given by Schumacher *et al.* (2001). Similar observations have also been made by Droste *et al.* (1999, 2002).

Unlike many previous statements of the Sharpened No Free Lunch Theorem, this statement emphasizes that the search algorithms which are being compared can be arbitrarily chosen. As we will see, one direction of the "if and only if" statement is not true if we compare the performance of a small number of algorithms (e.g., 2) and we can know in advance which algorithms are to be compared. Stated another way, the Sharpened No Free Lunch Theorem holds over all possible search algorithms, which then requires that all algorithms have exactly the same behavior. This precludes comparing specific algorithms.

For the permutation closure of a single function one can construct a proof using the adversarial argument. No matter which arbitrarily chosen search algorithms are compared, if they are compared on a set of functions that are closed under permutations, that set behaves like the grab bag:

any possible (remaining) value of the codomain can occur at the next time sample.

Schumacher *et al.* (2001) also note that the permutation closure has the following property.

$$P(F \cup F') = P(F) \cup P(F').$$

Given a function f and a different function g, where $g \notin P(f)$, we can then construct 3 permutation closures: $P(f), P(g), P(f \cup g)$. For example, this implies that NFL holds over the following sets of functions:

```
Set 1: {< 3, 0, 0 >,                          Set 3: {< 3, 0, 0 >,
        < 0, 3, 0 >,                                  < 0, 3, 0 >,
        < 0, 0, 3 >}                                  < 0, 0, 3 >,
                      Set 2: {< 1, 3, 2 >,            < 1, 3, 2 >,
                              < 2, 1, 3 >,            < 2, 1, 3 >,
                              < 2, 3, 1 >,            < 2, 3, 1 >,
                              < 3, 1, 2 >,            < 3, 1, 2 >,
                              < 3, 2, 1 >}            < 3, 2, 1 >}
```

In effect, the space of all possible discrete functions can be partitioned into permutation closures. Thus, No Free Lunch holds over all possible discrete functions because it holds over all possible permutation closures.

Because $P(f)$ is constructed from the codomain values of a single function, for NFL to hold we must insist that all members of $P(f)$ are considered with uniform probability. Otherwise, some functions are more likely to be sampled than others, and NFL breaks down. But, we can have a uniform sample over a permutation closure $P(g)$ and a (different) uniform sample over $P(f)$ and NFL still holds. Thus, sampling need not be uniform over $P(f \cup g)$. In general, No Free Lunch can still hold over non-uniform distributions of functions (Igel and Toussaint, 2004).

9.2.1. *No Free Lunch and Compressibility*

English (2000) first pointed out that NFL can hold over sets of functions such as needle-in-a-haystack functions. In the following example, NFL holds over just 3 functions.

$$f = \langle 0, 0, 3 \rangle \quad \Longrightarrow \quad P(f) = \{\langle 0, 0, 3 \rangle, \langle 0, 3, 0 \rangle, \langle 3, 0, 0 \rangle\}$$

Clearly, NFL does not just hold over sets that are uncompressible.

However, for functions that are bijections, compressibility is still an issue.

Lemma 9.2 (NFL and compressibility). *Let $P(f)$ represent the permutation closure of the function f. If f is a bijection, or if any fixed fraction of the codomain values of f are unique, then $|P(f)| = \mathcal{O}(N!)$ and the functions in $P(f)$ have a description length of $\mathcal{O}(N \log N)$ bits on average, where N is the number of points in the search space.*

A proof can be constructed based on the well known proof demonstrating that the best sorting algorithms have complexity $\mathcal{O}(N \log N)$. Assume that the function is a bijection and that $|P(f)| = N!$. We would like to "tag" each function in $P(f)$ with a bit string that uniquely identifies that function. We then make each of these tags a leaf in a binary tree. The "tag" acts as an address which tells us to go left or right at each point in the tree in order to reach a leaf node corresponding to that function. But the "tag" also uniquely identifies the function. The tree is constructed in a balanced fashion so that the height of the tree corresponds to the number of bits needed to tag each function. Since there are $N!$ leaves in the tree, the height of the tree must be $\mathcal{O}(\log(N!)) = \mathcal{O}(N \log N)$. Thus $\mathcal{O}(N \log N)$ bits are required to uniquely label each function. The bit string labels can be compressed somewhat but it can still be shown that the $\mathcal{O}(N \log N)$ complexity bound is a tight bound.

Note that the number of bits required to construct a full enumeration of any permutation of N elements is also $\mathcal{O}(N \log N)$ since there are N elements and $\lg N$ bits are needed to distinguish each element.

This is one of the major concerns about No Free Lunch results. Do "No Free Lunch" results really APPLY to sets of functions which are of practical interest? Yet this same concern is often overlooked when theoretical researchers wish to make other mathematical observations about search. For example, some proofs relating the number of expected optima over all possible functions (Rana and Whitley, 1998), or the expected path length to a local optimum over all possible functions (Tovey, 1985) under local search are computed with respect to the set of $N!$ functions.

Igel and Toussaint (2003) formalize the idea that if one considers all the possible ways that one can construct subsets over the set of all possible functions, then those subsets that are closed under permutation is a vanishing small percentage. However, the *a priori* probability of *any* subset of problems is vanishingly small–including any set of applications we might wish to consider.

On the other hand, Droste *et al.* have also shown that for any function for which a given algorithm is effective, there exist related functions for which performance of the same algorithm is substantially worse (Droste *et al.*, 2002). They describe this as the *Almost No Free Lunch* (ANFL) effect. Even search algorithms designed for specific problem classes might be subject to *Almost NFL* kinds of effects.

9.3. Representation and Local Optima

Local neighborhood search methods generally assume a given fixed representation and invariably become trapped in local optima during search. One hypothetical strategy for "escaping" local optima is to dynamically change representations based on the argument that a local optimum under one representation may not be a local optimum under another representation. However, Rana and Whitley (1998) show that random selection of an alternate representation is doomed to failure: high-quality points in the search space are likely to be local optima under most representations.

Consider an arbitrary discrete function f with a search space of size N. Assume the function is a bijection so that each point x in the search space has a unique evaluation $f(x)$. Given a search operator with a fixed neighborhood size k, a point i is a local optimum if its evaluation is better than that of all its k neighbors. We can sort the points in ascending order based on their evaluation under f to create a ranking $R = < r_1, r_2, ..., r_n >$, where r_1 and r_n respectively denote the best and worst points in the search space. We can then look at all the ways of constructing neighborhoods around point i. Using R, the probability $P(i)$ that a point x with rank i in R is a local optimum under an *arbitrary* representation is given by

$$P(i) = \frac{\binom{N-i}{k}}{\binom{N-1}{k}}, \tag{9.1}$$

where $1 \leq i \leq (N - k)$. Using Equation (9.1), we can then count the expected number of local optima for a function f with respect to all possible $N!$ representations. Let this be denoted by $\mathcal{E}(N, k)$. The expected number of times that a particular point x with rank i will be locally optimal is simply $N! \cdot P(i)$. The expected number of optima over the set of all representations with neighborhood size k is then given by

$$\mathcal{E}(N, k) = \sum_{i=1}^{N-k} P(i) \cdot N!. \tag{9.2}$$

Let $\mu(N,k)$ denote the average number of local optima over all neighborhood representations. (This can also be used as the expected value for a *single* randomly chosen representation.) $\mu(N,k)$ is obtained by dividing $\mathcal{E}(N,k)$ by $N!$, yielding

$$\mu(N,k) = \sum_{i=1}^{N-k} P(i). \tag{9.3}$$

Finally, using simple counting principles, it can be shown that

$$\mu(N,k) = \sum_{i=1}^{N-k} P(i) = \sum_{i=1}^{N-k} \frac{\binom{N-i}{k}}{\binom{N-1}{k}} = \frac{N}{k+1}. \tag{9.4}$$

These equations make it clear that highly fit (low ranked) points in the search space are local optima under almost all representations. In other words, to exploit dynamic representations, we cannot simply select a new representation at random.

9.3.1. *No Free Lunch over Representations*

Let $R(f)$ be a representation of f. We will assume that the representation does not change the codomain of f and that we can view the change in representation as a transformation of the objection function. Therefore

$$tr_A(R(f)) = \text{APPLY}(A, R(f), m)$$

when m is the size of the search space, and the trace must be in the permutation closure of function f.

Radcliffe and Surry (1995) first formalized the idea that we can also include representations under No Free Lunch. That is, when we consider all possible representations of a function, No Free Lunch still holds: No search algorithm is better than another when applied to all possible representations of a function. In effect, a change in representation just transforms one function into another.

One can look at a change in representation as a change in the algorithm used. The next section looks at this more closely by looking at Gray and Binary bit representations.

9.3.2. *Gray and Binary Representations*

Consider the set of functions whose domain consists of the integers 0 to $2^L - 1$ such that points can be represented using bit-strings of length L. Some functions with a different natural domain can still sometimes be mapped onto a bit representation.

Binary and Gray encodings are commonly used bit representations in evolutionary algorithms. There are in fact many Gray codes. A "Gray code" is any bit representation that has the property that adjacent integers are also adjacent neighbors in the bit-neighborhood hypercube graph defined over the set of bit strings of length L. Every Hamiltonian circuit in the hypercube can be used to form a Gray code by assigning the integers in sequence to the vertices along the Hamiltonian circuit. We can start the assignment at any arbitrary point along the circuit, so each circuit represents up to 2^L Gray codes under a simple shift operation which moves the first integer (e.g., 0) to some arbitrary point along the Hamiltonian circuit.

A Gray code representation can be constructed as follows. Let $\oplus$ denote a bitwise *exclusive-or* operator. Let b_i be the ith bit in the binary encoding, let g_i be the ith bit in the Gray encoding (technically, this is the *standard binary reflected Gray* encoding). Then $g_1 = b_1$ and $g_i = b_i \oplus b_{i-1}$.

The conversion from binary to standard binary reflected Gray code can also be performed using a matrix operation. There exists an $L \times L$ matrix G_L that maps a string of length L from Binary to to reflected Gray representation. There also exists a decoding matrix D_L that maps the reflected Gray back to its original representation.

The following are the G_4 and D_4 matrices.

$$G_4 = \begin{bmatrix} 1\,1\,0\,0 \\ 0\,1\,1\,0 \\ 0\,0\,1\,1 \\ 0\,0\,0\,1 \end{bmatrix} \qquad D_4 = \begin{bmatrix} 1\,1\,1\,1 \\ 0\,1\,1\,1 \\ 0\,0\,1\,1 \\ 0\,0\,0\,1 \end{bmatrix}.$$

In higher dimensions the G_L matrix continues to have 1 bits along the diagonal and the upper minor diagonal, and D_L has 1 bits in the diagonal and the upper triangle. Assuming the L-bit column vector x is a standard Binary representation, binary matrix multiplication (i.e., matrix multiplication mod 2), $x^T G_L$ produces a Gray coding.

For example, if $x^T = [1\ 0\ 0\ 1]$, then $(x^T G_4) = [1\ 1\ 0\ 1]$.

Assuming the L-bit column vector x is a standard reflected Gray representation and $x^T = [1\ 1\ 0\ 1]$, then $(x^T D_4) = [1\ 0\ 0\ 1]$ produces the corresponding standard Binary representation of the corresponding integer.

An interesting property of the G matrix is that all reorderings of the columns of the matrix produce a matrix that is also a Gray transformation corresponding to a different Hamiltonian circuit through the hypercube graph. Thus, there may be as many as $L! \cdot 2^L$ different Gray codes, since all 2^L shifts along each Hamiltonian circuit is also a Gray code. The exact number of possible Gray representations is an open question.

Empirically, Gray coding usually results in better performance than using a standard Binary coding when applying various forms of genetic algorithms and neighborhood search algorithm using common test functions (Caruana and Schaffer, 1988; Mathias and Whitley, 1994a,b). It has long been known that Gray coding removes *Hamming cliffs.* A Hamming cliff corresponds to adjacent integers whose bit representations are complementary. Thus in a four bit space, 7 and 8 are adjacent integers but their bit representations are 0111 and 1000.

It seems intuitive that one would like to preserve the adjacency of the original numeric/integer representation of the function in the bit representation. In 1-dimensional functions, this guarantees that the number of local optima under the Gray code is less than or equal to the number of local optima under the numeric/integer representation. This occurs because the numeric/integer neighborhood is embedded in the bit neighborhood.

The graph on the right in Figure 9.1 shows that all the adjacency relationships found in the numeric representation are preserved in Gray space. On the left, one can see that only half of the adjacent edges found in nu-

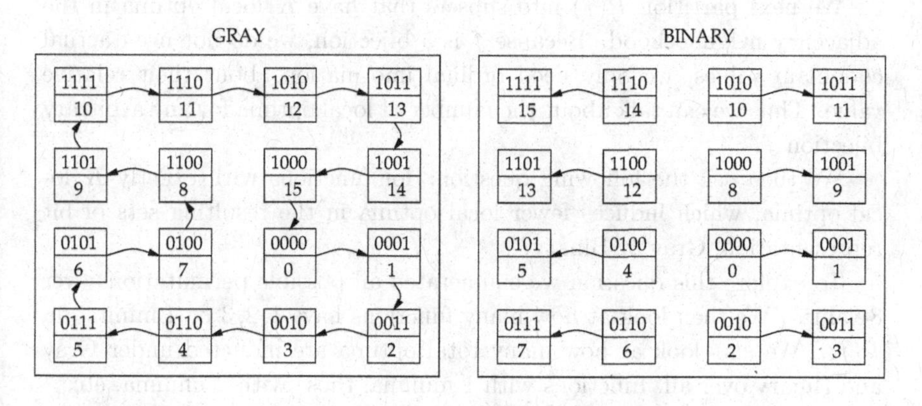

Fig. 9.1. Adjacency in 4-bit Hamming space for Gray and Binary encodings.

meric representations are preserved in Binary space; for representations of arbitrary length one can show by induction (Whitley *et al.*, 1996) that it continues to be true that half of the edges from the numeric representation are preserved under the Binary representation.

9.3.3. *Mini-Max Effects and NFL: Gray versus Binary*

Wolpert and Macready's original paper (Wolpert and Macready, 1995) also included another interesting idea: while two algorithms might have the same behavior when averaged over all functions, there can be "Mini-Max" differences in algorithm behaviors. Algorithm A_i could be *much* better than A_j by some metric $\mathcal{M}$ on a small subset of functions, while algorithm A_j is then *slightly* better than A_i on all other problems. Thus, A_j would be better than A_i over *most* functions, but over all functions (or over any set closed under permutation) there would be no difference between the algorithms as measured by $\mathcal{M}$.

A Mini-Max effect can be observed when comparing Gray and Binary representations. We start with some function f that is a bijection. We then generate the permutation closure of f, denoted by $P(f)$. Because f is a bijection, we can rank the codomain values of f. Next we define the *adjacency neighborhood* of a function in the permutation closure. We assume f is one dimensional, and for any point x_i its adjacent neighbors are x_{i-1} and x_{i+1}. We will assume the adjacency neighborhood wraps around (i.e., if there are 2^L points, x_0 and x_{2^L-1} are neighbors).

We next partition $P(f)$ into subsets that have K local optima in the adjacency neighborhood. Because f is a bijection, we do not need actual codomain values, we only need ordinal information about their relative value. Thus we can talk about the number of local optima for any arbitrary bijection f.

We then ask the following question: for functions with exactly K local optima, which induces fewer local optima in the resulting sets of bit representation, Gray or Binary?

To explore this question we enumerated all possible permutations over 8 values. We then look at how many functions have $1, 2, 3$ or 4 minima in $P(f)$. We also look at how many total optima are inducted under Gray and Binary over all functions with 1 minima, those with 2 minima, etc.

There are 8! or $40,320$ functions in this case. of minima for Gray or Binary representations must be No Free Lunch tells us is that the number

Table 9.1. This table counts the number of optima under Gray and Binary representations for all 3-bit bijective functions.

K	Functions with K optima	Number of optima with 2 neighbors	Number of optima with gray codes	Number of optima with binary codes
1	512	512	512	960
2	14,592	29,184	23,040	27,344
3	23,040	69,120	49,152	49,392
4	2,176	8,704	7,936	2,944
TOTAL	40,320	107,520	80,640	80,640

equal. We can use the formula

$$\mu(N, k) = \frac{N}{k+1} \tag{9.5}$$

to see that in a 3 bit neighborhood $k = 3$ and $\mu(N, k) = \frac{8}{4} = 2$, and $2 \cdot 8! = 80,640$. In general, given the permutation closure of a bijective function f using a bit representation, the Hamming neighborhood will result on average in $\mu(2^l, L) = \frac{2^L}{L+1}$ local optima. However, under the adjacency neighborhood $k = 2$ and $\frac{8}{3} \cdot 8! = 107,520$.

In Table 9.1, K represents the number of optima. The set of functions is then partitioned according to how many optima (minima in this case) the functions have using the adjacency neighborhood. Column two (Functions with K optima) counts the number of functions that have K optima under the adjacency neighborhood. The last two columns count the total number of optima induced by Gray and Binary encodings summed over all those functions that have K minima in the adjacency neighborhood.

This table makes it clear that Gray coding is more tightly coupled to the adjacency neighborhood representation of the function. As a side effect, Gray is biased toward producing fewer local minima for functions that have few minima in the adjacency neighborhood. This is true for *all* Gray codes. Gray codes are also biased toward producing a larger number of minima for functions that have a large number of minima in the adjacency neighborhood representation. Note that the adjacency neighborhood is "natural" in the sense that it captures the continuity of the function. This implies that (1) if we care about number of local minima and (2) if we are more likely to work with functions that have fewer local minima in the numeric (adjacent) function representation, then *on average* Gray coding is a better representation over this subset of functions if one is willing to define "better" in terms of inducing fewer optima in the corresponding bit representations.

Table 9.2. An illustration of the Mini-Max effect for Gray and Binary encodings.

K	Gray induces fewer optima	Binary induces fewer optima	Ties
1	448	0	64
2	6384	2176	6032
3	7088	6704	9248
4	0	2160	16
TOTAL=40,320	13,920	11,040	15,360

Over the set of functions with a single optimum is easy to prove that for all representations of functions in F defined over L bits, Gray always induces fewer minima than Binary (cf. Rana and Whitley (1997)). However, Binary is actually "better" than Gray over worst-case problems, where "worst-case" functions are those that have $N/2$ minima when N is the size of the search space (Whitley, 1999). When bit representations are involved, $N = 2^L$. Specifically in the above table, we consider "worst-case" functions to be those with $K = 4$ optima. It follows that Gray is therefore better than Binary over all remaining (i.e., not worst-case) functions. It is conjectured that one can select any bound K on the number of minima, and Gray codes will induce fewer optima than Binary on functions with less than K optima. However there are no general proofs of this.

We observe Wolpert and Macready's Mini-Max effect when examining Gray codes and Binary codes by doing the following comparison. For each function with K minima in the adjacency neighborhood, Table 9.2 counts how often Gray or Binary is better (induces fewer minima) relative to K. It ignores how much better one is relative to the other. Here, we find that Gray induces fewer minima on more functions than Binary, but it is still particularly "better" over functions with fewer minima. Of course, having fewer minima does not automatically translate into having a better algorithm. But the work of Christensen and Oppacher (2001) also suggests that number of local optima can be a reasonable way to classify functions according to complexity.

9.4. Focused No Free Lunch

Sharpened No Free Lunch theorems hold over sets that are closed under permutation. However, Focused No Free Lunch theorems can hold over sets that are not closed under permutation. We will refer to the set where a

Focused No Free Lunch result holds as a *focused set* (Whitley and Rowe, 2008).

Focused No Free Lunch effects can happen for two reasons.

The first reason that Focused No Free Lunch effects can occur is because practical search algorithms are limited to a small number of sample evaluations, denoted by m. It can be assumed that m is polynomial, while the search space is typically exponentially large. Thus if we are given A_i and f_j and A_k and we wish to find some f_l such that

$$\text{APPLY}(A_i, f_j, m) \equiv \text{APPLY}(A_k, f_l, m).$$

then there are many candidates for f_l because we only need to consider the first m steps of algorithm A_k. If f_j is a bijection, there are $(N - m)!$ functions (all with different reorderings of the unseen domain values) that might play the role of f_l. Because there is a choice in terms of which functions are added to the focus set, the focus set can be very small in some cases. In the smallest case, Focused No Free Lunch results can hold over two algorithms limited to m steps and the resulting *focused set* contains only two functions, both of which are guaranteed to be compressible.

One might object to the fact that the algorithms might have displayed different behaviors if search had continued beyond m steps. But search cannot explore more than a polynomial number of points in the search space, and m can be chosen to be sufficiently large to exploit this fact.

The second reason that Focused No Free Lunch effects can occur is because the behavior of two specific search algorithms induce a closure which is smaller than the permutation closure. This can happen even if $m = N$ and traces are complete permutations of the codomain values. This does not mean that the resulting focused sets are large. In some cases, this closure corresponds to the orbit of a permutation group. In this case, we leverage mathematical concepts from permutation groups to bound the maximal size of the focused closure.

We next look at an example from Gray and Binary representations.

9.4.1. *Focused NFL for Gray and Binary Representations*

Consider a benchmark test function, $f_b : \{0, 1, \ldots, 2^L - 1\} \rightarrow R$, where R is some finite subset of real numbers and L a positive integer. Let algorithm A_g be a search algorithm using a bit encoding with a Gray code representation. Let algorithm A_b use the same search algorithm except A_b

uses a Binary representation. (It does not matter what search algorithm is used, as long as it uses a bit representation of a parameter optimization problem.)

We now ask a normal No Free Lunch question: over what sets of functions do the two algorithms have the same performance?

The Sharpened No Free Lunch theorem proves that A_g and A_b will have the same performance when compared over the permutation closure of f_b.

However, because we are talking about two specific algorithms, not just any two arbitrary algorithms out of the space of all possible algorithms, there is a more focused result. Let string s_b be any binary string; we can also treat s_b as a column vector. There exists matrix G such that $(s_b)^T G = s_g$ where the addition is mod 2 and s_g is the Gray encoding corresponding to s_b. We can also "Gray" a string multiple times. Thus $(((s_b)^T G)G) = (s_b)^T G^2$ produces a string to which Gray coding has been applied twice.

It is straight forward to prove for strings of length L that there exists an integer $L \leq i < 2L$ such that $G^i = I$ where I is the identity matrix.

This means that for any function f_b there exists a *focused set* of functions with less than $2L$ members that can be generated by repeated application of Gray code. In contrast, the permutation closure can be exponential in size.

The following is an example for a 3 bit representation

s_b :	000	001	010	011	100	101	110	111
$(s_b)^T G^1$:	000	001	011	010	110	111	101	100
$(s_b)^T G^2$:	000	001	010	011	101	100	111	110
$(s_b)^T G^3$:	000	001	011	010	111	110	100	101
$(s_b)^T G^4$:	000	001	010	011	100	101	110	111

where $(s_b)^T G^4 = s_b$.

If we then interpret all of the bit strings as standard binary strings, then we can use the following permutation of indices into f_b to create a subset of 4 permutations.

P1:	0	1	2	3	4	5	6	7
P2:	0	1	3	2	6	7	5	4
P3:	0	1	2	3	5	4	7	6
P4:	0	1	3	2	7	6	4	5

We can convert these permutations into functions as follows. Let σ_i be the ith element in a permutation σ. A new function g is generated by the mapping $g(i) = f_b(\sigma_i)$.

This produces a focused subset of four functions, such that the performance of A_g and A_b will be identical when compared over all functions in the focused subset. Technically, a *group* has been defined that is closed with respect to the application of the graying matrix G to the bit representation.

This group has two cycles of length 1, a cycle of length 2, and a cycle of length 4, which can be denoted by:

$$(0)(1)(23)(4657), \quad \text{or more concisely by} \quad (23)(4657)$$

Closure is achieved when all of the cycles synchronize. The orbit of the permutation cycle defines both the size of the closure and the time required for the permutation cycles to synchronize. Since 2 and 1 are factors of 4, the orbit is 4.

Note that the group action just described can be viewed in three ways: a subset of functions is generated; a subset of different representations is generated, or a subset of different algorithms is generated. When this process is viewed as a change in representation, the Focused No Free Lunch result holds even if the search algorithms are nondeterministic, because from one point of view the algorithm does not change, only the representation changes. When this is the case, it does not matter if the algorithm is deterministic or nondeterministic.

This result can be extended to any pair of algorithms whose only difference is a choice of representation. Since a change in representation is the same as applying a bijection σ to the domain, this means that such a pair of algorithms A_i and A_j have the property that running A_i on f is the same as running A_j on σf, for all functions f. Then given any particular function f, we get a focused set $f, \sigma f, \sigma^2 f, \ldots$, whose size is the same as the order of σ (i.e., the smallest integer k such that σ^k is the identity). In the Gray and Binary case, σ is calculated using the G matrix, which induces an orbit of size $< 2L$.

When comparing two *specific* algorithms A_i and A_j, what we should care about is the not the permutation closure, but rather what other smaller closures or *focused sets* may exist.

9.4.2. *Focused NFL need not Require a Black Box*

The Focused No Free Lunch induced by the use of Gray and Binary encodings also makes it clear that Focused No Free Lunch is not limited to black box optimization. Given some parameter optimization function denoted by f, we might be given a great deal of information about f. We can be told how many parameters there are and the precision of the encoding for each parameter. We can also be given explicit information about which parameters interact, or it could be that f is a linear combination of lower complexity subfunctions and we might be able to do an exact Walsh analysis of f in polynomial time. Nevertheless the Focused No Free Lunch still holds even though the problem is not a black box optimization problem.

9.4.3. *More Focused Sets: Path-Search Algorithms*

Permutation groups are one mechanism that leads to focused sets. Permutation cycles can show up in various ways when comparing algorithms. We next look at this on a special class of algorithms which we call *path-search algorithms.*

Definition 9.3. A *path-search* algorithm searches a function using a predefined sequence of points in the domain X. The sequence of points determines a permutation of X.

This means a path-search algorithm does not use any information about the set of codomain values that are observed during search. Given N values in the domain of a function, there are $N!$ search path algorithm.

Path-search algorithms are strongly deterministic in as much as their behavior (and therefore their performance) is predetermined before search begins. This also means the path of a path-search algorithm and the resulting trace of the codomain values is strongly coupled as well.

Given two path-search algorithms, there is a unique set of cycles that describes the difference in the behaviors of the two path-search algorithms. This is not always true for other search algorithms. For now, this simplifies our discussion of permutation cycles.

We next construct an example using the domain $X = \{1, 2, 3, 4, 5, 6, 7\}$ and codomain $Y = \{a, b, c, d, e, f, g\}$ where a to g will represent any arbitrary real values. Consider the following path-search algorithms:

$$A_x = < 1\ 2\ 3\ 4\ 5\ 6\ 7 >$$
$$A_y = < 2\ 3\ 4\ 1\ 6\ 5\ 7 >$$

There again exists permutation cycles that define a closure: (1234)(56)(7).

Thus, there are three cycles in the two permutations that define the two path-search algorithms. Since all three cycles are synchronized after a period of 4, there exists an orbit of 4 functions that define a closure. Given any test function f_1 we can define three more functions f_2 and f_3 and f_4 such that the performance of algorithm A_x and A_y will be identical over these four functions.

For example, if $f_1 =< \mathrm{a\ b\ c\ d\ e\ f\ g} >$ then:

$$\begin{aligned} f_1 &=< \mathrm{a\ b\ c\ d\ e\ f\ g} > \\ f_2 &=< \mathrm{b\ c\ d\ a\ f\ e\ g} > \\ f_3 &=< \mathrm{c\ d\ a\ b\ e\ f\ g} > \\ f_4 &=< \mathrm{d\ a\ b\ c\ f\ e\ g} > \end{aligned}$$

Path-search algorithms A_x and A_y will have identical behaviors when compared over this set of functions.

We next formalize these ideas.

9.4.3.1. *Path Search and Closure*

Given finite sets X and Y, then Y^X is the set of all functions from X to Y. Given a deterministic black box search algorithm A for such functions then let $\mathrm{tr}_A(f)$ (the *trace* of A on f) be the sequence of codomain values generated as A searches through X. If we let T be the set of all possible sequences of codomain values, then one of the consequences of the NFL theorem is that $\mathrm{tr}_A : Y^X \to T$ is a bijection for all algorithms A. It follows that, given any two algorithms A and B that the function $(\mathrm{tr}_A)^{-1} \circ \mathrm{tr}_B$ is a permutation of the set of functions Y^X. There are, of course, many possible permutations of this set, and they form a *group* under function composition.

Definition 9.4. Let $\pi : X \to X$ be a bijection. Then the map $v_\pi(f) = f \circ \pi$ is a permutation of Y^X called a *value-permutation.*

The set of all value-permutations forms a subgroup of permutations of Y^X. This kind of permutation appears in the Sharpened NFL, which states that any set $S \subseteq Y^X$ is closed under value-permutations if and only if $\mathrm{tr}_A(S) = \mathrm{tr}_B(S)$ for all algorithms A, B. However, if we only consider some subset of algorithms, then S need not be closed under value-permutations for the result to hold.

Lemma 9.5. *Given a subset $\mathcal{A}$ of algorithms, let $\mathcal{G}$ be the group of permutations of Y^X generated by maps of the form $(\mathrm{tr}_A)^{-1} \circ \mathrm{tr}_B$, for pairs of algorithms $A, B \in \mathcal{A}$. Then for a given function $f : X \to Y$, the* orbit *of f under $\mathcal{G}$, defined by*

$$\mathcal{G}f = \{f' | f' = g(f) \text{ for some } g \in \mathcal{G}\}$$

has the property that $\mathrm{tr}_A(\mathcal{G}f) = \mathrm{tr}_B(\mathcal{G}f)$ for all $A, B \in \mathcal{A}$. Moreover, it is the smallest subset containing f for which this is the case.

Lemma 9.5 immediately from the definition of the group $\mathcal{G}$ that $\mathrm{tr}_A(\mathcal{G}f) = \mathrm{tr}_B(\mathcal{G}f)$. Now let $S \subseteq Y^X$ with $f \in S$ such that $\mathrm{tr}_A(S) = \mathrm{tr}_B(S)$ for all $A, B \in \mathcal{A}$. It follows that $S = g(S)$ for all $g \in \mathcal{G}$. Therefore $g(f) \in S$ for all $g \in \mathcal{G}$ which shows that $\mathcal{G}f \subseteq S$.

The orbit of a function f (corresponding to some set of algorithms $\mathcal{A}$) is the smallest set of functions containing f for which all the algorithms in $\mathcal{A}$ have equal performance. The orbits corresponding to different functions partition the set Y^X. The union of two such orbits is another set for which the algorithms have equal performance. We therefore get a lattice (under set inclusion) of sets of functions for which NFL results hold for a given set of algorithms.

Of course, the orbit $\mathcal{G}f$ must be contained within the value-permutation closure of f. But it is simple to verify this is the case. Let $g = (\mathrm{tr}_A)^{-1} \circ \mathrm{tr}_B(f)$. Then the trace of A on g is the same as the trace of B on f and g and f are value-permutations drawn from the same codomains values.

From this result, the Sharpened No Free Lunch theorem follows.

Corollary 9.6 (Sharpened NFL). *If $\mathcal{G}$ is the group generated by all deterministic black box algorithms on Y^X and V is the group of value-permutations, then $\mathcal{G}(F) = V(F)$ for any set of functions $F \subseteq Y^X$. Thus, No Free Lunch holds over arbitrary search algorithms and over a set of functions if and only if the set of functions is closed under permutations.*

It may well be the case that the orbit is by no means the whole of the value-permutation closure. Consider the example where there are just two algorithms $A, B \in \mathcal{A}$. Then the group $\mathcal{G}$ is generated by the single element $z = (\mathrm{tr}_A)^{-1} \circ \mathrm{tr}_B$ and is a cyclic group. If f is a one-to-one function then the size of the orbit of f is the order of the element z. That is, it is the smallest integer k for which z^k is the identity. Examples can be given where this is considerably smaller than the size of the value-permutation closure.

9.4.4. *Algorithms Limited to m Steps*

What happens when deterministic algorithms are limited to m steps? Given the pragmatic constraints on search, we will assume that for an arbitrarily chosen problem the search space is exponentially large relative to the problem representation size and that m is polynomial.

For example, given $A1, A2$ and f_0 there exists at least two functions f_1 and f_2 such that:

$$\text{APPLY}(A1, f_0, m) \equiv \text{APPLY}(A2, f_1, m).$$

$$\text{APPLY}(A2, f_0, m) \equiv \text{APPLY}(A1, f_2, m).$$

If by coincidence $f_1 = f_2$ then this subset of functions is closed under the APPLY operator when $A1, A2$ and f_0 are fixed. Thus if $A1$ is better than $A2$ on function f_0 then $A2$ has identical but opposite performance relative to $A1$ on f_1. If $m < N$, then we only require that the traces are the same for the first m steps. We then ask under what circumstances this does happen?

9.4.5. *Benignly Interacting Algorithms*

Let $A1$ and $A2$ be two deterministic algorithms that are executed on function f_1. We assume a method exists for indexing the domain. Assume each algorithm samples m points, with each domain sample indexed by $i = 1$ to m.

For the current discussion, a key aspect of deterministic algorithms is that the search paths explored by these algorithms may "intersect" in the sense of sampling the same domain values. However, given that m is dramatically smaller than the size of the search space, it is also possible that two algorithm do not "intersect" after m steps. What does this mean for the comparison of the two algorithms?

It is possible to show that when 2 algorithms are executed on a single function, there are conditions under which a Focused No Free Lunch holds over just 2 functions. In other words, if $A1$ is better than $A2$ on test function f_1, there can exist a second function f_2 where the performance is exactly the reverse. This induced a Focused No Free Lunch result which holds over exactly 2 functions. In the following discussion it is critical that the search algorithms only examine a small sample from the entire search space. The following explanation is also constructive in nature: given $A1$

and $A2$ and f_1 we construct a function f_2. In fact, since only a small sample of the search space is examined, there can be exponentially many ways to construct a function f_2 with the required properties.

We first construct two arrays: Array D stores the domain values $D_{i,j}$ sampled by algorithm j at step i. Array V stores the codomain values $V_{i,j}$ sampled by algorithm j at step i.

We next construct two new arrays, D^* and V^*. Note that V and V^* contain information about *two* traces constructed from the codomain values. Our goal is to combined the information contained in V^* into a single potential function. We construct V^* as follows.

$$\forall i \in \{1, \ldots, m\}, \qquad V^*_{i,2} = V_{i,1} \quad \text{and} \quad V^*_{i,1} = V_{i,2}.$$

To assign values to D^*, we define a function denoted by XDT to extend the domain trace. The function $XDT(i, j, D^*, V^*)$ executes algorithm A_j, using the provided domain values and codomain values from steps 1 to $i-1$. It returns the domain value that algorithm A_j will visit at step i. In effect, $XDT(i, j, D^*, V^*)$ simulates algorithm A_j; it uses the $i - 1$ values in D^* and V^* to determine $D^*_{i,j}$.

We then iteratively construct D^* as follows:

$$\text{for } i = 1 \text{ to } m, \; D^*_{i,j} = XDT(i, j, D^*, V^*)$$

Note that D and D^* stores m elements for each algorithm. We will treat the elements of D and D^* as sets, and apply intersection and union operations.

$$D_{A1} = \bigcup D_{i,1} \quad \text{and} \quad D^*_{A2} = \bigcup D_{i,2}$$

$$V_{A1} = \bigcup V_{i,1} \quad \text{and} \quad V^*_{A2} = \bigcup V_{i,2}$$

For example, this denotes that D_{A1} is the union of the m domain values sampled by $A1$ at each time step i.

Definition 9.7. Two algorithms $A1$ and $A2$ *benignly interact* with respect to a function f_1 and the construction of V^* if one of the following conditions hold:

1) $(D^*_{A1} \cap D^*_{A2}) = \emptyset$.
2) if ($d \in (D^*_{A1} \cap D^*_{A2})$ and $d = D^*_{x,1} = D^*_{y,2}$)
   then $V^*_{x,1} = V^*_{y,2}$.

Note that condition 2 for benign interaction can occur in two ways. If f_1 is a bijection then

$$((D^*_{A1} \cap D^*_{A2}) \neq \emptyset \text{ and } V^*_{x,1} = V^*_{y,2}) \iff x = y.$$

Thus, if the function is a bijection and both algorithms sample the same domain value, they must visit that domain value at time step $x = y$ in order to produce the same codomain value at the same time in the traces V and V^*. This produces a permutation cycle of one element in the two trace V and V^*.

If f_1 is not a bijection and $x \neq y$, then as long as $V^*_{x,1} = V^*_{y,2}$ when $(D^*_{A1} \cap D^*_{A2}) \neq \emptyset$ then the same codomain values will still occur at the same time step in V and V^*.

We will use D^* and V^* to construct a new function f_2. The notion of benign interaction is a way of guaranteeing that either the algorithms $A1$ and $A2$ do not visit the same domain values, or if they do visit the same domain values, they also find the same codomain values at those locations.

Lemma 9.8. *Assume two algorithms $A1$ and $A2$ are executed for a polynomial number of time steps denoted by m on function f_1. If $A1$ and $A2$ "benignly interact" there exists a function f_2 such that $A1$ and $A2$ have two equal but opposite traces when executed for m steps on f_1 and f_2; if f_1 is compressible, then there exists at least one function f_2 which is compressible. In general, when f_1 is a bijection and N is the size of the search space, there are $(N-m)!$ ways to construct f_2.*

To see if this is true, note that array V defines the behavior of algorithms $A1$ and $A2$ on function f_1. Using D^* and V^* we construct a second function f_2; when D^* does not contain a domain value we can make an assignment to that domain value randomly or we use f_1 to make the assignment.

By construction $V^*_{i,2} = V_{i,1}$ and $V^*_{i,1} = V_{i,2}$.

The construction is feasible and V^* captures the correct traces of $A1$ and $A2$ executed on f_2 given that $A1$ and $A2$ benignly interact on function f_1. If $(D^*_{A1} \cap D^*_{A2}) = \emptyset$ then we are free to assign values to f_2 at all domain values using V^* without conflict. If $(D^*_{A1} \cap D^*_{A2}) \neq \emptyset$ but $V^*_{x,1} = V^*_{y,2}$ because $A1$ and $A1$ benignly interact, then there is no conflict in the construction since the codomain values that appear in both V and V^* occur at exactly the same time steps (and thus act as a permutation cycle of length 1).

In the construction of f_2 if domain points that are not in D^* are assigned codomain values randomly using codomain values that are not in V^*, then for a search space of size N, there are $(N-m)!$ possible constructions when f_1 is a bijection.

In the construction of f_2 if f_1 is used to make the assignment to domain points not in D^*, f_2 is compressible when f_1 is compressible, since we only need a copy of f_1 and the arrays D^* and V^* which are of length m to construct f_2.

One objection to only looking at m steps is the following. If we construct a focused set based on m time steps, and then run algorithms $A1$ and $A2$ for $2m$ time steps on these functions, their performance will almost certainly not still be the same after $2m$ time steps. This is true. But either we don't have that access to that information, or we can construct another new focused set using the $2m$ time steps.

This might seem like a flaw. But in practice, we can never have complete traces for any algorithm on any non-trivial (exponentially large) problem. Also consider the following. There are many examples of reasonable search algorithms where the performance of $A1$ is better than $A2$ after $100,000$ evaluations, but $A2$ is better than $A1$ after $200,000$ evaluations. It is not just a matter of one algorithm being better than another, but rather being better than another given some amount of effort such as number of function evaluations. Thus, arguments about the traces an algorithm would have produced if it had exhaustively explored the search space are not practical given that all comparisons of algorithms are based on a limited view of the function based on m samples.

9.4.6. *A Focused NFL Theorem*

We will state a more general version Focused No Free Lunch result. We will again us the APPLY function to generate a full or partial *trace* when algorithm A is executed on function f.

Theorem 9.9 (A Focused No Free Lunch theorem). *Let $A1$ and $A2$ be two search algorithms and let $\mathcal{F}$ be a set of functions such that for all $f \in \mathcal{F}$, there exists $f_2, f_3 \in \mathcal{F}$ such that*

$$\text{APPLY}(A1, f_1, m) = \text{APPLY}(A2, f_2, m)$$

and

$$\text{APPLY}(A2, f_1, m) = \text{APPLY}(A1, f_3, m)$$

Algorithms A1 and A2 will generate identical sets of traces when applied to all the functions in $\mathcal{F}$. The set $\mathcal{F}$ need not be closed under permutation, and is referred to as a focused set.

Given a seed function f_1 we can compute $\mathcal{F}$ iteratively by generating f_2 and f_3. Note the construction of $\mathcal{F}$ yields a closure with respect to the APPLY function. Thus, by construction the set of traces which are produced when algorithm A_1 is executed on all the functions in $\mathcal{F}$ must be exactly the same as the set of traces which are produced when algorithm A_2 is executed on all the functions in $\mathcal{F}$. That $\mathcal{F}$ need not be closed under permutation can be illustrated by example: different path search algorithms are one example. Also, if we create two versions of an algorithm such that one uses a Gray representation and the other a Binary representation, we have another example. In both of these cases, the set $\mathcal{F}$ corresponds to the orbit of a group. Also in both of these cases, the number of steps m can be as large as the size of the search space.

The set $\mathcal{F}$ need not corresponds to the orbit of a group (or unions of such sets). When m is polynomial and the size of the search space is exponential, there are exponentially many functions that satisfy the constraints outlined in the above Focused No Free Lunch theorem 9.9 and which could serve as f_2 or f_3. These means there are many ways in which a set $\mathcal{F}$ could be constructed. However, when constructing the set $\mathcal{F}$ there may be way to minimize the size of the set. For example, if we are constructing the set $\mathcal{F}$ iteratively and f_i is the most recent element to be added to $\mathcal{F}$ and there already exists a function $f_j \in \mathcal{F}$ such that

$$\text{APPLY}(A1, f_i, m) = \text{APPLY}(A2, f_j, m)$$

then a new function need not be added to $\mathcal{F}$. Note that the same partial trace can be generated multiple times by distinct functions in $\mathcal{F}$. However, we must insure that every partial trace produced by the two algorithms being compared is generated an equal number of times by each algorithm.

Consider the following example. Label this set of functions $\mathcal{F}_1$.

	f1	f2	f4	f6	f8	f3	f5	f7	f9
A1	t1	t2	t4	t2	t4	t3	t5	t2	t9
A2	t3	t1	t2	t4	t2	t5	t2	t9	t4

In this table, the rows are indexed by the algorithms and the columns are indexed by functions. Each entry is a trace that is produced when a particular algorithm is executed on a particular function.

Both algorithms generate the same nine traces: $t1$, $t2$, $t2$, $t2$, $t3$, $t4$, $t4$, $t5$, $t9$. This set of functions can also be broken into two subsets: subset $\mathcal{F}_1 - \{f6, f8\}$ and subset $\{f6, f8\}$. The algorithms have identical behaviors over both subsets. Thus, both subsets are also focused sets.

The nine functions form a set, but the traces can be a multiset. Thus, algorithm $A1$ produces the same trace on functions $f2, f6$ and $f7$.

When comparing two algorithms there can be multiple sets contained in the permutation closure where the algorithms generate exactly the same sets of traces. Given multiple focused sets, it is clear that these sets can be unioned if the intersection of the set of functions is empty. Focused No Free Lunch would still hold over the unioned sets. However, if there are two focused sets and the intersection is not empty, these sets cannot always be unioned to form a new focused set.

For example, consider a second focused set, which we will label $\mathcal{F}_2$ and the partial traces that are produced by the algorithms.

```
    |   f6  f10  f11  f7  f8
====|=========================
A1  |   t2  t10  t9   t2  t4
A2  |   t4   t2  t10  t9  t2
```

Both algorithms produce the same traces: $t2, t2, t4, t9, t10$.

Note that the functions $f6, f7$ and $f8$ are in the intersection of $\mathcal{F}_1$ and $\mathcal{F}_2$.

But if we union the set of functions $\mathcal{F}_1$ and set $\mathcal{F}_2$ the two algorithms do not generate the same set of traces: this happens because the set the traces associated with the functions in the intersection of $\mathcal{F}_1$ and $\mathcal{F}_2$ are not the same for both algorithms.

In general, if sets $\mathcal{F}_i$ and $\mathcal{F}_j$ are focused sets with respect to algorithms $A1$ and $A2$, but the intersection of sets $\mathcal{F}_i$ and $\mathcal{F}_j$ is not a focused sets with respect to algorithms $A1$ and $A2$, then the union of $\mathcal{F}_i$ and $\mathcal{F}_j$ is not a focused sets with respect to algorithms $A1$ and $A2$.

9.5. Additional Reading

Many papers related to No Free Lunch theorems have been published. The following additional references are not meant to be exhaustive.

A recent paper by Rowe *et al.* (2009) present a generalization of the No Free Lunch theorem for deterministic non-repeating in set-theoretic terms. This is in constrast to the original work of Wolpert and Macready and well as work by Köppen (2000) and Köppen *et al.* (2001) that use probabilistic proofs to describe No Free Lunch. Auger and Teylaud (2007) shows conditions where No Free Lunch fails to hold for continuous domains using the classical probabilistic interpretation of No Free Lunch. Corne and Knowles (2003) have studied NFL effects for multi-objective optimization.

9.6. Conclusions

Ultimately, No Free Lunch theorems are concerned with how researchers and practitioners use and compare search algorithms. If algorithm $A1$ is better than algorithm $A2$ on a benchmark β, the Sharpened No Free Lunch the theorem tells us that we can construct another set of function β' with the same number of functions where algorithm $A2$ is better than algorithm $A1$.

Or, if we compute the permutation closure of β (denoted by $P(\beta)$), then algorithm $A2$ is equally better than $A1$ on the set $P(\beta)-\beta$ in the aggregate. The problem is that usually the size of β is small and $P(\beta)-\beta$ is enormous and there average behaviors would be different.

Focused No Free Lunch results show that there can exists a *focused set* denoted by $C(\beta)$ such that when $A1$ is better than $A2$ on β, then $A2$ is better than $A1$ on $C(\beta)-\beta$. In some cases, $C(\beta)$ can be quite small.

These results also address some of the concerns raised by Igel and Toussaint (2003) and Streeter (1989) who show many broad classes of problems (for example, consider ONEMAX, MAXSAT, trap functions, or N-K Landscapes) are not closed under permutation. Focused No Free Lunch theorems demonstrate that a subset of algorithms can also display identical performance over a focused set that is not closed under permutation.

Acknowledgments

This research was sponsored by the Air Force Office of Scientific Research, Air Force Materiel Command, USAF, under grant number FA9550-07-1-0403. The U.S. Government is authorized to reproduce and distribute reprints for Governmental purposes notwithstanding any copyright notation thereon. We would also like to thank Dagstuhl for giving us the chance to meet and exchange ideas.

References

Auger, A. and Teylaud, O. (2007). Continuous lunches are free! in *Genetic and Evolutionary Computation Conference (GECCO 2007)* (ACM Press), pp. 916–922.

Caruana, R. and Schaffer, J. (1988). Representation and Hidden Bias: Gray vs. Binary Coding for Genetic Algorithms, in *Proc. of the 5th Int'l. Conf. on Machine Learning* (Morgan Kaufmann), pp. 152–161.

Christensen, S. and Oppacher, F. (2001). What can we learn from No Free Lunch? in *Genetic and Evolutionary Computation Conference (GECCO 2001)* (Morgan Kaufmann), pp. 1219–1226.

Corne, D. and Knowles, J. (2003). No free lunch and free leftover theorems for multi-objective optimization problems, in *Evolutionary Multi-Criterion Optimization (EMO 2003), LNCS*, Vol. 2632 (Springer), pp. 327–341.

Culberson, J. (1998). On the Futility of Blind Search, *Evolutionary Computation* **6**, pp. 109–127.

Droste, S., Janson, T. and Wegener, I. (1999). Perhaps not a free lunch, but at least a free appetizer, in *Genetic and Evolutionary Computation Conference (GECCO 1999)* (Morgan Kaufmann), pp. 833–839.

Droste, S., Jansen, T. and Wegener, I. (2002). Optimization with randomized search heuristics - the (a)nfl theorem, realistic scenarios, and difficult functions, *Theoretical Computer Science* **287**, pp. 131–144.

English, T. (2000). Practical Implications of New Results in Conservation of Optimizer Performance, in *Parallel Problem Solving from Nature VI (PPSN 2000)* (Springer), pp. 69–78.

Igel, C. and Toussaint, M. (2003). On classes of functions for which No Free Lunch results hold, *Information Processing Letters.*

Igel, C. and Toussaint, M. (2004). A No-Free-Lunch Theorem for non-uniform distribution of target functions, *Journal of Mathematical Modelling and Algorithms* **3**, pp. 313–322.

Köppen, M. (2000). Some technical remarks on the proof of the no free lunch theorem, in *Proceedings of the Joint Conference on Information Sciences*, pp. 1020–1024.

Köppen, M., Wolpert, D. and Macready, W. (2001). Remarks on a recent paper on the "no free lunch" theorems, *IEEE Transactions on Evolutionary Computation* **5**, pp. 295–296.

Mathias, K. E. and Whitley, L. D. (1994a). Changing Representations During Search: A Comparative Study of Delta Coding, *Journal of Evolutionary Computation* **2**, pp. 249–278.

Mathias, K. E. and Whitley, L. D. (1994b). Initial Performance Comparisons for the Delta Coding Algorithm, in *IEEE Int'l. Conf. on Evolutionary Computation* (IEEE Service Center), pp. 433–438.

Radcliffe, N. and Surry, P. (1995). Fundamental limitations on search algorithms: Evolutionary computing in perspective, in *Computer Science Today, Lecture Notes in Computer Science*, Vol. 1000 (Springer), pp. 275–291.

Rana, S. and Whitley, D. (1997). Representations, Search and Local Optima, in *Proceedings of the 14th National Conference on Artificial Intelligence AAAI-97* (MIT Press), pp. 497–502.

Rana, S. and Whitley, D. (1998). Search, representation and counting optima, in *Proc. IMA Workshop on Evolutionary Algorithms*, pp. 177–189.

Rowe, J., Vose, M. and Wright, A. (2009). Reinterpreting No Free Lunch, *Evolutionary Computation* **17**, pp. 117–129.

Schumacher, C. (2000). *Fundamental Limitations of Search*, Ph.D. thesis, University of Tennessee, Department of Computer Sciences, Knoxville, TN.

Schumacher, C., Vose, M. and Whitley, D. (2001). The No Free Lunch and Problem Description Length, in *Genetic and Evolutionary Computation Conference (GECCO 2001)* (Morgan Kaufmann), pp. 565–570.

Streeter, M. (1989). Two broad classes of functions for which a No Free Lunch result does not hold, in *Genetic and Evolutionary Computation Conference (GECCO 2003)* (Morgan Kaufmann), pp. 1418–1430.

Tovey, C. A. (1985). Hill climbing and multiple local optima, *SIAM Journal on Algebraic and Discrete Methods* **6**, pp. 384–393.

Whitley, D. (1999). A Free Lunch Proof for Gray versus Binary Encodings, in *Genetic and Evolutionary Computation Conference (GECCO 1999)* (Morgan Kaufmann), pp. 726–733.

Whitley, D. (2000). Functions as Permutations: Regarding No Free Lunch, Walsh Analysis and Summary Statistics, in *Parallel Problem Solving from Nature VI (PPSN 2000)* (Springer), pp. 169–178.

Whitley, D., Mathias, K., Rana, S. and Dzubera, J. (1996). Evaluating Evolutionary Algorithms, *Artificial Intelligence Journal* **85**, pp. 1–32.

Whitley, D., Rana, S. and Heckendorn, R. (1997). Representation Issues in Neighborhood Search and Evolutionary Algorithms, in *Genetic Algorithms and Evolution Strategies in in Engineering and Computer Science* (John Wiley and Sons), pp. 39–57.

Whitley, D. and Rowe, J. (2008). Focused No Free Lunch theorems, in *Genetic and Evolutionary Computation Conference (GECCO 2008)* (ACM Press), pp. 811–818.

Wolpert, D. H. and Macready, W. G. (1995). No free lunch theorems for search, Tech. Rep. SFI-TR-95-02-010, Santa Fe Institute.

Wolpert, D. H. and Macready, W. G. (1997). No free lunch theorems for optimization, *IEEE Transactions on Evolutionary Computation* **4**, pp. 67–82.

Chapter 10

Theory of Evolution Strategies: A New Perspective

Anne Auger and Nikolaus Hansen
TAO Team, INRIA Saclay — Île-de-France, Orsay, France
anne.auger@inria.fr
nikolaus.hansen@inria.fr

Evolution Strategies (ESs) are stochastic optimization algorithms recognized as powerful algorithms for difficult optimization problems in a black-box scenario. Together with other stochastic search algorithms for continuous domain (like Differential Evolution, Estimation of Distribution Algorithms, Particle Swarm Optimization, Simulated Annealing...) they are so-called *global optimization algorithms*, as opposed to gradient-based algorithms usually referred to as local search algorithms. Many theoretical works on stochastic optimization algorithms focus on investigating convergence to the global optimum with probability one, under very mild assumptions on the objective functions. On the other hand, the theory of Evolution Strategies has been restricted for a long time to the so-called *progress rate theory*, analyzing the one-step progress of ESs on unimodal, possibly noisy functions. This chapter covers global convergence results, revealing slow convergence rates on a wide class of functions, and fast convergence results on more restricted function classes. After reviewing the important components of ESs algorithms, we illustrate how global convergence with probability one can be proven easily. We recall two important classes of convergence, namely sub-linear and linear convergence, corresponding to the convergence class of the pure random search and to the optimal convergence class for rank-based algorithms respectively. We review different lower and upper bounds for adaptive ESs, and explain the link between lower bounds and the progress rate theory. In the last part, we focus on recent results on linear convergence of adaptive ESs for the class of spherical and ellipsoidal functions, we explain how almost sure linear convergence can be proven using different laws of large numbers (LLN).

Contents

10.1. Introduction

In this chapter we focus on numerical optimization where the functions to be optimized are mapping $\mathcal{D}$, a subset of the euclidian space $\mathbb{R}^d$ equipped with the euclidian norm $\|.\|$, into $\mathbb{R}$ and without loss of generality we assume minimization. Moreover, we consider derivative free optimization algorithms, i.e., algorithms that do not use derivatives of the function to optimize. Often, those derivatives do not exist or are too costly to evaluate. For instance, the function f can be given by an expensive numerical simulation, a frequent situation in many real-world optimization problems. Such a context is called black-box optimization. The objective function $f : \mathcal{D} \subset \mathbb{R}^d \to \mathbb{R}$ is modeled as a black-box that is able to compute, for a given $x \in \mathbb{R}^d$, the associated objective function value, $f(x)$. The algorithms considered in this chapter are Evolution Strategies (ESs), known as robust and efficient algorithms for black-box optimization without derivative with numerous successful applications to scientific and industrial problems. ESs are just an instance of adaptive stochastic search algorithms where the search space is explored by sampling probe points according to a (continuous) search distribution that can change (be adapted) during the search process.

We assume for the moment that the optimization goal is to approach, with the least search cost, i.e., the least number of function evaluations, a solution x^* such that $f(x^*) \leq f(x)$ for all $x \in \mathcal{D}$ and refer to Section 10.2.1 for a more suitable definition of x^*. The solution x^* will be called global optimum.

The simplest stochastic search algorithm for black-box optimization is the pure random search (PRS), proposed by Brooks (1958) that samples solutions independently with the same probability distribution defined over $\mathcal{D}$ and takes as approximation, or estimate, of the optimal solution, the best solution visited so far, where best refers to the solution having the smallest objective function value. In PRS, the samples are independent and identically distributed, therefore the exploration is "blind"—no feedback from the objective function values observed is taken into account for guiding the next steps. However, in general, the functions to optimize have an underlying structure that can be exploited by optimization algorithms. This fact was soon recognized after the introduction of the PRS in 1958 and research on the algorithmic point of view focused on finding techniques to adapt the search distribution according to the already observed solutions.

10.1.1. *Adaptive Search Algorithms: A Tour d'Horizon*

Adaptive search algorithms refer here to algorithms where the sampling distribution—as opposed to PRS—is adapted along the search. We denote by X_n the best estimate of the (global) minimum after n iterations and by Y_n a *probe point* or *candidate solution.* In PRS, $Y_0, \ldots, Y_n, \ldots$ are independent and identically distributed (i.i.d.) over $\mathcal{D}$, $X_0 = Y_0$ and

$$\mathrm{X}_{n+1} = \begin{cases} \mathrm{Y}_n \text{ if } f(\mathrm{Y}_n) \leq f(\mathrm{X}_n) \ , \\ \mathrm{X}_n \text{ otherwise.} \end{cases} \tag{10.1}$$

The probably most important concept in an *adaptive* algorithm is to probe preferably in the neighborhood of the current solution since for non pathological functions good solutions are usually close to even better ones. If $W_0, \ldots, W_n, \ldots$ are i.i.d. with a probability measure usually centered in zero, independent of the initial estimate X_0, then $\mathrm{X}_n + W_n$ is the new probe point and the update is given by

$$\mathrm{X}_{n+1} = \begin{cases} \mathrm{X}_n + W_n & \text{if } f(\mathrm{X}_n + W_n) \leq f(\mathrm{X}_n) \ , \\ \mathrm{X}_n & \text{otherwise.} \end{cases} \tag{10.2}$$

This algorithm has different names like *local random search* (Devroye and Krzyżak, 2002) or *Markov monotonous search* (Zhigljavsky and Žilinskas, 2008) and will be referred here—following the terminology for evolutionary algorithms—as (1+1)-Evolution Strategy. The "(1+1)" notation comes from the fact that one point X_n (the first "1" in (1+1)) is used to generate another point $X_n + W_n$ (the second "1" in (1+1)) and both are compared to achieve the point X_{n+1} (the best among X_n plus $X_n + W_n$ is kept). In ESs the random vectors $(W_k)_{k \in \mathbb{N}}$ follow typically a multivariate normal distribution with zero mean. The pioneers of Evolution Strategies are I. Rechenberg (Rechenberg, 1973) and H.-P. Schwefel (Schwefel, 1981).

The term *local random search* suggests that a close neighborhood of X_n is explored. The size of this neighborhood is determined by the dispersion (or variance if it exists) of the distribution of W_n which is fixed once for all in the local random search algorithm. However, it seems natural that this dispersion needs to be adapted as well: in the first iterations, exploration of the search space and thus a large dispersion should be preferred and in the last iterations smaller dispersions are needed so as to converge. The question of *how to adapt the dispersion* has been central in the field of stochastic optimization. Consequently, many different methods have been proposed to address this issue and, already in 1971, a survey of different techniques was written by White (1971).

Before to review some important steps in this respect, we set a new framework, restricting the distribution for the new probe point to a spherical multivariate normal distribution. Let $N_0, \ldots, N_n, \ldots$ be i.i.d. multivariate normal random vectors where for each n, each coordinate of N_n follows a standard normal distribution independent of the other coordinates. A new probe point at iteration n is given by $X_n + \sigma_n N_n$, where σ_n is a strictly positive parameter, called step-size. Since the coordinates of N_n are standard normally distributed, the parameter σ_n corresponds to the standard deviation of each coordinate of N_n. The update for a so-called (1+1)-ES with adaptive step-size reads

$$X_{n+1} = \begin{cases} X_n + \sigma_n N_n & \text{if } f(X_n + \sigma_n N_n) \leq f(X_n) \ , \\ X_n & \text{otherwise,} \end{cases} \tag{10.3}$$

for the update of X_n plus an update equation for the step-size σ_n, $(X_0, \sigma_0) \in \mathcal{D} \times \mathbb{R}^+$.

One basic principle behind step-size adaptive algorithms is to try bigger steps if an improvement is observed, and smaller steps if a probe is unsuccessful (Matyas, 1965). Schumer and Steiglitz (1968) propose to maintain a constant probability of success of around 1/5, where probability of success is defined as the probability that the new probe point has an objective function value smaller than the current solution. This idea can also be found in Devroye (1972) and Rechenberg (1973) and is known in the domain of evolutionary algorithms under the name of 1/5th-success rule[a].

Adaptivity of random search algorithms is not limited to a single parameter, like the step-size. The multivariate normal distribution used to sample the new probe point has, besides its mean value, $(d^2 + d)/2$ variation parameters—variances and covariances. These parameters reflect a quadratic model. More recently, a surprisingly effective method has been introduced to adapt all variances and covariances in the *covariance matrix adaptation* (CMA) evolution strategy (Hansen and Ostermeier, 2001). Three main ideas are exploited: (1) the search path (evolution path) over a backward time horizon of about d iterations is pursued and its length and direction are analyzed, (2) new probe points are favored in directions where previously probe points were successful, (3) invariance properties are maintained and the method is in particular invariant under the choice of the coordinate system.

The methods outlined above sample only one new probe point. An important ingredient of *evolutionary algorithms* however is to sample a *population* of probe points. ESs loop over the following steps: (1) sample new solutions from a multivariate normal distribution, (2) evaluate the objective function value of those solutions and (3) adapt the parameters of the multivariate normal distribution (mean vector, step-size and/or covariance matrix) using the observed data. The last step is a crucial step for an ES to converge faster than random search as we will see later.

Following the terminology used for evolutionary algorithms, we might, in the sequel, call *parent* the current search point X_n and *offspring* the new probe points generated from X_n.

10.1.2. *What to Analyze Theoretically?*

Given an optimization algorithm, the first question usually investigated is the one of convergence that can be formulated as: will the algorithm, when time grows to infinity, get arbitrarily close to an optimum of the optimization problem?

[a] A simple implementation can be found in (Kern *et al.*, 2004): the step-size is multiplied by $\alpha > 1$ in case of success and divided by $\alpha^{1/4}$ otherwise. This algorithm will be analyzed in Section 10.4.3.2.

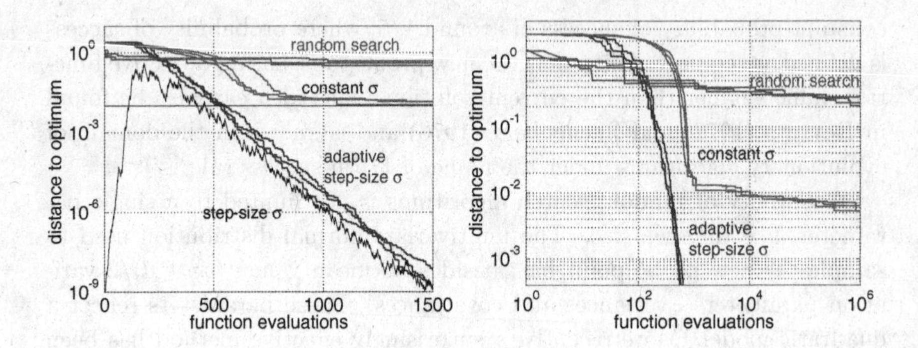

Fig. 10.1. Time evolution of the distance to the global minimum $\|X_n - x^*\|$ of three different (1+1)-algorithms on the function $f : x \mapsto g(\|x - x^*\|)$, where g is strictly monotonically increasing. For each algorithm, three trials on the left and three on the right are shown. On the left, the jagged black line starting at 10^{-9} shows additionally the step-size of one trial with adaptive step-size. The constant horizontal line shows the step-size of the constant step-size algorithm, 10^{-2}. The initial step-size of the adaptive step-size algorithm was intentionally chosen much smaller than 10^{-2} in order to observe the effect of the adaptation. On the right, the trials are shown in a log-log plot and the adaptive step-size algorithm has an initial step-size equal to 10^{-2} to have a fair comparison with the algorithm with constant step-size. From the right plot, we estimate that the algorithm with constant step-size reaches a distance to the optimum of 10^{-3} in at least 10^{12} function evaluations and is therefore more than 1 billion times slower than the adaptive step-size algorithm.

Convergence is illustrated in Figure 10.1. The objective function to be minimized is the sphere function defined as $f(x) = \|x\|$. The global minimum is here unique and equals 0. The optimization goal is thus to approach 0 as fast as possible. Shown are 18 realizations (or runs) from three different algorithms. The y-axis depicts the objective function value of the best solution reached so far (note the log-scale) plotted against the number of objective function evaluations (x-axis) regarded as execution time[b]. The three algorithms converge to the optimum: they approach zero arbitrarily close when the time grows to infinity.

Global convergence can either refer to convergence to a local optimum independently of the starting point or convergence to a global optimum. The former is usually meant in the "deterministic" optimization community, where it is often formulated as convergence to zero of the sequence of

[b]We assume that the search cost (and thus the overall execution time) is mainly due to the objective function evaluations and not to the internal operations of the algorithm. This assumption is reasonable for many problems that are handled with evolutionary algorithms.

gradients associated to the candidate solutions. The latter is usually meant in the "stochastic" optimization community.

Convergence results alone are of little interest from a practical viewpoint because they do not tell us how long it takes to approach a solution. The three algorithms presented in Figure 10.1 all converge with probability one to the optimum. However, we clearly observe that they do not converge at the same speed. The (1+1) with constant step-size is estimated to be 1 billion times slower than the (1+1) with adaptive step-size to reach a distance to the optimum of 10^{-3}.

Therefore, it is important to investigate, together with convergence, the speed of convergence or convergence rate of an algorithm, i.e., to study how fast the algorithm approaches the global minimum. The question of speed of convergence can be tackled in different ways.

(1) We can study the speed at which the distance to x^*, $\|X_n - x^*\|$, decreases to zero.
(2) We can study the *hitting time* τ_ϵ of an ϵ-ball $B_\epsilon(x^*)$ around the optimum, $\tau_\epsilon = \inf\{n : X_n \in B_\epsilon(x^*)\}$ or the hitting time of a sublevel set $\{x|f(x) \leq f(x^*) + \epsilon\}$. Assume $E[\tau_\epsilon]$ is finite, studying the rate of convergence amounts to studying how $E[\tau_\epsilon]$ depends on ϵ when ϵ goes to zero.

Convergence speeds are classified into classes (quadratic convergence, linear convergence, sublinear convergence, ...) that we will define more precisely in Section 10.2.3. For instance, the fastest algorithm depicted in Figure 10.1, left, is in the linear convergence class (the logarithm of the distance to the optimum decreases linearly with the number of function evaluations). Within a class, constants, usually referred to as convergence rates, determine more precisely the speed of convergence.

After investigating for a given class of functions to which convergence class an algorithm belongs to, one is usually interested in estimating the constants (convergence rates). Determining the convergence rate will allow to compare the speed of convergence of two algorithms from the same convergence class. For instance, for the fastest algorithm depicted in Figure 10.1, left, for which the logarithm of the distance to the optimum decreases linearly, the convergence rate corresponds to the coefficient for the linear decrease, i.e., the averaged slope of the three curves represented in Figure 10.1. However, determining the constants analytically is in general much more difficult than finding the convergence class (Nekrutkin and

Tikhomirov, 1993) and requires further assumptions. Related to this last question, one is interested in the scaling of the constants with respect to the dimension of the search space.

An analysis in discrete search spaces is different in several respects. In a discrete space one can consider the hitting time of the optimum, $\tau_{x^*} = \inf\{n : \mathrm{X}_n = x^*\}$. Then, only the scaling of $E[\tau_{x^*}]$ with respect to the size of the search space is investigated. For instance, if the domain is the set of bit-strings of length d, $\mathcal{D} = \{0,1\}^d$, one investigates the scaling of $E[\tau_{x^*}]$ with respect to d.

10.1.3. *Notations and Structure of the Chapter*

We denote $\mathbb{N}$ the set of non-negative integers $\{0, 1, \ldots\}$, denote $\mathbb{R}^+$ the set $(0, +\infty)$, denote $B_\epsilon(x)$ the ball in $\mathbb{R}^d$ of center x and radius ϵ for the euclidian norm, i.e., $B_\epsilon(x) = \{y \in \mathbb{R}^d, \|y - x\| \leq \epsilon\}$ and denote $\mathrm{V}ol(.)$ the volume in $\mathbb{R}^d$ for the Lebesgue measure, i.e., $\mathrm{V}ol(A) = \int_A dx$. The volume of the ball $B_\epsilon(x)$ for all x is, for a fixed dimension d, proportional to ϵ^d, more precisely

$$\mathrm{V}ol(B_\epsilon(x)) = \frac{\pi^{d/2}\epsilon^d}{\Gamma(\frac{d}{2}+1)}, \tag{10.4}$$

where $\Gamma(.)$ is the gamma function. The Borel σ-algebra on $\mathbb{R}^d$ is denoted $\mathcal{B}(\mathbb{R}^d)$. The binary operator $\wedge$ denotes either the logical conjunction or the minimum between two real numbers. Technically, when a is a real number and b a random variable, $b : \Omega \to \mathbb{R}$ then $a \wedge b$ is a random variable such that for each $\omega \in \Omega$, $(a \wedge b)(\omega) = \min\{a, b(\omega)\}$. A vector distributed according to a multivariate normal distribution with zero mean vector and identity as covariance matrix is said to be distributed as $N(0, I_d)$. The set of strictly increasing transformations on $\mathbb{R}$ is denoted $\mathcal{M}$, namely $\mathcal{M} = \{g : \mathbb{R} \to \mathbb{R}, \forall x, y \text{ such that } x < y, g(x) < g(y)\}$.

For a real-valued function $x \mapsto h(x)$, we introduce its positive part $h^+(x) := \max\{0, h(x)\}$ and negative part $h^- = (-h)^+$. In other words $h = h^+ - h^-$ and $|h| = h^+ + h^-$. In the sequel, we denote by e_1 a unitary vector in $\mathbb{R}^d$ and w.l.o.g. $e_1 = (1, 0, \ldots, 0)$.

The organization of the chapter is the following. In Section 10.2 we start by giving a rigorous definition of the optimization goal; define different modes of convergence for sequences of random vectors; define linear convergence and sub-linear convergence; illustrate simple proofs for convergence (without convergence rate) for the (1+1)-ES and explain the invariance to strictly increasing transformations of ESs together with its consequences. In Section 10.3 we analyze in detail the convergence class of the pure random search and review lower and upper bounds of non-adaptive ESs. In Section 10.4, we present tight lower bounds for step-size adaptive ESs and link those results with the progress rate theory. We then explain how linear convergence of adaptive ESs can be proven.

10.2. Preliminary Definitions and Results

In this section we come back on the mathematical formulation of a search problem and define different modes of convergence needed throughout the chapter, as well as important convergence classes for optimization. We illustrate convergence proofs for the simple (1+1)-ES with fixed sample distribution. We also explain the invariance of ESs to order preserving transformations of the objective function and its consequences.

10.2.1. *Mathematical Formulation of the Search Problem*

The goal in numerical optimization is to approximate the global minimum of a real valued function f defined on a subset $\mathcal{D}$ of $\mathbb{R}^d$. Without further assumptions on f, this problem may have no solution. First of all, f may have no minimum in $\mathcal{D}$, take for instance $f(x) = x$ for $x \in \mathcal{D} = (0, 1)$ or $\mathcal{D} = \mathbb{R}$, or the minimum may not be unique as for $f(x) = \sin(x)$ in $\mathcal{D} = \mathbb{R}$. However, even if f admits a unique global minimum on $\mathcal{D}$, this minimum can be impossible to approach in practice: take for instance $f(x) = x^2$ for all $x \in \mathbb{R}\backslash\{1\}$ and $f(1) = -1$. Then the global minimum of f is located in 1, however, it would be impossible to approach it in a black-box scenario. To circumvent this, we take the approach from measure theory considering classes of functions (instead of functions) where two functions belong to the same class if and only if they are equal except on a set of measure zero. Let ν be a measure on $\mathcal{D}$, and $f, g : \mathcal{D} \to \mathbb{R}$, then g belongs to the class of f, denoted $[f]$, if g and f are equal almost everywhere, that is $\nu\{x, f(x) \neq g(x)\} = 0$. The generalization of the minimum of a function to classes of functions is the so-called essential infimum defined for a function f as

$$m_\nu(f) = \operatorname{ess\,inf} f = \sup\{b \in [-\infty, \infty], \nu(\{x \in \mathcal{D} : f(x) < b\}) = 0\}, \tag{10.5}$$

and which is constant for all g in $[f]$. When the context is clear, we write m_ν instead of $m_\nu(f)$ and we will not differentiate between the minimum of the function class $[f]$ and of any $g \in [f]$.

If X is a random variable with probability measure ν, then $\Pr[f(X) < m_\nu] = 0$ and $\Pr[f(X) < m_\nu + \epsilon] > 0$ for all $\epsilon > 0$. The value m_ν depends on ν but will be the same for equivalent measures—measures having the same null sets. We denote in the following with m the essential infimum with respect to the Lebesgue measure on $\mathcal{D}$ or, equivalently, with respect to any probability measure with a strictly positive density everywhere on $\mathcal{D}$.

From now on, we assume that there exists a unique point x^* in the domain closure $\overline{\mathcal{D}}$, such that $\nu\{x \in B_\epsilon(x^*) : f(x) < m + \delta\} > 0$ for all $\epsilon, \delta > 0$. The definition of x^* is independent of the choice of a function in $[f]$. Then, for any $g \in [f]$, the well-defined optimization goal is to approach the unique *global optimum* $x^* \in \overline{\mathcal{D}}$.

With the above definitions we obtain $x^* = 0$ for $f(x) = x$ and $\mathcal{D} = (0, 1)$, and $x^* = 0$ for $f(x) = x^2$ for $x \in \mathbb{R}\backslash\{1\}$ and $f(1) = -1$. However, not all functions admit a global optimum according to our definition, take for instance $f(x) = x$ and $\mathcal{D} = \mathbb{R}$. For solving such a function in practice, one expects an algorithm to generate X_n with $\lim_{n\to\infty} X_n = -\infty$.

10.2.2. *Different Modes of Convergence*

If $(x_n)_{n\in\mathbb{N}}$ is a deterministic sequence of $\mathbb{R}^d$, all possible definitions of convergence are equivalent to the following: x_n converges to x^* if for all $\epsilon > 0$, there exists n_1 such that for all $n \geq n_1 \in \mathbb{N}$, $\|x_n - x^*\| \leq \epsilon$. Notations used are $\lim_{n\to\infty} \|x_n - x^*\| = 0$ or $\lim_{n\to\infty} x_n = x^*$.

However, for a sequence $(\mathrm{X}_n)_{n\in\mathbb{N}}$ of random vectors there exist different modes of convergence inducing different definitions of convergence which are not equivalent.

Definition 10.1 (Almost sure convergence). *The sequence X_n converges to a random variable X almost surely (a.s.) or with probability one if*

$$\Pr[\lim_{n\to\infty} X_n = X] = 1 .$$

Except on an event of probability zero, all single instances of the sequence $(\mathrm{X}_n)_n$ converge to X. Both notations $\lim_{n\to\infty} \|X_n - X\| = 0$ a.s. or $\lim_{n\to\infty} X_n = X$ a.s. will be used, where a.s. stands for almost surely.

A weaker type of convergence is convergence in probability.

Definition 10.2 (Convergence in probability). *The sequence X_n converges in probability towards X if for all $\epsilon > 0$*

$$\lim_{n\to\infty} \Pr\left[\|X_n - X\| \geq \varepsilon\right] = 0.$$

Almost sure convergence implies convergence in probability.

Definition 10.3 (Convergence in p-mean or in L_p). *The sequence X_n converges towards X in p-mean or in L_p for $p \geq 1$ if $E[\|X_n\|^p] < \infty$ and*

$$\lim_{n\to\infty} E[\|X_n - X\|^p] = 0\ .$$

If $p = 1$ we say simply that X_n converges in mean or in expectation towards X.

10.2.3. *Convergence Order of Deterministic Sequences*

Let a deterministic sequence $(z_n)_{n\in\mathbb{N}}$ converge to $z \in \mathbb{R}^d$ with $z_n \neq z$ for all n, and let

$$\lim_{n\to\infty} \frac{\|z_{n+1} - z\|}{\|z_n - z\|^q} = \mu, \text{ with } q \geq 1 \text{ and } \mu \in (0,1)\ . \tag{10.6}$$

Depending on q and μ we define different *orders of convergence.*

Super-linear convergence, if $\mu = 0$ or $q > 1$. If $\mu > 0$, we speak about convergence with order $q > 1$ and about quadratic convergence if $q = 2$.

Linear convergence, if $q = 1$ and $\mu \in (0,1)$, where we have consequently

$$\lim_{n\to\infty} \frac{\|z_{n+1} - z\|}{\|z_n - z\|} = \mu \in (0,1)\ . \tag{10.7}$$

Sub-linear convergence, if $q = 1$ and $\mu = 1$. Furthermore, the sequence $(z_n)_n$ converges sub-linear *with degree* $p > 0$, if

$$\frac{\|z_{n+1} - z\|}{\|z_n - z\|} = 1 - c_n\|z_n - z\|^{1/p} \text{ and } c_n \to c > 0\ . \tag{10.8}$$

The distance to z is reduced in each iteration by a factor that approaches one[c]. For $p = \infty$ and $c < 1$, linear convergence is recovered.

[c]This definition does not allow to characterize the convergence of all sequences converging sub-linearly, for instance $1/(n\log(n))$ converges sub-linearly to zero but lies in between convergence of degree 1 and $1 + \epsilon$ for all $\epsilon > 0$.

Sub-linear convergence with degree p implies that

$$\|z_n - z\| \sim \left(\frac{p}{c}\right)^p \frac{1}{n^p}. \tag{10.9}$$

See Stewart (1995).

10.2.4. *Log- and Sub-linear Convergence of Random Variables*

Since the limit of $\frac{\|z_{n+1}-z\|}{\|z_n-z\|}$ is a random variable, defining convergence like with Equations (10.6) or (10.7) is in general not appropriate. We could use the deterministic sequence $E[\|z_n - z\|]$ instead of $\|z_n - z\|$. However, we rather use a weaker definition of linear convergence implied by Equation (10.7): we say that a sequence of random variables converges log-linearly, if there exits $c < 0$ such that

$$\lim_n \frac{1}{n} \ln \frac{\|z_n - z\|}{\|z_0 - z\|} = c \ . \tag{10.10}$$

The definition of linear convergence in Equation (10.10) is implied by Equation (10.7) since Equations (10.7) is equivalent to $\lim_k \ln(\|z_{k+1} - z\|/\|z_k - z\|) = \ln(\mu)$ and by the Cesàro mean result we obtain $\lim_n \frac{1}{n}\sum_{k=0}^{n-1} \ln(\|z_{k+1} - z\|/\|z_k - z\|) = \ln(\mu)$ which after simplification gives Equation (10.10) where $c = \ln(\mu)$. Equation (10.10) implies also that the logarithm of $\|z_n - z\|$ converges to $-\infty$ like cn, suggesting the name *log-linear convergence.* We will say that log-linear convergence or linear convergence holds almost surely if there exists $c < 0$ such that Equation (10.10) holds almost surely. We will say that log-linear convergence or linear convergence holds in mean or expectation if there exists $c < 0$ such that

$$\lim_n \frac{1}{n} E\left[\ln \frac{\|z_n - z\|}{\|z_0 - z\|}\right] = c \ . \tag{10.11}$$

Log-linear convergence is illustrated in Figure 10.2.

Following Equation (10.9), we will say that a sequence of random variables converges sub-linearly with degree p if $\|z_n - z\|$ converges to zero like $\frac{1}{n^p}$. Sub-linear convergence is illustrated in Figure 10.3.

As explained in Section 10.1.2, speed of convergence and thus linear and sublinear convergence can be alternatively defined with respect to the hitting time of an ϵ-ball around the optimum, $E[\tau_{B_\epsilon}]$. Linear convergence corresponds to $E[\tau_{B_\epsilon}]$ increasing like $K(-\ln(\epsilon)) = K\ln(1/\epsilon)$ where $K > 0$. The constants $-1/K$ and c play the same role and will typically decrease like

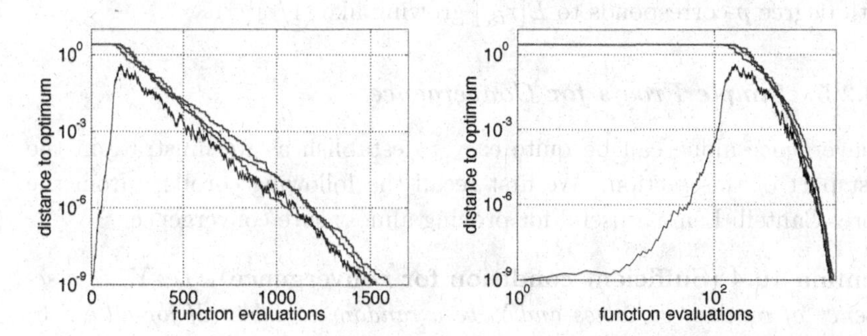

Fig. 10.2. Illustration of the (log)-linear convergence: Time evolution of the distance to the optimum of the (1+1)-ES with one-fifth success rule, three trials respectively, on the function $f : x \in \mathbb{R}^{10} \mapsto g(\|x\|)$, where g is strictly monotonically increasing. The jagged line starting at 10^{-9} shows additionally the step-size of one of the trials with adaptive step-size. The initial step-size has been chosen very small (10^{-9}) compared to the initial search points to show the capability to increase the step-size (during the first 100 evaluations). **Left:** log scale on y-axis: after about 100 evaluations, the logarithm of the distance to the optimum decreases linearly, the slope of the curves corresponds to the convergence rate c defined in Equation (10.10). **Right:** log scale on y-axis and x-axis.

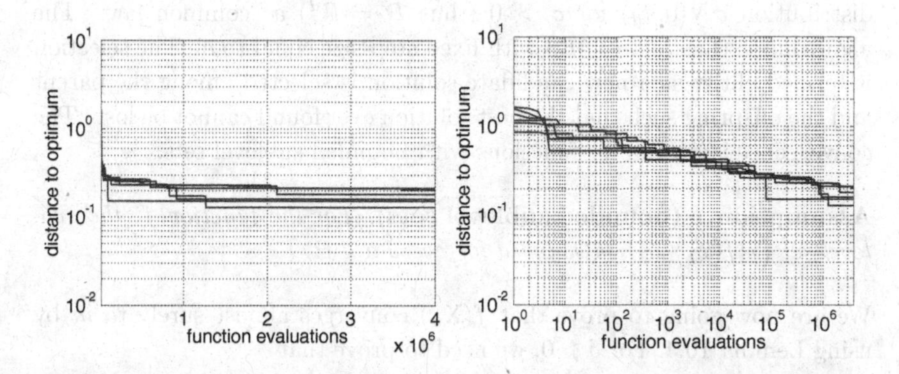

Fig. 10.3. Illustration of sub-linear convergence with degree p: Time evolution of the distance to the optimum from six runs of pure random search in a log-linear plot (left) and a log-log plot (right). Each probe point is sampled uniformly at random in $[-0.2, 0.8]^{10}$ on the function $f : x \mapsto g(\|x\|)$, where g is strictly monotonically increasing. Apart from the scale of the x-axis both plots are identical. In the log-log plot the graphs shape up as linear with only stochastic deviations. According to Theorem 10.8, $p = 1/d$ and here $d = 10$.

$1/d$, i.e., K will grow like d for evolution strategies. Sub-linear convergence with degree p corresponds to $E[\tau_{B_\epsilon}]$ growing like $(1/\epsilon)^{1/p}$.

10.2.5. *Simple Proofs for Convergence*

Convergence alone can be quite easy to establish as we illustrate in the first part of this section. We first recall the following corollary from the Borel-Cantelli Lemma useful for proving almost sure convergence.

Lemma 10.4 (Sufficient condition for convergence). *Let Y_n be a sequence of random variables and Y be a random variable. If for all $\epsilon > 0$, $\sum_n \Pr[|Y_n - Y| > \epsilon] < \infty$, then Y_n converges almost surely to Y.*

Proof. The proof can be found in probability textbooks, see for instance (Karr, 1993, p. 138). □

The previous lemma is implicitly or explicitly used to investigate convergence of evolution strategies in (Baba, 1981; Rudolph, 2001; Greenwood and Zhu, 2001).

In the sequel, we will prove convergence for the local random search given in Equation (10.2) where $(W_k)_{k\in\mathbb{N}}$ are i.i.d. with multivariate normal distribution $\sigma N(0, I_d)$ for $\sigma > 0$ (thus $\mathcal{D} = \mathbb{R}^d$) as common law. The algorithm is thus a (1+1)-ES with fixed step-size equal to σ. The selection is termed elitist as a new candidate solution is selected among the parent and the offspring such that the best solution ever found cannot be lost. The convergence is proven for functions with bounded sublevel sets.

Assumption 1 (bounded sublevel sets). *For the function f, the sublevel set $\{x|f(x) \leq \alpha\}$ is bounded for every $\alpha \in \mathbb{R}$.*

We are now going to prove that $f(\mathrm{X}_n)$ converges almost surely to m by using Lemma 10.4. For $\delta > 0$, we need to prove that

$$\sum_n \Pr[|f(\mathrm{X}_n) - m| > \delta] < \infty \ . \tag{10.12}$$

We denote A_δ the set $\{x \in \mathbb{R}^d, f(x) \leq m+\delta\}$, its complementary A_δ^c satisfies thus $A_\delta^c = \{x \in \mathbb{R}^d, f(x) > m + \delta\}$. Rewriting the condition (10.12), we need to show that

$$\sum_n \Pr[\mathrm{X}_n \in A_\delta^c] < \infty \ . \tag{10.13}$$

Equation (10.13) will follow from (a) a constant lower bound for the probability to hit A_δ at any step and (b) elitist selection that prevents to escape A_δ once it has been hit. First, we sketch the proof idea for (a).

Lemma 10.5. *For any $R > 0$, there exists $\gamma > 0$ such that for all $x \in B_R(0)$, $\Pr[x + W_0 \in A_\delta] \geq \gamma$.*

Proof. We sketch the idea of the proof and leave the technical details to the reader. The definition of m (essential infimum of f) guarantees that $Vol(A_\delta)$ is strictly positive. The probability that $x + W_0$ hits A_δ is given by the integral over A_δ of the density of a multivariate normal distribution centered in x. This integral will be always larger or equal to the integral when x is placed in $B_R(0)$ and at the largest distance from A_δ. The constant γ will be this latter integral. □

The previous lemma together with elitist selection implies the following lemma:

Lemma 10.6. *Let $R > 0$ such that $\{x, f(x) \leq f(X_1)\} \subset B_R(0)$ and γ the constant from Lemma 10.5, then $\Pr[\mathrm{X}_n \in A_\delta^c] \leq (1-\gamma)^n$.*

Proof. Note first that R does exist because of Assumption 1 and that the elitist selection ensures that for all n, $\mathrm{X}_n \in B_R(0)$. Therefore, $\Pr[\mathrm{X}_n \in A_\delta^c] = \Pr[\mathrm{X}_n \in A_\delta^c \cap B_R(0)]$. Because of the elitist selection, once X_n enters A_δ, the next iterates stay in A_δ such that if X_n does not belong to A_δ it means that X_{n-1} did not belong to A_δ and thus $\Pr[\mathrm{X}_n \in A_\delta^c \cap B_R(0)] = \Pr[\mathrm{X}_n \in A_\delta^c \cap B_R(0), X_{n-1} \in A_\delta^c \cap B_R(0)]$. By the Bayes formula for conditional probabilities, we can write $\Pr[\mathrm{X}_n \in A_\delta^c \cap B_R(0), X_{n-1} \in A_\delta^c \cap B_R(0)]$ as $\Pr[\mathrm{X}_n \in A_\delta^c \cap B_R(0) | X_{n-1} \in A_\delta^c \cap B_R(0)]$ times $\Pr[X_{n-1} \in A_\delta^c \cap B_R(0)]$. However, by Lemma 10.5, $\Pr[\mathrm{X}_n \in A_\delta^c \cap B_R(0) | X_{n-1} \in A_\delta^c \cap B_R(0)] \leq 1 - \gamma$ such that we have now that

$$\Pr[\mathrm{X}_n \in A_\delta^c \cap B_R(0)] \leq (1-\gamma) \Pr[X_{n-1} \in A_\delta^c \cap B_R(0)]$$

and then by induction we obtain that $\Pr[\mathrm{X}_n \in A_\delta^c] = \Pr[\mathrm{X}_n \in A_\delta^c \cap B_R(0)] \leq (1-\gamma)^n$. □

A direct consequence of Lemma 10.6 is the convergence of $f(\mathrm{X}_n)$ in probability since we have that $\Pr[\mathrm{X}_n \in A_\delta^c]$ converges to zero when n grows to infinity. However, using now Lemma 10.6, we can also obtain the stronger convergence, that is almost sure convergence. Indeed, we deduce from

Lemma 10.6 that

$$\sum_{n=1}^{\infty} \Pr[X_n \in A_\delta^c] \leq \sum_{n=1}^{\infty} (1-\gamma)^n = \frac{1}{\gamma}$$

and thus applying Lemma 10.4, the almost sure convergence of $f(X_n)$ to m holds. In conclusion, we have proven the following theorem:

Theorem 10.7 (Almost sure convergence of (1+1)-ES). *For f satisfying Assumption 1, the (1+1)-ES with constant step-size $\sigma > 0$ converges almost surely to m the essential infimum of f in the sense that*

$$f(X_n) \to m \;\; a.s.$$

The techniques illustrated in this section can be also used for disproving convergence. Almost sure convergence implies convergence in probability such that a necessary condition for convergence with probability one is

$$\Pr[X_n \in A_\delta^c] \to 0 \; .$$

Rudolph is using this fact to disprove convergence with probability one of the (1+1)-ES with one-fifth success rule on a multi-modal function (Rudolph, 2001).

Over a compact set and without further assumptions on f, convergence towards x^* holds if $\sigma_n \sqrt{\ln(n)} \to \infty$ (Devroye and Krzyżak, 2002), i.e., if σ_n decreases not too fast, global convergence is preserved. We will see later that for step-size adaptive ESs, the step-size σ_n converges much faster to 0, more precisely $\lim_{n\to\infty} \sigma_n \alpha^n < \infty$ for some $\alpha > 1$.

10.2.6. *Invariance to Order Preserving Transformations*

One important aspect of evolution strategies is their invariance with respect to monotonically increasing transformations of the objective function. Remind that $\mathcal{M}$ denotes the set of strictly increasing transformations on $\mathbb{R}$. For an ES, optimizing f or $g \circ f$ for $g \in \mathcal{M}$ is exactly the same, in that the exact same sequence $(X_n)_{n\in\mathbb{N}}$ is constructed when optimizing f or $g \circ f$, provided the independent random numbers needed to sample the probe points are the same. This is due to the fact that all the updates are *only* based on the ranking of new solutions (which is preserved if $g \circ f$ is considered instead of f) and not on their absolute objective function value. This property is not true for all stochastic search algorithms, in particular not for the simulated annealing algorithm where the selection of a new probe depends on the fitness difference between the current solution and the probe point or for genetic algorithms with fitness proportionate selection.

Invariance to strictly increasing transformations is illustrated in Figure 10.4 where the function $f : x \mapsto \|x\|^2$ is plotted (in dimension 1 for the sake of illustration) on the left together with two functions $x \mapsto g(f(x))$ for $g \in \mathcal{M}$ (middle and right). Though the left function seems to be easier to optimize (the function being convex and quadratic), the behavior of ESs on the three functions will be identical, in the sense that the sequences $(\mathrm{X}_n)_n$ will be indistinguishable.

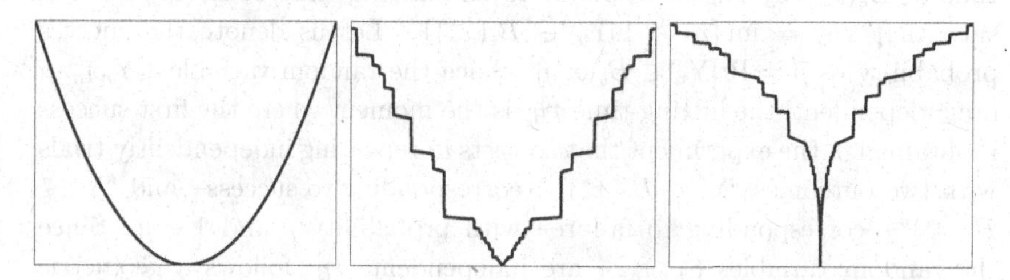

Fig. 10.4. **Left:** sphere function $f : x \mapsto \|x\|^2$. **Middle and right:** $x \mapsto g(f(x))$ for two different $g \in \mathcal{M}$ and $d = 1$. ESs are invariant to $\mathcal{M}$ in the sense that $(\mathrm{X}_n)_n$ generated when optimizing $g \circ f$ for any $g \in \mathcal{M}$ will be the same.

10.3. Rate of Convergence of Non-adaptive Algorithms

In this section, *non-adaptive* means that dispersion parameters of the sampling distribution, like its width and shape, remain constant, however, the sampling distribution can be shifted during the search. The local random search algorithm described in Equation (10.2) is non-adaptive in this sense. The (1+1)-ES described in Equation (10.3) is adaptive, in case σ_n changes over time. Rates of convergence are investigated for the pure random search first and for local random search algorithms afterwards.

10.3.1. *Pure Random Search*

We investigate the pure random search from Equation (10.1). We denote μ the probability measure of the random vectors $(\mathrm{Y}_n)_{n\in\mathbb{N}}$. The probability measure μ characterizes the probability to hit (Borel) sets of $\mathcal{D}$: for a set A included in $\mathcal{D}$, $\Pr[\mathrm{Y}_n \in A] = \mu(A)$. We start by looking at a simple case:

Theorem 10.8 (Case of uniform sampling on the unit cube). *Let f be the spherical function $f(x) = g(\|x - x^*\|)$ where $g \in \mathcal{M}$ and with $x^* \in \mathcal{D} =]0,1[^d$. Let μ be the uniform distribution on $\mathcal{D}$, then for ϵ such that $B_\epsilon(x^*) \subset \mathcal{D}$, the first hitting time of $B_\epsilon(x^*)$ by X_n, i.e., $\tau_{B_\epsilon} = \inf\{n \geq 1 \,|X_n \in B_\epsilon(x^*)\}$ satisfies*

$$E[\tau_{B_\epsilon}] = \frac{\Gamma(\frac{d}{2}+1)}{\pi^{d/2}} \frac{1}{\epsilon^d}. \tag{10.14}$$

Proof. Since the objective function is the sphere function, the hitting time of $B_\epsilon(x^*)$ by X_n corresponds to the hitting time of $B_\epsilon(x^*)$ by Y_n, such that $\tau_{B_\epsilon} = \inf\{n \geq 1 | \mathrm{Y}_n \in B_\epsilon(x^*)\}$. Let us denote the success probability as $p = \Pr[\mathrm{Y}_n \in B_\epsilon(x^*)]$. Since the random variables $(\mathrm{Y}_n)_{n\in\mathbb{N}}$ are independent, the hitting time τ_{B_ϵ} is the moment where the first success is obtained in the experiment that consists in repeating independently trials with two outcomes "$\mathrm{Y}_n \in B_\epsilon(x^*)$"—corresponding to success—and "$\mathrm{Y}_n \notin B_\epsilon(x^*)$"—corresponding to failure—with probability p and $1-p$. Since the random variables $(\mathrm{Y}_n)_{n\in\mathbb{N}}$ are independent, τ_{B_ϵ} follows a geometric distribution with parameter p and $E[\tau_{B_\epsilon}] = 1/p$.

Since μ is the uniform distribution on $\mathcal{D}$ and $B_\epsilon(x^*) \subset \mathcal{D}$, the probability $p = \Pr[\mathrm{Y}_n \in B_\epsilon(x^*)] = \frac{\mathrm{V}ol(B_\epsilon(x^*))}{\mathrm{V}ol(\mathcal{D})}$. Since $\mathrm{V}ol(\mathcal{D}) = 1$, with Equation (10.4) we obtain the result. □

The arguments used in Theorem 10.8 transfer to more general objective functions and sampling distributions if we consider the hitting time of the set $A_\delta = \{x, f(x) \leq m + \delta\}$, i.e.,

$$\tau_{A_\delta} = \inf\{n, \mathrm{X}_n \in A_\delta\} = \inf\{n, f(\mathrm{X}_n) \leq m + \delta\} \ . \tag{10.15}$$

We now assume that the search domain $\mathcal{D}$ is bounded and the sampling distribution μ admits a density $p_\mu(x)$ with respect to the Lebesgue measure that satisfies the following assumption.

Assumption 2 (Bounded sample distribution). *There exist $c_2, c_1 > 0$ such that $c_1 \leq p_\mu(x) \leq c_2$ for all $x \in \mathcal{D}$.*

In order to derive the dependence in ϵ and d of the expected hitting time, we also need to make an assumption on the objective function:

Assumption 3 (Bounded volume of A_δ). *There exists $\delta_0 > 0$ such that for all $\delta > 0$ and $\delta \leq \delta_0$, there exists two constants K_1 and K_2 such that*

$$K_1 \epsilon^d \leq \mathrm{V}ol(A_\delta) \leq K_2 \epsilon^d \ .$$

We can now state the following theorem for PRS:

Theorem 10.9. *Let f satisfy A 3 and the sampling distribution of PRS satisfy A 2, then $E[\tau_{A_\delta}] = \Theta(1/\epsilon^d)$. Specifically, the expected hitting time of A_δ satisfies*

$$\frac{1}{K_2 c_2} \frac{1}{\epsilon^d} \leq E[\tau_{A_\delta}] \leq \frac{1}{K_1 c_1} \frac{1}{\epsilon^d} . \tag{10.16}$$

Proof. The hitting time τ_{A_δ} defined in Equation (10.15) can be expressed as the hitting time of the variable Y_n, i.e., $\tau_{A_\delta} = \inf\{n, \mathrm{Y}_n \in A_\delta\}$. Note that this would not have been the case if we would have considered the hitting time of $B_\epsilon(x^*)$. The same arguments used in Theorem 10.8 hold now here: at each iteration, Y_n hits A_δ with probability $p = \Pr[\mathrm{Y}_n \in A_\delta]$. Because the Y_n are independent, the expectation of τ_{A_δ} equals $1/p$. It remains now to estimate p. Since $p = \int_{A_\delta} p_\mu(x)dx$, using A 2, we have that $c_1 \int_{A_\delta} dx \leq p \leq c_2 \int_{A_\delta} dx$, in other words $c_1 \mathrm{V}ol(A_\delta) \leq p \leq c_2 \mathrm{V}ol(A_\delta)$. Using the bounds of $\mathrm{V}ol(A_\delta)$ from A 3 we obtain the result. □

Both theorems state sub-linear convergence with degree $1/d$ of the pure random search algorithm which is illustrated for the spherical function in Figure 10.3.

10.3.2. *Lower and Upper Bounds for Local Random Search*

The local random search defined via Equation (10.2) includes a (1+1)-ES with constant step-size and is a particular case of the so-called Markov monotonous search algorithms.

10.3.2.1. *A Detour Through Markov Monotonous Search (Zhigljavsky and Zilinskas, 2008)*

Definition 10.10 (Markov montonous search). *A Markov chain sequence $(\mathrm{X}_n)_{n\in\mathbb{N}}$ that moreover satisfies $f(\mathrm{X}_{n+1}) \leq f(\mathrm{X}_n)$ with probability one is called a Markov monotonous search sequence.*

Informally, a sequence $(\mathrm{X}_n)_n$ is a Markov chain if the value of the n-th variable depends on the past variables only through the immediate predecessor. A transition kernel can be associated to a Markov chain:

Definition 10.11 (Transition kernel). *For a Markov chain $(\mathrm{X}_n)_n$ the transition kernels are the collection of $P_n(.,.) : \mathbb{R}^d \times \mathcal{B}(\mathbb{R}^d) \to [0,1]$ where for each n, for each $A \in \mathcal{B}(\mathbb{R}^d)$, $P_n(.,A)$ is a measurable mapping and for all*

$x \in \mathbb{R}^d$, $P_n(x,.)$ is a probability measure that characterizes the distribution of the sequence: $P_n(x,A) = \Pr[\mathrm{X}_{n+1} \in A | \mathrm{X}_n = x]$ for all x, A.

When P_n is independent of n, the Markov chain is homogeneous and thus the local random search is an *homogeneous* Markov monotonous search algorithm. However, if the random variable W_n is, for example, scaled using a cooling schedule, say $1/(n+1)$, i.e., a new probe point is $\mathrm{X}_n + \frac{1}{n+1} W_n$ the algorithm is not homogeneous.

Let Q be the transition kernel $Q(x,A) = \Pr[x + W_0 \in A]$ representing the probability that $x + W_0$ belongs to A. The transition kernel for the local random search can be written as

$$P(x,A) = \delta_x(A) Q(x, B_f^c(x)) + Q(x, A \cap B_f(x)) \,, \tag{10.17}$$

where $B_f(x) = \{y \in \mathbb{R}^d, f(y) \leq f(x)\}$ and $B_f^c(x)$ denote its complement and $\delta_x(.)$ is the probability measure concentrated at x, i.e., $\delta_x(A) = 1$ if $x \in A$ and 0 otherwise.

10.3.2.2. *Upper Bounds for Convergence Rates of Certain Homogeneous Markov Monotonous Search*

The question of how to choose the distribution $Q(x,.)$ to converge "fast" has been addressed by Nekrutkin and Tikhomirov (1993) and Tikhomirov (2006). The main result is that for an appropriate choice of $Q(x,.)$ one can upper bound the expected hitting time of M_ϵ defined as

$$M_\epsilon = \{y \in B_\epsilon(x^*) : f(y) < f(z) \text{ for every } z \in B_\epsilon^c(x^*)\}$$

by $O(\ln^2(1/\epsilon))$ under mild assumptions on the objective function. The result holds for a fixed dimension and thus the constant hidden in the O notation depends on d. We now prove that the density associated to the distribution $Q(x,.)$ needs to have a singularity in 0.

Theorem 10.12. *The density associated to the distribution $Q(x,.)$ needs to have a singularity in zero.*

Proof. If the density of the sampling distribution is upper bounded, non-adaptive algorithms cannot be faster than random search, because there exists a PRS with uniform density larger than the upper bound over M_ϵ (for small enough ϵ). This PRS has in each iteration a larger probability for hitting M_ϵ and consequently a smaller expected hitting time. Therefore, if the density is upper bounded, the upper bound of the expected hitting time cannot be essentially faster than $1/\epsilon^d$, i.e., cannot belong to $o(1/\epsilon^d)$. □

The results by Tikhomirov (2006) have been established in the context of homogeneous Markov symmetric search where Q admits a density $q(x, y)$ that can be written as $q(x, y) = g(\|x - y\|)$ where g is a non-increasing, non-negative left-continuous function defined on $\mathbb{R}^+$ and normalized such that $q(0, .)$ is a density. More precisely for a fixed precision ϵ, Tikhomirov (2006) proves that there exists a function g_ϵ that depends on the precision ϵ such that for the associated Markov Monotonous search algorithm, the hitting time of M_ϵ satisfies on average $E_x[\tau_\epsilon] \leq O(\ln^2(1/\epsilon))$. A slightly worse upper bound with a function g independent of the precision ϵ can be obtained (Zhigljavsky, 1991).

10.4. Rate of Convergence of Adaptive ESs

We focus now on the rate of convergence of *adaptive* ESs where the dispersion of the sampling distribution is adapted during the course of the search process. Various results show that adaptive ESs cannot converge faster than linear, with a convergence rate decreasing like $1/d$ when d goes to infinity. This has been in particular proven by Nekrutkin and Tikhomirov (1993) in the context of Markov Monotonous search (i.e., for a (1+1)-ES) without showing the dependency in $1/d$ for the convergence rate though, by Jägersküpper (2008) for (1+λ)-ES with isotropic sampling distributions and by Teytaud and Gelly (2006) for general rank-based algorithms.

10.4.1. *Tight Lower Bounds for Adaptive ESs*

In this section, we establish *tight constants for the lower bounds* associated to the (1+1)-ES with adaptive step-size defined in Equation (10.3) and explain how the results generalize to the (1,λ)-ES with adaptive step-size. The parent and step-size of an adaptive (1+1) or (1,λ)-ES are denoted (X_n, σ_n). Both algorithms sample new points by adding to X_n random vectors following a *spherical multivariate normal distribution* scaled by σ_n. Whereas a single new probe point is created for the (1+1)-ES, λ *independent* probe points are created in the case of the (1,λ)-ES where the best among the λ points becomes subsequently the new parent.

We start by defining a specific artificial step-size adaptation rule where the step-size is proportional to the distance to the optimum of the function to optimize. This algorithm is essentially relevant for spherical functions as we will illustrate, however it can be defined for an arbitrary function f:

Definition 10.13 (Scale-invariant step-size). *Let x^* be the optimum of a function f to optimize and (X_n, σ_n) be the parent and step-size at iteration n of a (1,λ) or (1+1)-ES. If $\sigma_n = \sigma \|X_n - x^*\|$ for $\sigma > 0$, the step-size is called scale-invariant.*

We now define the expected log-progress towards a solution x^*.

Definition 10.14 (Expected log-progress). *Let (X_n, σ_n) be the parent and step-size of an adaptive ES, we define the expected conditional log-progress towards a solution x^* at iteration n as*

$$\varphi_{\ln}(X_n, \sigma_n) := E\left[\ln \frac{\|X_n - x^*\|}{\|X_{n+1} - x^*\|} \Big| X_n, \sigma_n\right] . \qquad (10.18)$$

The previous definition implicitly assumes that (X_n, σ_n) is a Markov Chain but can be adapted to more general settings by replacing the conditioning with respect to (X_n, σ_n) by conditioning with respect to the past in Equation (10.18). In the next lemma, we define the function $F_{(1+1)}$. We will then give its interpretation in terms of expected log-progress.

Lemma 10.15 (Jebalia *et al.* (2008)). *Let $\mathcal{N}$ be a random vector of distribution $N(0, I_d)$. The map $F_{(1+1)} : [0, +\infty] \to [0, +\infty]$ defined by $F_{(1+1)}(\sigma) := E\left[\ln^-(\|e_1 + \sigma\mathcal{N}\|)\right]$, $F_{(1+1)}(+\infty) := 0$ and that can be written as*

$$F_{(1+1)}(\sigma) = \frac{1}{(2\pi)^{d/2}} \int_{\mathbb{R}^d} \ln^- \|e_1 + \sigma x\| e^{-\frac{\|x\|^2}{2}} dx\,, \qquad (10.19)$$

otherwise, is continuous on $[0, +\infty]$ (endowed with the usual compact topology), finite valued and strictly positive on $]0, \infty[$.

We will now prove that for $\sigma \in [0, +\infty[$, $F_{(1+1)}(\sigma)$ is equal to (i) the expected log-progress achieved on the spherical functions $g(\|x\|), g \in \mathcal{M}$ by the (1+1)-ES starting from $e_1 = (1, 0, \ldots)$ and with step-size σ; (ii) the expected log-progress of the (1+1)-ES with scale-invariant step-size. We formalize and prove those results in the next lemma.

Lemma 10.16. *Let $\sigma \in [0, +\infty[$, on the class of spherical functions $f(x) = g(\|x\|), g \in \mathcal{M}$, $F_{(1+1)}(\sigma)$ coincides with (i) the expected log-progress of a (1+1)-ES starting from e_1 and with step-size σ, i.e.,*

$$E\left[\ln \frac{\|X_n\|}{\|X_n + \sigma_n \mathcal{N}\| \wedge 1} \Big| X_n = e_1, \sigma_n = \sigma\right] = F_{(1+1)}(\sigma) \ ,$$

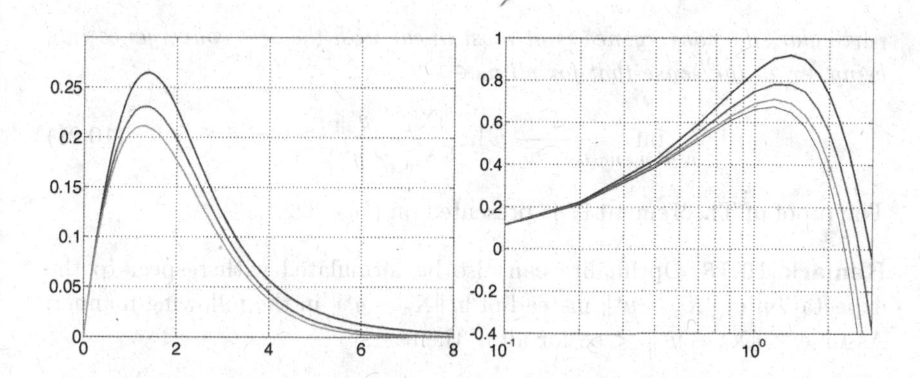

Fig. 10.5. **Left:** Plot of $\sigma \mapsto dF_{(1+1)}(\sigma/d)$ where $F_{(1+1)}$ is defined in Equation (10.19) for $d = 5, 10, 30$ (top to bottom). **Right:** Plot of $\sigma \mapsto dF_{(1,5)}(\sigma/d)$ where $F_{(1,5)}$ is defined in Equation (10.25) for $d = 5, 10, 30$, top to bottom. The lowest line is the limit of $\sigma \mapsto dF_{(1,5)}(\sigma/d)$ for d to infinity given in Equation (10.26).

(ii) the expected log-progress of the (1+1)-ES with scale-invariant step-size ($\sigma_n = \sigma\|\mathrm{X}_n\|$) at any iteration n, i.e., for all $n \in \mathbb{N}$

$$\varphi_{\ln}(\mathrm{X}_n, \sigma_n) = F_{(1+1)}(\sigma) .$$

Proof. Starting from $X_n = e_1$, a new search point sampled with a step-size σ denoted $e_1 + \sigma\mathcal{N}$ is accepted if its objective function $g(\|e_1 + \sigma\mathcal{N}\|)$ is smaller than the objective function of e_1, which is equal to $g(1)$. Thus, $\|\mathrm{X}_{n+1}\|$ is the minimum between $g(\|e_1 + \sigma\mathcal{N}\|)$ and $g(1)$. The expected log-progress will therefore be equal to $\ln\|e_1\| - E\left[\ln\left(\|e_1 + \sigma\mathcal{N}\| \wedge 1\right)\right]$ which simplifies to $E[\ln^- \|e_1 + \sigma\mathcal{N}\|]$ with $\ln^-(x) := \max(0, -\ln(x))$. The second point (ii) will be proven together with Theorem 10.17 (see Equation (10.43)). □

Plots of the function $F_{(1+1)}$ for different dimensions are given in Figure 10.5. For a given dimension d, *minus the maximum of the function $F_{(1+1)}$ is a lower bound for the convergence rate of step-size adaptive (1+1)-ESs* as stated in the following theorem.

Theorem 10.17 (Convergence at most linear). *Let $(\mathrm{X}_n, \sigma_n)_n$ be generated by a (1+1)-ES with adaptive step-size, on any function f. Let y^* be any vector in $\mathbb{R}^d$ and $E[|\ln\|\mathrm{X}_n - y^*\||] < +\infty$ for all n. Then*

$$E\left[\ln\|\mathrm{X}_{n+1} - y^*\|\right] \geq E\left[\ln\|\mathrm{X}_n - y^*\|\right] - \tau , \tag{10.20}$$

where τ is a strictly positive constant defined as $\tau = \sup F_{(1+1)}([0, +\infty])$ where $F_{(1+1)}$ is the real valued function defined in Equation (10.19). In *particular, the convergence is at most linear with the best convergence rate being $-\tau$ in the sense that for all $n_0 \in \mathbb{N}$*

$$\inf_{n\in\mathbb{N},n>n_0} \frac{1}{n-n_0} E \ln \frac{\|X_n - y^*\|}{\|X_{n_0} - y^*\|} \geq -\tau \ . \tag{10.21}$$

The proof of Theorem 10.17 is presented on page 322.

Remark 10.18. Optimality can also be formulated with respect to the expectation of $\|X_n - y^*\|$ instead of $\ln \|X_n - y^*\|$ in the following manner: Assume $E[\|X_n - y^*\|] < \infty$ for all n, then

$$E\|X_{n+1} - y^*\| \geq E[\|X_n - y^*\|] \ \tau' \ , \tag{10.22}$$

where $\tau' = \min \tilde{F}_{(1+1)}([0, +\infty])$ with $\tilde{F}_{(1+1)}(\sigma) = E[\|e_1 + \sigma\mathcal{N}\| \wedge 1]$.

The two formulations are not equivalent. The constants $-\tau$ and $\ln(\tau')$ play the same role but are not equal due to Jensen's inequality that implies for all σ

$$-F_{(1+1)}(\sigma) = E[\ln(\|e_1 + \sigma\mathcal{N}\| \wedge 1)] < \ln E[\|e_1 + \sigma\mathcal{N}\| \wedge 1] = \ln(\tilde{F}_{(1+1)}(\sigma)) \ .$$

The formulation with the logarithm inside the expectation is compatible with almost sure convergence (Auger and Hansen, 2006).

We will now prove that the lower bound given in Equation (10.21) is reached on spherical functions by the (1+1)-ES with scale-invariant step-size and an appropriate choice of σ. However before to state this result we formulate the linear convergence in expectation of the (1+1)-ES with scale-invariant step-size.

Proposition 10.19. *On spherical functions, $f(x) = g(\|x - x^*\|), g \in \mathcal{M}$, the (1+1)-ES with scale-invariant step-size ($\sigma_n = \sigma\|X_n - x^*\|$) converges linearly in expectation, moreover for all $n_0 \in \mathbb{N}$ and for all $n > n_0$*

$$\frac{1}{n-n_0} E\left[\ln \frac{\|X_n - x^*\|}{\|X_{n_0} - x^*\|}\right] = -F_{(1+1)}(\sigma) \ . \tag{10.23}$$

The proof of Proposition 10.19 is presented on page 323. As a consequence, lower bounds are reached for the (1+1)-ES with scale-invariant step-size and σ chosen to maximize $F_{(1+1)}$.

Corollary 10.20 (Lower bound reached for ES with scale-invariant step-size). *The lower bound in Equation (10.20) is reached on spherical functions $f(x) = g(\|x - x^*\|)$ with $g \in \mathcal{M}$ for the scale-invariant* *step-size rule where at each $n \in \mathbb{N}$, $\sigma_n = \sigma_{(1+1)} \|X_n - x^*\|$ with $\sigma_{(1+1)} > 0$ such that $F_{(1+1)}(\sigma_{(1+1)}) = \tau$. Moreover, for all $n_0 \in \mathbb{N}$ and for all $n > n_0$*

$$\frac{1}{n - n_0} E\left[\ln \frac{\|X_n - x^*\|}{\|X_{n_0} - x^*\|}\right] = -\tau \ . \tag{10.24}$$

The proof is presented together with the proof of Proposition 10.19 on page 323.

Equations (10.23) and (10.24) imply linear convergence in expectation as defined in Equation (10.11). However it does *not only* hold *asymptotically* but also for any *finite* n and n_0. We will prove in addition later that almost sure linear convergence also holds.

The constant $-\tau$ corresponds thus to the convergence rate on spherical functions of the (1+1)-ES with scale-invariant step-size where at each iteration n, $\sigma_n = \sigma_{(1+1)}\|X_n - x^*\|$. Because the convergence rate is reached, the lower bound $-\tau$ is tight. Those results were presented in (Jebalia *et al.*, 2008). They hold for the (1+1)-ES, but the same analysis can be applied to a (1,λ)-ES, resulting in optimality of the scale-invariant step-size (1,λ)-ES where $\sigma_{(1,\lambda)}$ realizes the maximum of $F_{(1,\lambda)}$ defined as the expected log-progress of a (1,λ)-ES

$$F_{(1,\lambda)}(\sigma) = -E\left[\ln \min_{1 \leq i \leq \lambda} \|e_1 + \sigma \mathcal{N}_i\|\right] , \tag{10.25}$$

where $\mathcal{N}_i$ are λ independent random vectors distributed as $N(0, I_d)$ (Auger and Hansen, 2006).

10.4.2. *Link with Progress Rate Theory*

Developments of ESs are closely related to the so-called progress-rate theory (Rechenberg, 1973) that constitutes the main core of the book called "The Theory of Evolution Strategies" (Beyer, 2001). In this section we explain the progress rate approach and its connexions with the convergence of scale-invariant step-size ESs.

The progress rate is defined as a one-step expected progress towards the optimal solution. Assuming w.l.o.g. that the optimum is $x^* = 0$, we can define the normalized expected progress φ^* as

$$\varphi^* = d\, E\left[\frac{\|X_n\| - \|X_{n+1}\|}{\|X_n\|}\right] = d\left(1 - E\left[\frac{\|X_{n+1}\|}{\|X_n\|}\right]\right) .$$

The normalized log-progress can also be considered

$$\varphi^*_{\ln} = d\left(E\left[\ln \frac{\|X_n\|}{\|X_{n+1}\|}\right]\right) .$$

It coincides with the expectation of Equation (10.18) times d. For an ES with a spherical search distribution and on the sphere function, we define additionally

$$\sigma^* = d\sigma_n / \|X_n\| \ .$$

As we will see in Section 10.4.3.2, the sequence $(\sigma_n / \|X_n\|)_n$ is an homogeneous Markov chain. To take out the dependency of σ^* in n, it is in addition assumed that σ^* *is constant*, and thus that the step-size is scale-invariant with $\sigma_n = \sigma^* \|X_n\| / d$. Consequently, the normalized progress φ^* and $\varphi^*_{\ln}$ are functions of σ^* that are independent of n (the proof of this fact is similar to the proof of Lemma 10.16) and of further initial values. Moreover $\varphi^*_{\ln}$ *equals the convergence rate of ESs with scale-invariant step-size multiplied by d*, for example for the (1,λ), for all σ^*

$$dF_{(1,\lambda)}(\sigma^*/d) = \varphi^*_{\ln}(\sigma^*) \ ,$$

or see Proposition 10.19 for the case of the (1+1)-ES. The function $\varphi^*_{\ln}$ as a function of σ^* was plotted in Figure 10.5. Progress rate derivations are in general asymptotic for d to infinity so as to provide comprehensive quantitative estimates for convergence rates. In general, in the limit for $d \to \infty$, φ^* and $\varphi^*_{\ln}$ coincide (Auger and Hansen, 2006). As an example of a simple asymptotic formula, we give the asymptotic progress φ^* (or $\varphi^*_{\ln}$) on the sphere function for the (1,λ)-ES,

$$\lim_{d \to \infty} \varphi^*(\sigma^*) = c_{1,\lambda}\sigma^* - \frac{\sigma^{*2}}{2} \ , \tag{10.26}$$

where $c_{1,\lambda}$ is the expected value of the maximum of λ standard normal distributions and usually is in the order of one.

10.4.3. *Linear Convergence of Adaptive ESs*

We have seen that convergence of adaptive ESs is at most linear and have proven, on spherical functions, the linear convergence *in expectation* of the artificial scale-invariant step-size (1+1)-ES where the step-size is scaled to the distance to the optimum. In this section, we explain how the linear convergence of real adaptation schemes can be analyzed. We assume that f is a spherical function, $f(x) = g(\|x\|)$ for g in $\mathcal{M}$. We will first present the proof of *almost sure* convergence of the (1+1)-ES with scale-invariant step-size and will illustrate afterwards that the convergence of real step-size adaptation schemes is the natural extension of this result.

10.4.3.1. *Almost Sure Linear Convergence of the (1+1)-ES Scale-invariant Step-size*

We remind that for the (1+1)-ES with scale-invariant step-size on $g(\|x\|), g \in \mathcal{M}$, for each $n \in \mathbb{N}$ we have $\sigma_n = \sigma\|\mathrm{X}_n\|$ with $\sigma \geq 0$. A new probe point $\mathrm{X}_n + \sigma\|\mathrm{X}_n\|N_n$ is accepted if $\|\mathrm{X}_n + \sigma\|\mathrm{X}_n\|N_n\| \leq \|\mathrm{X}_n\|$ or normalizing by $\|\mathrm{X}_n\|$ if $\left\|\frac{\mathrm{X}_n}{\|\mathrm{X}_n\|} + \sigma N_n\right\| \leq 1$. Therefore $\|\mathrm{X}_{n+1}\|/\|\mathrm{X}_n\|$ satisfies

$$\frac{\|\mathrm{X}_{n+1}\|}{\|\mathrm{X}_n\|} = \left\|\frac{\mathrm{X}_n}{\|\mathrm{X}_n\|} + \sigma N_n 1_{\{\|\frac{\mathrm{X}_n}{\|\mathrm{X}_n\|} + \sigma N_n\| \leq 1\}}\right\|, \tag{10.27}$$

where $1_{\{\|\frac{\mathrm{X}_n}{\|\mathrm{X}_n\|} + \sigma N_n\| \leq 1\}} = 1$ if $\left\|\frac{\mathrm{X}_n}{\|\mathrm{X}_n\|} + \sigma N_n\right\| \leq 1$ and zero otherwise.

We connect now linear convergence and law of large numbers by stating the following technical lemma.

Lemma 10.21. *For $n \geq 2$, the following holds*

$$\frac{1}{n} \ln \frac{\|\mathrm{X}_n\|}{\|X_0\|} = \frac{1}{n} \sum_{k=0}^{n-1} \ln \frac{\|\mathrm{X}_{k+1}\|}{\|\mathrm{X}_k\|}, \; a.s. \tag{10.28}$$

The proof of the lemma is trivial: using the property $\ln(a) + \ln(b) = \ln(ab)$ for all $a, b > 0$ we find that both sides equal $n^{-1} \ln \prod_{k=0}^{n-1} \|\mathrm{X}_{k+1}\|/\|\mathrm{X}_k\|$. Linear convergence defined in Equation (10.10) means that the left-hand side (and thus the RHS) of Equation (10.28) converges to a constant. In order to prove linear convergence, we exploit the fact that the right-hand side is the sum of n random variables divided by n, suggesting the use of a Law of Large Numbers (LLN):

Lemma 10.22 (LLN for independent random variables). *Let $(Y_n)_{n\in\mathrm{N}}$ be a sequence of independent, identically distributed, integrable ($E[|Y_0|] < +\infty$) random variables. Then*

$$\frac{1}{n} \sum_{k=0}^{n-1} Y_k \xrightarrow[n\to\infty]{} E[Y_0] \; a.s.$$

In order to apply Lemma 10.22 to the (1+1)-ES with scale-invariant step-size, it remains to be shown that the summands in the right-hand side of Equation (10.28) are i.i.d. and integrable random variables:

Proposition 10.23. *For the (1+1)-ES with scale-invariant step-size, $(\ln(\|\mathrm{X}_{n+1}\|/\|\mathrm{X}_n\|) : n \in \mathbb{N})$ are independent identically distributed as*

$\ln^-(\|e_1 + \sigma\mathcal{N}\|)$ *where* $\mathcal{N}$ *is a random vector following the distribution* $N(0, I_d)$.

For the technical details of the proof we refer to (Jebalia *et al.*, 2009, Lemma 7) where the result was proven in a slightly different setting where the objective function include noises. A weaker result stating that the random variables are orthogonal was proven in (Jebalia *et al.*, 2008). Moreover, we have seen in Lemma 10.15 that $\ln^- \|e_1 + \sigma\mathcal{N}\|$ is integrable such that we can apply Lemma 10.22 and together with Lemma 10.21 obtain the almost sure linear convergence of $\frac{1}{n}\ln\|\mathrm{X}_n\|/\|X_0\|$:

Theorem 10.24. *The (1+1)-ES with scale-invariant step-size converges linearly almost surely on the sphere function:*

$$\lim_{n\to\infty}\frac{1}{n}\ln\frac{\|\mathrm{X}_n\|}{\|X_0\|} = E[\ln^- \|e_1 + \sigma\mathcal{N}\|] = F_{(1+1)}(\sigma), a.s.$$

The idea of using Laws of Large Numbers for analyzing the convergence of evolution strategies was introduced in (Bienvenüe and François, 2003) and used for analyzing ESs with scale-invariant step-size in (Auger and Hansen, 2006; Jebalia *et al.*, 2008, 2009).

10.4.3.2. *How to Analyze Linear Convergence of Real Step-size Adaptation Schemes?*

The linear convergence of real adaptation schemes will also follow from applying a Law of Large Numbers. However, contrary to the scale-invariant step-size case, the sequence $(\ln(\|\mathrm{X}_{n+1}\|/\|\mathrm{X}_n\|), n \in \mathbb{N})$ is not independent and a different LLN needs thus to be apply. We will illustrate the different steps of the analysis exemplary for the one-fifth success rule that we define now precisely on spherical functions $g(\|x\|)$, $g \in \mathcal{M}$. Assume (X_n, σ_n) are given, the next iterate $(\mathrm{X}_{n+1}, \sigma_{n+1})$ is constructed in the following manner:

$$\mathrm{X}_{n+1} = \mathrm{X}_n + \sigma_n N_n 1_{\{\|\mathrm{X}_n + \sigma_n N_n\| \leq \|\mathrm{X}_n\|\}}, \tag{10.29}$$

$$\sigma_{n+1} = \sigma_n \left(\alpha^{-1/4} + (\alpha - \alpha^{-1/4}) 1_{\{\|\mathrm{X}_n + \sigma_n N_n\| \leq \|\mathrm{X}_n\|\}}\right), \tag{10.30}$$

where $\alpha > 1$, i.e., the step-size is multiplied by α in case of success and divided by $\alpha^{1/4}$ otherwise such that Equation (10.30) can be rewritten

$$\sigma_{n+1} = \begin{cases} \alpha\sigma_n & \text{if } \|\mathrm{X}_n + \sigma_n N_n\| \leq \|\mathrm{X}_n\| , \\ \alpha^{-1/4}\sigma_n & \text{otherwise,} \end{cases} \tag{10.31}$$

and $(X_0, \sigma_0) \in \mathbb{R}^d \times \mathbb{R}^+$. The equivalent of Equation (10.27) is now

$$\frac{\|\mathrm{X}_{n+1}\|}{\|\mathrm{X}_n\|} = \left\| \frac{\mathrm{X}_n}{\|\mathrm{X}_n\|} + \frac{\sigma_n}{\|\mathrm{X}_n\|} N_n 1_{\{\| \frac{\mathrm{X}_n}{\|\mathrm{X}_n\|} + \frac{\sigma_n}{\|\mathrm{X}_n\|} N_n \| \leq 1\}} \right\| , \tag{10.32}$$

where σ is replaced by $\sigma_n/\|\mathrm{X}_n\|$. Because Equation (10.32) depends on the random variable $\sigma_n/\|\mathrm{X}_n\|$, the random sequence $(\ln(\|\mathrm{X}_{n+1}\|/\|\mathrm{X}_n\|), n \in \mathbb{N})$ will not be independent and, thus, the LLN for independent random variables cannot be applied. However, $\|\mathrm{X}_n\|/\sigma_n$ is a Markov chain whose distribution can be defined in a simple way. Let $Z_0 = \|X_0\|/\sigma_0$, and define

$$Z_{n+1} = \frac{1}{\alpha^*} \| Z_n e_1 + N_n 1_{\{\|Z_n e_1 + N_n\| \leq Z_n\}} \| , \tag{10.33}$$

where $\alpha^* = \alpha^{-1/4} + (\alpha - \alpha^{-1/4}) 1_{\{\|Z_n e_1 + N_n\| \leq Z_n\}}$, i.e., corresponding to the multiplicative factor in Equation (10.30). Then it is clear that Z_n is a Markov Chain and not difficult to show that Z_n follows the same distribution as $\|\mathrm{X}_n\|/\sigma_n$. The Markov chain Z_n can be exploited to prove linear convergence thanks to the following lemma.

Lemma 10.25. *The following equality holds in distribution*

$$\frac{1}{n} \ln \frac{\|\mathrm{X}_n\|}{\|X_0\|} = \frac{1}{n} \sum_{k=0}^{n-1} \ln \frac{\| Z_k e_1 + N_k 1_{\{\|Z_k e_1 + N_k\| \leq Z_k\}} \|}{Z_k} . \tag{10.34}$$

The summands of the right-hand side of Equation(10.34) correspond to replacing in the right-hand side of Equation (10.32), $\sigma_n/\|\mathrm{X}_n\|$ by $1/Z_n$ and $\mathrm{X}_n/\|\mathrm{X}_n\|$ by e_1. The proof of this lemma is similar to the proof of Lemma 3 in (Jebalia *et al.*, 2009). Its main ingredients are the isotropy of the sampling distribution and of spherical functions. In addition, with Equation (10.33) we have $\alpha^* Z_{k+1} = \| Z_k e_1 + N_k 1_{\{\|Z_k e_1 + N_k\| \leq Z_k\}} \|$ and thus in distribution

$$\frac{1}{n} \ln \frac{\|\mathrm{X}_n\|}{\|X_0\|} = \frac{1}{n} \sum_{k=0}^{n-1} \ln \frac{\alpha^* Z_{k+1}}{Z_k} . \tag{10.35}$$

Since (Z_n) is a Markov chain, Equations (10.34) or (10.35) suggest to apply a LLN for Markov chains. However, not all Markov chains satisfy a LLN. The properties needed to satisfy a LLN are so-called stability properties, namely φ-irreducibility, Harris recurrence and positivity that are explained in the next following paragraphs.

Given a homogeneous Markov chain $(Z_n)_n \subset \mathbb{R}$, with transition kernel $P(.,.)$ and denoting $\mathcal{B}(\mathbb{R})$ the Borel sigma-algebra on $\mathbb{R}$, $(Z_n)_n$ is φ-irreducible if there exists a measure φ such that:

$$\forall (x, A) \in \mathbb{R} \times \mathcal{B}(\mathbb{R}), \varphi(A) > 0, \exists\, n_0 \geq 0 \text{ such that } P^{n_0}(x, A) > 0\,, \tag{10.36}$$

where $P^{n_0}(x, A)$ equals $\Pr[Z_{n_0} \in A | Z_0 = x]$. Another equivalent definition for the φ-irreducibility of the Markov chain $(Z_n)_n$ is: for all $x \in \mathbb{R}$ and for all $A \in \mathcal{B}(\mathbb{R})$ such that $\varphi(A) > 0$, $\Pr[\tau_A < +\infty | Z_0 = x] > 0$, where τ_A is the hitting time of Z_n on A, i.e.,

$$\tau_A = \min\{n \geq 1 \text{ such that } Z_n \in A\}.$$

If the last term of Equation (10.36) is equal to one, the chain is *recurrent.* A φ-irreducible chain $(Z_n)_n$ is *Harris recurrent* if:

$$\forall A \in \mathcal{B}(\mathbb{R}) \text{ such that } \varphi(A) > 0; \Pr[\eta_A = \infty | Z_0 = x] = 1, x \in \mathbb{R}\,,$$

where η_A is the occupation time of A defined as $\eta_A = \sum_{n=1}^{\infty} 1_{\{Z_n \in A\}}$. A chain $(Z_n)_n$ which is Harris-recurrent admits an *invariant measure*, i.e., a measure π on $\mathcal{B}(\mathbb{R})$ satisfying:

$$\pi(A) = \int_{\mathbb{R}} P(x, A) \mathrm{d}\pi(x), A \in \mathcal{B}(\mathbb{R})\,.$$

If in addition this measure is a probability measure, the chain is called *positive.* Positive, Harris-recurrent chains satisfy a LLN as stated in (Meyn and Tweedie, 1993, Theorem 17.0.1) and recalled here.

Theorem 10.26 (LLN for Harris positive chains). *Suppose that* $(Z_n)_n$ *is a positive Harris chain with invariant probability measure* π, *then for any function* G, *satisfying* $\pi(|G|) := \int |G(x)| \mathrm{d}\pi(x) < \infty$, *holds*

$$\lim_{n\to\infty} \frac{1}{n} \sum_{k=0}^{n-1} G(Z_k) = \pi(G)\,. \tag{10.37}$$

Therefore, in order to prove linear convergence of the (1+1)-ES with one-fifth success rule, one can investigate the stability of Z_n and prove that Theorem 10.26 applies to the right-hand side of Equation (10.34) deducing thus the convergence of $\frac{1}{n} \ln (\|\mathrm{X}_n\| / \|X_0\|)$[d].

---

[d]In fact, the right-hand side of Equation (10.34) can be written $\frac{1}{n} \sum_k G(Z_k, N_k)$ with $G(z, N_n) = \ln \|z e_1 + N_n 1_{\{\|z e_1 + N_n\| \leq z\}}\| / z$ such that one needs to study the stability of the Markov chain (Z_k, N_k). However, the stability of (Z_k, N_k) is a direct corollary of the stability of Z_k since N_k is independent of Z_k.

Theorem 10.27. *If Z_n is φ-irreducible, Harris-recurrent and positive with invariant probability measure π and*

$$\int_{\mathbb{R}} E\left[|\ln \|ze_1 + N_0 1_{\{\|ze_1+N_0\|\leq z\}}\|/z|\right] d\pi(z) < \infty \ ,$$

then the (1+1)-ES with one-fifth success rule converges linearly, more precisely

$$\frac{1}{n}\ln\frac{\|X_n\|}{\|X_0\|} \xrightarrow[n\to\infty]{} \int_{\mathbb{R}} E\left[\ln \|ze_1 + N_0 1_{\{\|ze_1+N_0\|\leq z\}}\|/z\right] d\pi(z) \ . \quad (10.38)$$

Assuming that $\|X_0\|/\sigma_0$ is distributed according to π, we can formulate a non-asymptotic linear convergence result (the equivalent of Equation (10.24)):

Theorem 10.28. *If Z_n is φ-irreducible, Harris-recurrent and positive with invariant probability measure π,*

$$\int_{\mathbb{R}} E[|\ln \|ze_1 + N_0 1_{\{\|ze_1+N_0\|\leq z\}}\|/z|] d\pi(z) < \infty,$$

and $Z_0 \sim \pi$, then for all $n_0 \in \mathbb{N}$ and for $n > n_0$

$$\frac{1}{n-n_0} E\left[\ln\frac{\|X_n\|}{\|X_{n_0}\|}\right] = \int_{\mathbb{R}} E\left[\ln \|ze_1 + N_0 1_{\{\|ze_1+N_0\|\leq z\}}\|/z\right] d\pi(z).$$

Proving the stability is in general the most difficult part in the analysis and has been achieved for the (1,λ)-ES with self-adaptation for $d = 1$ using drift conditions (Auger, 2005). The convergence rate in Equation (10.38) is expressed implicitly by means of the invariant distribution of the chain Z_n. However, it is also possible to derive a central limit theorem to characterize the convergence of Equation (10.38) and, then derive confidence intervals for a Monte Carlo estimation of the convergence rate. For this, a stronger stability property needs to be satisfied, namely the geometric ergodicity. Moreover, using the Fubini theorem, it is possible to prove that $\frac{1}{n}\ln\sigma_n$ converges to the same limit than $\frac{1}{n}\ln\frac{\|X_n\|}{\|X_0\|}$ and to prove an alternative expression for the convergence rate, namely

$$\int_{\mathbb{R}} E\left[\ln \|ze_1 + N_0 1_{\{\|ze_1+N_0\|\leq z\}}\|/z\right] d\pi(z) = \int_{\mathbb{R}} E[\ln(\alpha^*(z))] d\pi(z)$$

where $\alpha^*(z)$ is the multiplicative factor for the step-size change, i.e., $\alpha^*(z) = \alpha^{-1/4} + (\alpha - \alpha^{-1/4})1_{\{\|ze_1+N_0\|\leq z\}}$ (Auger, 2005). The fact that both σ_n and $\|X_n\|$ converge (log)-linearly at the same rate can be observed on the left plot in Figure 10.2 where we observe the same rate for the decrease of $\ln \|X_n\|$ and $\ln \sigma_n$.

The link between stability of the normalized chain $\|X_n\|/\sigma_n$ and linear convergence or divergence of ESs was first pointed out in Bienvenüe and François (2003) and exploited in Auger (2005). Beyond the fact that stability of Z_n implies linear convergence, it is interesting to note that the stability is a natural generalization of the scale-invariant step-size updaté rule. Indeed, stability implies that after a transition phase, the distribution of $\|X_n\|/\sigma_n$ will be close to the invariant distribution π, i.e., $\|X_n\|/\sigma_n \approx \pi$ whereas for the algorithm with scale-invariant step-size we have $\|X_n\|/\sigma_n = \sigma$. In other words, the scale-invariant step-size rule approximates π by a constant. The benefit of this simplification is the possibility to derive explicit formulae for the convergence rates for d to infinity. The transition phase is illustrated in Figure 10.2 where the experiment was started in the tail of the invariant distribution: the step-size was chosen very small equal to 10^{-9} such that Z_0 is very large. The adaptation stage lasts up to the iteration 150. Afterwards $Z_n = \|X_n\|/\sigma_n$ "looks" stable as both $\|X_n\|$ and σ_n decrease at the same rate.

Other approaches to investigate linear convergence have been used on the sphere and certain convex quadratic functions by Jägersküpper (2007, 2005) who derives lower and upper bounds on the time needed to halve the distance to the optimum for a special one-fifth success rule algorithm. With such an approach, it is possible to derive the dependence in the dimension of the convergence rate. However, the approach seems less general in terms of step-size update rules that can be tackled (Jägersküpper and Preuss, 2008).

10.5. Discussion and Conclusion

Stochastic optimization algorithms for numerical optimization are studied in different communities taking different viewpoints. Showing (or disproving) global convergence on a broad class of functions is often a comparatively easy task. In contrast, proving an associated *rate* of convergence, or convergence speed, is often much more intricate. In particular fast, i.e., linear convergence, with running times proportional to $d \log 1/\epsilon$, can only be proven on comparatively restricted classes of functions or in the vicinity of a well-shaped optimum. Here, d is the search space dimension and ϵ is a precision to reach. Linear convergence is a general lower bound: rank-based algorithms (and thus ESs) cannot be faster than linear with a convergence rate decreasing like $1/d$ (see also Chapter 11).

We believe that global convergence is per se rather meaningless in practice. The (1+1)-ES with fixed step-size as well as the pure random search converge with probability one to the global optimum of functions belonging to a broad class, where the main assumption is that a neighbourhood of the global optimum should be reachable by the search distribution with a positive probability. However, the convergence rate is sub-linear with degree $1/d$, therefore, the running time is proportional to $(1/\epsilon)^d$. Even for moderate dimension, e.g., $d = 10$, this is prohibitively slow in practice. More promising upper bounds for the convergence rate can be achieved for non-adaptive algorithms, when the sampling distribution admits a singularity. For a sampling distribution chosen depending on ϵ, the bound is $O(\ln^2(1/\epsilon))$, and it is slightly worse if we relax the dependency in ϵ. Corresponding practical algorithms have yet to be implemented.

Adaptive ESs, however, do not converge to the global optimum with probability one on such broad classes of functions, because they might never recover from converging to a local optimum (Rudolph, 2001). Instead, adaptive ESs have been shown to achieve linear convergence on restricted function classes. For example, Jägersküpper (2007) lower and upper bounds the time to halve the distance to the optimum with the (1+1)-ES with one-fifth success rule on special ellipsoidal functions. Linear convergence can also be investigated using the Markov chain $\|\mathrm{X}_n\|/\sigma_n$. We have illustrated the corresponding proof techniques in the context of evolution strategies.

One might argue that linear convergence results on convex quadratic functions are weak results because the class of functions is rather limited. However, much slower convergence results are rather irrelevant in practice and linear convergence is not limited to convex quadratic functions: (1) as pointed out in this chapter, the invariance of ESs to strictly monotonic transformations implies the generalization of the result to the class of functions $\{g \circ f, g \in \mathcal{M}, f \text{ convex quadratic}\}$, that contains non-convex, non-smooth functions; (2) linear convergence with a positive probability (on a large class of functions) will imply linear convergence with probability one of a restart version of the algorithm, where a constant distribution is sampled simultaneously and the restart is conducted when a superior solution is sampled[e]; (3) robustness of the linear convergence in presence of noise has been proven when using a scale-invariant-constant step-size

[e]This idea was suggested to the authors first by Günter Rudolph.

(Jebalia *et al.*, 2009); (4) adaptation has been recognized as a key of the success of evolution strategies also in practice.

10.6. Appendix

10.6.1. *Proof of Theorem 10.17*

We prove now Theorem 10.17 that was stated page 311.

Proof. We fix n and assume that we are at the iteration n of a (1+1)-ES with adaptive step-size such that (X_n, σ_n) is known.

Maximal progress towards x^* in one step: The next iterate X_{n+1} either equals the sampled offspring $\mathrm{X}_n + \sigma_n N_n$ or the parent X_n (depending on what is the best according to f) and thus the distance between X_{n+1} and y^* is always larger or equal than the minimum between the distance between the offspring and y^* and the parent and y^*:

$$\|\mathrm{X}_{n+1} - y^*\| \geq \min\{\|\mathrm{X}_n - y^*\|, \|\mathrm{X}_n + \sigma_n N_n - y^*\|\} \ . \tag{10.39}$$

If $a > 0$, the minimum of (a, b) equals $a \min(1, b/a)$ such that

$$\|\mathrm{X}_{n+1} - y^*\| \geq \|\mathrm{X}_n - y^*\| \min\{1, \|\frac{\mathrm{X}_n - y^*}{\|\mathrm{X}_n - y^*\|} + \frac{\sigma_n}{\|\mathrm{X}_n - y^*\|} N_n\|\} \ . \tag{10.40}$$

Taking the logarithm of the previous equation we obtain

$$\ln \|\mathrm{X}_{n+1} - y^*\| \geq \ln \|\mathrm{X}_n - y^*\| + \ln\left[\min\left\{1, \left\|\frac{\mathrm{X}_n - y^*}{\|\mathrm{X}_n - y^*\|} + \frac{\sigma_n}{\|\mathrm{X}_n - y^*\|} N_n\right\|\right\}\right] \ . \tag{10.41}$$

We assume that $E[\ln \|\mathrm{X}_n - y^*\|] < +\infty$ for all n such that we can take the expectation in Equation (10.41) condition to (X_n, σ_n). We use the notation $\ln^-(x) = \max(0, -\ln(x))$ such that $\ln(\min(1, h(x))) = -\ln^-(h(x))$. By linearity of the expectation we obtain that

$$E[\ln \|\mathrm{X}_{n+1} - y^*\| | \mathrm{X}_n, \sigma_n] \geq \ln \|\mathrm{X}_n - y^*\| - E\left[\ln^- \left\|\frac{\mathrm{X}_n - y^*}{\|\mathrm{X}_n - y^*\|} + \frac{\sigma_n}{\|\mathrm{X}_n - y^*\|} N_n\right\| \Big| \mathrm{X}_n, \sigma_n\right] \ . \tag{10.42}$$

The offspring distribution $N(0, I_d)$ being spherical, i.e., the direction of $N(0, I_d)$ is uniformly distributed, it does not matter where the parent inducing the offspring is located on the unit hypersphere and thus

$$E\left[\ln^- \left\|\frac{\mathrm{X}_n - y^*}{\|\mathrm{X}_n - y^*\|} + \frac{\sigma_n}{\|\mathrm{X}_n - y^*\|} N_n\right\| \Big| \mathrm{X}_n, \sigma_n\right] = F_{(1+1)}\left(\frac{\sigma_n}{\|\mathrm{X}_n - y^*\|}\right), \tag{10.43}$$

where $F_{(1+1)}$ is defined in Lemma 10.15. Using the same lemma, we know that $F_{(1+1)}$ is continuous, the supremum $\tau := \sup F_{(1+1)}([0, +\infty])$ is reached and the step-size σ_F such that $F_{(1+1)}(\sigma_F) = \tau$ exists. Injecting this in Equation (10.42) we obtain $E[\ln \|\mathrm{X}_{n+1} - y^*\| | \mathrm{X}_n, \sigma_n] \geq \ln \|\mathrm{X}_n - y^*\| - \tau$ and consequently $E[\ln \|\mathrm{X}_{n+1} - y^*\|] \geq E[\ln \|\mathrm{X}_n - y^*\|] - \tau$. □

10.6.2. *Proof of Proposition 10.19 and Corollary 10.20*

We prove Proposition 10.19 and Corollary 10.20 stated page 312.

Proof. If $f(x) = g(\|x - x^*\|)$, Equation (10.42) with $y^* = x^*$ is an equality. If $\sigma_n = \sigma\|\mathrm{X}_n - x^*\|$, we obtain $E[\ln \|\mathrm{X}_{n+1} - x^*\|] = E[\ln \|\mathrm{X}_n - x^*\|] - F_{(1+1)}(\sigma)$ or $E[\ln \|\mathrm{X}_{n+1} - x^*\|] - E[\ln \|\mathrm{X}_n - x^*\|] = -F_{(1+1)}(\sigma)$. Summing from $n = n_0, \ldots, N$, we obtain that $E[\ln \|X_N - x^*\|] - E[\ln \|X_{n_0} - x^*\|] = -(N - n_0)F_{(1+1)}(\sigma)$. Dividing by N we obtain Equation (10.23). If $\sigma_n = \sigma_F \|\mathrm{X}_n - x^*\|$ where $F(\sigma_F) = \tau$, we obtain Equation (10.24). □

References

Auger, A. (2005). Convergence results for (1,λ)-SA-ES using the theory of φ-irreducible markov chains, *Theoretical Computer Science* **334**, pp. 35–69.

Auger, A. and Hansen, N. (2006). Reconsidering the progress rate theory for evolution strategies in finite dimensions, in A. Press (ed.), *Proceedings of the Genetic and Evolutionary Computation Conference (GECCO 2006)*, pp. 445–452.

Baba, N. (1981). Convergence of random optimization methods for constrained optimization methods, *Journal of Optimization Theory and Applications.*

Beyer, H.-G. (2001). *The Theory of Evolution Strategies*, Natural Computing Series (Springer-Verlag).

Bienvenüe, A. and François, O. (2003). Global convergence for evolution strategies in spherical problems: some simple proofs and difficulties, *Theoretical Computer Science* **306**, 1-3, pp. 269–289.

Brooks, S. (1958). A discussion of random methods for seeking maxima, *The computer journal* **6**, 2.

Devroye, L. (1972). The compound random search, in *International Symposium on Systems Engineering and Analysis* (Purdue University), pp. 195–110.

Devroye, L. and Krzyżak, A. (2002). Random search under additive noise, in P. L. M. Dror and F. Szidarovsky (eds.), *Modeling Uncertainty* (Kluwer Academic Publishers), pp. 383–418.

Greenwood, G. W. and Zhu, Q. J. (2001). Convergence in evolutionary programs with self-adaptation, *Evolutionary Computation* **9**, 2, pp. 147–157.

Hansen, N. and Ostermeier, A. (2001). Completely derandomized self-adaptation in evolution strategies, *Evolutionary Computation* **9**, 2, pp. 159–195.

Jägersküpper, J. (2005). Rigorous runtime analysis of the (1+1)-ES: 1/5-rule and ellipsoidal fitness landscapes, in LNCS (ed.), *Foundations of Genetic Algorithms: 8th International Workshop, FoGA 2005*, Vol. 3469, pp. 260–281.

Jägersküpper, J. (2007). Analysis of a simple evolutionary algorithm for minimization in euclidean spaces, *Theoretical Computer Science* **379**, 3, pp. 329–347.

Jägersküpper, J. (2008). Lower bounds for randomized direct search with isotropic sampling, *Operations research letters* **36**, 3, pp. 327–332.

Jägersküpper, J. and Preuss, M. (2008). Aiming for a theoretically tractable CSA variant by means of empirical investigations, in *Proceedings of the 2008 Conference on Genetic and Evolutionary Computation*, pp. 503–510.

Jebalia, M., Auger, A. and Hansen, N. (2009). Log-linear convergence and divergence of the scale-invariant (1+1)-ES in noisy environments, *Algorithmica* In press.

Jebalia, M., Auger, A. and Liardet, P. (2008). Log-linear convergence and optimal bounds for the (1+1)-ES, in N. Monmarché and al. (eds.), *Proceedings of Evolution Artificielle (EA'07), LNCS*, Vol. 4926 (Springer), pp. 207–218.

Karr, A. F. (1993). *Probability*, Springer Texts in Statistics (Springer-Verlag).

Kern, S., Müller, S., Hansen, N., Büche, D., Ocenasek, J. and Koumoutsakos, P. (2004). Learning Probability Distributions in Continuous Evolutionary Algorithms - A Comparative Review, *Natural Computing* **3**, 1, pp. 77–112.

Matyas, J. (1965). Random optimization, *Automation and Remote control* **26**, 2.

Meyn, S. and Tweedie, R. (1993). *Markov Chains and Stochastic Stability* (Springer-Verlag, New York).

Nekrutkin, V. and Tikhomirov, A. (1993). Speed of convergence as a function of given accuracy for random search methods, *Acta Applicandae Mathematicae* **33**, pp. 89–108.

Rechenberg, I. (1973). *Evolutionstrategie: Optimierung Technisher Systeme nach Prinzipien des Biologischen Evolution* (Frommann-Holzboog Verlag, Stuttgart).

Rudolph, G. U. (2001). Self-adaptive mutations may lead to premature convergence, *IEEE Transactions on Evolutionary Computation* **5**, pp. 410–414.

Schumer, M. and Steiglitz, K. (1968). Adaptive step size random search, *Automatic Control, IEEE Transactions on* **13**, pp. 270–276.

Schwefel, H.-P. (1981). *Numerical Optimization of Computer Models* (John Wiley & Sons, Inc., New York, NY, USA).

Stewart, G. W. (1995). On sublinear convergence, Tech. Rep. CS-TR-3534, University of Maryland.

Teytaud, O. and Gelly, S. (2006). General lower bounds for evolutionary algorithms, in 10th *International Conference on Parallel Problem Solving from Nature (PPSN 2006)*, Vol. 4193 (Springer), pp. 21–31.

Tikhomirov, A. S. (2006). On the markov homogeneous optimization method, *Computational Mathematics and Mathematical Physics* **46**, 3, pp. 361–375.

White, R. (1971). A survey of random methods for parameter optimization, *SIMULATION* **17**, pp. 197–205.

Zhigljavsky, A. A. (1991). *Theory of Global Random Search* (Kluwer Academic Publishers).

Zhigljavsky, A. A. and Zilinskas, A. (2008). *Stochastic global optimization, Springer Optimization and its applications*, Vol. 1 (Springer).

Chapter 11

Lower Bounds for Evolution Strategies

Olivier Teytaud

TAO (INRIA), LRI, Umr 8623 (CNRS, Univ. Paris-Sud), Bât. 490, Univ. Paris-Sud, 941405 Orsay, France
teytaud@lri.fr

The mathematical analysis of optimization algorithms involves upper and lower bounds; we here focus on the second case. Whereas other chapters will consider black box complexity, we will here consider complexity based on the key assumption that the only information available on the fitness values is the rank of individuals — we will not make use of the exact fitness values. Such a reduced information is known efficient in terms of robustness (Gelly *et al.*, 2007), what gives a solid theoretical foundation to the robustness of evolution strategies, which is often argued without mathematical rigor — and we here show the implications of this reduced information on convergence rates. In particular, our bounds are proved without infinite dimension assumption, and they have been used since that time for designing algorithms with better performance in the parallel setting.

Contents

11.1. Introduction

Optimization methods in continuous domains (we will consider also discrete cases in this chapter) can be classified in a few categories:

- Order 2 methods, computing fitness values, gradients and Hessians;
- Order 1 methods, computing fitness values and gradients only;
- Order 0 methods, computing fitness values only.

However, the order of the optimization algorithm sometimes refers to the convergence rate of the algorithm for some family of fitness functions: order $q > 1$ means that, with y_k the fitness value of the best evaluated point after k steps,

$$(y_{k+1} - y^*)/(y_k - y^*)^q = O(1).$$

Order $q = 1$ means that

$$(y_{k+1} - y^*)/(y_k - y^*) \leqslant c \text{ for some } c < 1.$$

These two definitions do not coincide, for several reasons:

- Newton methods are of order 2 for both definitions (at least mathematically, as limited precision often reduce the "real world" order of convergence).
- Quasi-Newton methods have a superlinear convergence; therefore, they are "a bit more" than order 1 in spite of the fact that they use only gradients (they approximate the Hessian thanks to successive gradients, but they never compute it explicitly).
- Algorithms with no gradient can nonetheless be linear. This is the case for most evolution strategies (Auger, 2005). Also, thanks to approximation of the objective function built on previously visited points, algorithms with no gradient can be superlinear; (Powell, 1974), cited by Conn *et al.* (1997), shows that, in dimension 1, the secant model has average order $1.618\cdots = \frac{1}{2}(1+\sqrt{5})$, i.e., more than the Newton-Raphson algorithm with finite differences which reaches $\sqrt{2}$.

Evolution strategies (ES) are often termed "order 0" methods as they do not use gradients or Hessians. Interestingly, most ES are in fact a special case of order 0 methods: in addition to not using gradients, they only use comparisons between fitness values and not the fitness values themselves - *i.e.*, they have only access to a discrete information, with finitely many possible values. Using comparisons only makes an important difference since it is known that, even without gradient, a super-linear convergence can be obtained when using fitness values (as done by (Auger *et al.*, 2005), using surrogate models for a super-linear convergence rate), whereas it is not possible with comparison-based non-elitist algorithms. Getting rid of the gradient or Hessian is easy to justify: often, the gradient is quite expensive, and the Hessian is even much more expensive - sometimes, they both just do not exist. But is there a good reason for considering algorithms using only comparisons and discarding the other information provided by the fitness values?

A first answer lies in the optimality of this comparison-based principle for some robustness criterion; this robustness was suggested by Bäck *et al.* (1991); Whitley (1989); Baker (1987) and was formally established by Gelly *et al.* (2007). The robustness criterion is the supremum over compositions with monotonic functions: instead of

$$\underbrace{E_{f\in F}}_{\text{sup. on fitness}} \quad E\|x_{N,f} - x^*(f)\|^2$$

where $x_{N,f}$ is the approximation of the optimum of f provided by the Alg. after N visited points and $x^*(f)$ is the optimum of f, the robust criterion considered by (Gelly *et al.*, 2007) is

$$\underbrace{E_{f\in F}}_{\text{sup. on fitness}} \quad \sup_{g \text{ monotonic}} \quad E\|x_{N,g\circ f} - x^*(f)\|^2$$

and for this criterion, there is an optimal algorithm which is comparison-based. A second answer is natural in the case of problems in which you have only comparisons between parametrizations, but no fitness values: for example, when tuning a strategy in transitive games, or in interactive optimization when the comparison is provided by a human.

In many cases, comparisons between individuals are used in order to select a subset of the population; these selected individuals are the parent of the next generation. However, they can also be used for ranking these selected individuals or even the whole population. Examples of algorithms using more information than just the selection of a subset are the roulette-wheel with rank-based fitness assignment (stochastic sampling (Baker, 1987), rank-based fitness assignment (Bäck *et al.*, 1991; Whitley, 1989)), weighted recombination (Hansen and Ostermeier, 2001; Arnold, 2005) or BREDA (Gelly *et al.*, 2007).

Hence, the wide family of order 0 methods can be divided, from the more general to the more specific case, into (i) algorithms using fitness values; (ii) algorithms using the full ranking of all the population; (iii) algorithms using the full ranking of selected points; (iv) algorithms using only the selection information (this will be formalized in Section 11.2). In cases (ii) to (iv), the branching factor of the algorithm, i.e., the number of possible outcomes for the information extracted from the population in one iteration, is finite and bounded. This simple fact has been used by Teytaud and Gelly (2006) in order to provide lower bounds that match some upper bounds known for evolutionary algorithms (Droste, 2005; Auger, 2005; Rudolph, 1997); this simple technique extends previous lower bounds shown by Witt and Jägersküpper (2005).

We will first introduce formal notations in Section 11.2 and background in Section 11.3, so that we can formalize the main results. Lower bounds on $(\mu \overset{+}{,} \lambda)$-ES* based on the branching factor, obtained by Teytaud and Gelly (2006), are then recalled (Section 11.4). However, improvements under some mild assumptions were proved by Fournier and Teytaud (2008); these results, based on VC-dimension, are presented in Section 11.6. The traditional case of the sphere function is studied in Section 11.7. We also present a simple algorithm based on full ranking, which allows to obtain an almost linear speed-up† when the size of the offspring is linear in the dimension. Conclusions and elements for further research are presented in Section 11.8.

Notations. In all the chapter, $\log(x)$ denotes the logarithm with basis 2, i.e., $\log(2) = 1$. The set of integers $\{1, 2, \ldots, n\}$ is denoted by $[[1, n]]$. The notation $|\cdot|$ is used to denote both the cardinal of a set and the length of a vector.

---

*$(\mu \overset{+}{,} \lambda)$-ES stands for either $(\mu + \lambda)$-ES or (μ, λ)-ES.

†The speed-up is the ratio between the new convergence ratio and the old convergence ratio when a modification is performed; here, we compare the convergence ratio of the algorithm with λ linear in the dimension and the convergence ratio with λ constant.

11.2. Evolution Strategies of Type $(\mu \overset{+}{,} \lambda)$

This section is devoted to a formal definition of evolution strategies of type $(\mu \overset{+}{,} \lambda)$ (denotes $(\mu \overset{+}{,} \lambda)$-ES). The aim of $(\mu \overset{+}{,} \lambda)$-ES is to find the minimum of a real-valued function on a domain D ($f : D \to \mathbb{R}$); f is termed the fitness function. We will here focus on the classical case of $(\mu \overset{+}{,} \lambda)$-ES which are not allowed to have access to the fitness value, but only to comparisons between fitness values: given x and y in D, the algorithm can request the sign of $f(x) - f(y)$. More precisely, given some points $z_1, \ldots, z_p \in D$ (usually termed population), these algorithms are given (possibly partial) information on the signs of $f(z_i) - f(z_j)$, for all $1 \leqslant i < j \leqslant p$: all the signs in the case of complete ranking of the population, only a subset of these signs when the algorithms only uses a selection information (these notions will be formalized below). Of course these algorithms are not required to work for only one fitness function, but for a whole family of fitness functions. In the following, we denote by $\mathcal{F}$ the set of fitness functions considered, and we will consider worst cases on functions in $\mathcal{F}$.

In the rest of the chapter, otherwise explicitly stated, we assume that equality of fitness values $f(x) = f(y)$, for two points x and y generated at any iteration, never occurs. This is a reasonable hypothesis in the continuous setting (e.g., when $D = [0, 1]^d$) where this assumption almost surely holds for many algorithms and many fitness functions; the case with equality is developped by Fournier and Teytaud (2010).

Let λ and μ be two integers. The first case is the case of Selection Based evolution strategies (SB-$(\mu \overset{+}{,} \lambda)$-ES). In this case, there is a set $\mathcal{I}$ of internal states. The algorithm starts in the initial state I_0 returned by the function *initial_state*. At each iteration, the algorithm follows these three successive steps. First generate a set of λ points, called the *offspring*. Then select only the μ best ones, i.e., the μ points with lowest fitness values; in the case of an SB-(μ, λ)-ES (termed non-elitist), points generated at previous stages are forgotten and this selection is performed only among the offspring, while an algorithm of type SB-$(\mu + \lambda)$-ES (termed elitist) selects the μ best points among the offspring *and* the points selected at the previous step[‡]. Finally, the internal state is updated: this might include for example cumulative step-length adaptation (CSA) by Hansen and Ostermeier (2001), covariance matrix adaptation (CMA) by Hansen and Ostermeier (2001), mutative self-adaptation (SA) by Rechenberg (1973); Schwefel (1974), covariance matrix

[‡]As a consequence, in elitist cases, these μ selected points are always the μ points with lowest fitness values found so far.

adaptation by SA (CMSA) by Beyer and Sendhoff (2008), the one-fifth rule for step-size adaptation (Rechenberg, 1973; Beyer, 2001).

Classical ES could be cast into our framework; in the continuous case, just use Gaussian mutations with parameters encoded in the internal state. Please notice that the framework is in fact much more general than that: it includes adaptation by cross-entropy methods (CEM by de Boer *et al.* (2005)) or Estimation of Distribution Algorithms like e.g., UMDA (Mühlenbein and Mahnig, 2002), Compact Genetic Algorithm (Harik *et al.*, 1999), Population-Based Incremental Learning (Baluja, 1994), Relative Entropy (Mühlenbein and Höns, 2005) and Estimation of Multivariate Normal Algorithms (EMNA by Larranaga and Lozano (2001)); this similarity points out the strong similarity between evolution strategies, cross-entropy methods, estimation of distribution algorithms which are all in the scope of this paper. We can even include so-called direct search methods (Conn *et al.*, 1997), including the Nelder-Mead algorithm (Nelder and Mead, 1965), the Hooke&Jeeves algorithm (Hooke and Jeeves, 1961).

Notice that $\mu \leqslant \lambda$ must hold in the case of an SB-(μ, λ)-ES.

Algorithm 20 (Fournier and Teytaud, 2008) Selection Based (μ, λ)-ES (resp. Selection Based $(\mu + \lambda)$-ES). Framework for evolution strategies based on selection, working on a fitness function f. The random variable ω is a random seed. An algorithm matching this framework is obtained by specifying the distribution of ω, the space of states, and the functions *initial_state*, *generate*, *update* and *proposal*.

Initialization: $I_0 \leftarrow \text{initial\_state}(\omega)$; $S_0 \leftarrow \emptyset$; $n \leftarrow 0$
while true **do**
  $n \leftarrow n + 1$
  Generation step (generate an offspring O_n of λ distinct points): $O_n \leftarrow \text{generate}(I_{n-1})$
  $E_n \leftarrow O_n$ (resp. $E_n \leftarrow O_n \oplus S_{n-1}$)
  $\ell \leftarrow \min(\mu, |E_n|)$
  Selection step: $v_n \leftarrow \text{select}(E_n, f)$, i.e.:
    The vector $v_n = (i_1, \ldots, i_\ell)$ is defined by:

$$\begin{cases} 1 \leqslant i_1 < i_2 < \cdots < i_\ell \leqslant |E_n| \\ \text{for all } j \text{ and } k, \text{ if } j \in v_n \text{ and } k \notin v_n, \text{ then } f(E_{n,j}) < f(E_{n,k}) \end{cases}$$

  Update the internal state: $I_n \leftarrow \text{update}(I_{n-1}, E_n, v_n)$
  $S_n \leftarrow (E_{n,i_1}, \ldots, E_{n,i_\ell})$
  $x_{\omega,n}^{(f)} \leftarrow \text{proposal}(I_n)$
end while

We present SB-(μ, λ)-algorithms (resp. SB-$(\mu + \lambda)$-algorithms) in Algorithm 20. In this algorithm (also in Algorithm 21), the concatenation of the two vectors $x = (x_1, \dots, x_k)$ and $x' = (x'_1, \dots, x'_\ell)$ is denoted by $x \oplus x' = (x_1, \dots, x_k, x'_1, \dots, x'_\ell)$; we also use the shortcut $v \in (x_1, \dots, x_k)$ to express that there exists $i \in [[1, k]]$ such that $x_i = v$.

Let us detail how the selection step is performed. If $E_n = (z_1, \dots, z_p)$ is the vector of points considered at step n (either $E_n = O_n$ in the case of SB-(μ, λ)-ES or $E_n = O_n \oplus S_{n-1}$ in the case of SB-$(\mu+\lambda)$-ES), the function *select* returns a vector of μ distinct integers $v_n = (i_1, \dots, i_\mu)$ such that:

$$\begin{cases} i_1 < \dots < i_\mu \\ \text{for all } j \in \{i_1, \dots, i_\mu\} \text{ and } k \in [[1, p]] \setminus \{i_1, \dots, i_\mu\},\ f(z_j) < f(z_k) \end{cases}$$

Notice that the length of the vector v_n is equal to μ except maybe during the first iterations of the algorithm in the case $\lambda < \mu$: this explains the use of the auxiliary variable ℓ in Algorithm 20.

Algorithms with the "+" are usually termed *elitist*; this means that we always keep the best individuals. Algorithms with the "," are termed *non-elitist.*

Finally, we would like to explain a generalization of SB-$(\mu \overset{+}{,} \lambda)$-ES, called Full Ranking $(\mu \overset{+}{,} \lambda)$-ES (FR-$(\mu \overset{+}{,} \lambda)$-ES). Instead of just giving the best μ points (i.e., the μ points with the lowest fitness values), we can consider a selection procedure which returns the best μ points *ordered with respect to their fitness.*

The outline of these algorithms is summarized in Algorithm 21. More precisely, the selection step described in this algorithm works as follows. Given the vector of points $E_n = (z_1, \dots, z_p)$ considered at step n, the function *select* returns a vector of μ distinct integers $v_n = (i_1, \dots, i_\mu)$ such that:

$$\begin{cases} f(z_{i_1}) < \dots < f(z_{i_\mu}) \quad \text{(the difference is here)} \\ \text{for all } j \in [[1, p]] \setminus \{i_1, \dots, i_\mu\},\ f(z_{i_\mu}) < f(z_j). \end{cases}$$

(Once again, the length of the vector v_n may not be equal to μ at the beginning of the algorithm.)

Note that both Algorithms 20 and 21 define a class of algorithms: in order to obtain an algorithm, one has to specify the distribution of ω, how the offspring is generated (function *generate*), the space of states $\mathcal{I}$ as well as the functions *initial_state* and *update*, and finally the function *proposal.* Throughout the chapter, we assume that all functions involved in these algorithms are measurable.

Algorithm 21 (Fournier and Teytaud, 2008) Full Ranking (μ, λ)-ES (resp. Full Ranking $(\mu + \lambda)$-ES). Framework for evolution strategies based on full ranking, working on a fitness function f. The random variable ω is a random seed. Compared to Algorithm 20, the vector of integers v_n obtained at the selection step is now ordered with respect to the fitness values of points from E_n; this framework is thus more general.

Initialization: $I_0 \leftarrow$ initial_state(ω); $S_0 \leftarrow \emptyset$; $n \leftarrow 0$
while true **do**
  $n \leftarrow n + 1$
  Generation step (generate an offspring O_n of λ distinct points): $O_n \leftarrow$ generate(I_{n-1})
  $E_n \leftarrow O_n$ (resp. $E_n \leftarrow O_n \oplus S_{n-1}$)
  $\ell \leftarrow \min(\mu, |E_n|)$
  Selection step: $v_n \leftarrow$ select(E_n, f), i.e.:
    The vector $v_n = (i_1, \ldots, i_\ell)$ is defined by:

$$\begin{cases} i_1, \ldots, i_\ell \in [[1, |E_n|]] \\ f(E_{n,i_1}) < f(E_{n,i_2}) < \cdots < f(E_{n,i_\ell}) \\ \text{for all } j \notin v_n,\ f(E_{n,i_\ell}) < f(E_{n,j}) \end{cases}$$

  Update the internal state: $I_n \leftarrow$ update(I_{n-1}, E_n, v_n)
  $S_n \leftarrow (E_{n,i_1}, \ldots, E_{n,i_\ell})$
  $x^{(f)}_{\omega,n} \leftarrow$ proposal(I_n)
end while

Importantly, in spite of the formulation in Algorithms 20 and 21, the formalization is not restricted to deterministic algorithms. In Algorithms 20 and 21, all steps are deterministic, but there is an initial source of randomization, ω, which can be as large as required, e.g., it might be an infinite sequence of realizations of a random Gaussian variable or random uniform variables. So this is not a restriction, the algorithm must just be rephrased accordingly, so that ω is reported in the internal state and used in, *e.g.*, the "proposal" function.

11.3. Informal Section: How to Define Convergence Rates?

There are algorithms which are termed quadratic, superlinear, or linear. Roughly speaking, quadratic means that the number of exact digits doubles at each iteration; superlinear means that the number of exact digits increases by a number which goes to infinity with iterations (but does not necessarily doubles); linear means that the number of exact digits increases by a (on average) constant number per iteration. The convergence order is 2 in the quadratic case, 1 in the linear case.

In the linear case, the convergence rate is the minimum constant c such that the distance to the optimum is upper bounded by $O(c^n)$, where n is the number of iterations. The convergence rate 0 is reserved for superlinear algorithms (e.g., BFGS).

These definitions are not so easy as it seems at first view. For example, what is an iteration? Should we consider each iterate? This would imply that we are not allowed to have an exploration step sometimes; each point far from the optimum "kills" the convergence rate. If the algorithm is very fast, and uses one random exploration step (uniform in the search space) once per 10 steps, then it is not even linear because of these random diversifications. Definitions can be modified in order to include such cases: we can distinguish between the points at which we compute the fitness, and the points which are proposed to the user as approximations of the optimum; only the latter is considered when considering the distance to the optimum, and only the former for counting the number of iterations.

Other troubles remain: for which families of fitness functions does the convergence holds? In order to make these points clear, a simple solution is to define the supremum, over all fitness functions in a given family, of the number n of visited points such that the algorithm finds the optimum with precision ε.

However, many algorithms use random numbers. Therefore, we should consider the number n of points such that the algorithm finds the optimum with precision ε and with probability at least $1 - \delta$; this number will be termed hitting time and denoted by $n_{\epsilon,\delta}$.

Normalizations matter. In many cases, $n_{\epsilon,\delta}$ depends linearly in the dimension; therefore, we will consider normalizations so that a measure of convergence is the same for any dimensionality. This is the reason for our definition or convergence ratio, below. Incidentally, this definition will have the advantage that it can be used both for discrete and for continuous domains. We will see that recovering the convergence rate from the convergence ratio is not difficult.

11.4. Branching Factor and Convergence Speed

We consider a bounded domain $D \subset \mathbb{R}^d$ and a norm $\|\cdot\|$ on $\mathbb{R}^d$. For $\varepsilon > 0$, we define $N(\varepsilon)$ to be the maximum integer n such that there exist n distinct points $x_1, \ldots, x_n \in D$ with $\|x_i - x_j\| \geqslant 2\varepsilon$ for all $i \neq j$. $N(\varepsilon)$ is termed the packing number, and quantifies the "size" and "dimensionality" of the domain as explained below. In particular, $N(\varepsilon) = |D|$ when ε is small enough in the case of a finite domain D, and $\log N(\varepsilon) \sim d\log(1/\varepsilon)$ when $\varepsilon \to 0$ if the domain D is bounded with non-empty interior (Kolmogorov and Tikhomirov, 1961).

Please notice that these definitions are consistent both for continuous domain $D \subset \mathbb{R}^d$ and discrete domain $D \subset \mathbb{R}^d$.

If each function $f \in \mathcal{F}$ has one and only one optimum $x^*(f) \in D$, for any given optimization algorithm as in Algorithm 21, and for $\varepsilon > 0$ and $\delta > 0$, we define $n_{\varepsilon,\delta}$ be the minimum number n of iterations such that with probability at least $1 - \delta$, an optimum is found at the n-th iteration within distance ε; i.e., $n_{\varepsilon,\delta}$ is minimal such that for all $n \geqslant n_{\varepsilon,\delta}$ and for all $f \in \mathcal{F}$,

$$\Pr_\omega[\|x_{\omega,n}^{(f)} - x^*(f)\| < \varepsilon] \geqslant 1 - \delta.$$

In order to state lower bounds on the convergence speed of evolution strategies obtained by Fournier and Teytaud (2008); Teytaud and Gelly (2006), we first need to introduce a couple of definitions. Consider an algorithm of type $(\mu \dagger \lambda)$-ES working over a set of fitness functions $\mathcal{F}$. Let us define $L_n(\omega)$, the number of different paths followed by the algorithm on the random seed ω after n steps of computation when the function f runs over $\mathcal{F}$. More precisely,

$$L_n(\omega) = |\{(I_0, I_1, \ldots, I_n) \ : \ f \in \mathcal{F}\}|,$$

where the states I_i in the sequence above implicitly depend on both the function f and the random seed ω.

The *branching factor* K of this algorithm is defined as

$$K = \sup_E |\{\text{select}(E, f) \ : \ f \in \mathcal{F}\}|,$$

where the supremum holds for:

- E any set of λ in the case of SB-(μ, λ)-ES or Full Ranking (μ, λ)-ES;
- E any set of $\mu + \lambda$ in the case of SB-$(\mu + \lambda)$-ES or Full Ranking $(\mu + \lambda)$-ES.

The crucial idea around the concept of branching factor is that the number of different paths followed by some algorithm can be bounded in terms of the branching factor as follows:

$$L_n(\omega) \leqslant K \cdot L_{n-1}(\omega)$$

for all $n > 0$. Since the number of different points proposed after n steps of computation – denoted by $x_{w,n}^{(f)}$ in Algorithms 20 and 21 – is bounded by $L_n(\omega)$, the algorithm should converge slowly if $L_n(\omega)$ is small. This is quantified by the following result from Teytaud and Gelly (2006), restricted here to our purpose, relating the convergence speed to the branching factor of a $(\mu \overset{+}{,} \lambda)$-algorithm.

Theorem 11.1 (Lower bound on the hitting time of $(\mu \overset{+}{,} \lambda)$-ES). *Consider a Full Ranking (μ, λ)-ES or $(\mu+\lambda)$-ES, as defined in Algorithm 21, and a set $\mathcal{F}$ of fitness functions on domain D, i.e., F is a set of functions from D to $\mathbb{R}$, i.e., $\mathcal{F} \subset \mathbb{R}^D$, such that any fitness function $f \in \mathcal{F}$ has only one min-argument $x^*(f)$, and such that $\{x^*(f) \,:\, f \in \mathcal{F}\} = D$. Let $\varepsilon > 0$ and $\delta \in]0,1[$. Let $L_n(\omega)$ be the number of different paths (when the function f runs over $\mathcal{F}$) followed by the algorithm on the random seed ω. Then*

$$E_\omega[L_{n_{\varepsilon,\delta}}(\omega)] \geqslant (1-\delta)N(\varepsilon).$$

In particular, if K denotes the branching factor of the algorithm, then

$$n_{\varepsilon,\delta} \geqslant \frac{\log(1-\delta)}{\log K} + \frac{\log N(\varepsilon)}{\log K}.$$

We point out that the assumptions on $\mathcal{F}$ are very mild; essentially we only assume that any point in the domain is a possible optimum.

We now define a quantity capturing the convergence speed of evolution strategies. Our aim is to have a unifying notion for both discrete and continuous domains, with a quantity analogous to the bounds of the form $O(1/d)$ established by Witt and Jägersküpper (2005) when $D \subset [0,1]^d$. We call *convergence ratio* of an algorithm for precision ε and parameters (μ, λ) the following:

$$\mathrm{CR}_\varepsilon^{(\mu,\lambda)} = \frac{\log N(\varepsilon)}{dn_{\varepsilon,\frac{1}{2}}}.$$

This quantity is very related to convergence rate - however, it is defined also in discrete cases, in which convergence rate does not make sense. People who are familiar with convergence rate will recover it by Formula 11.1 when both quantities exist.

$$-\log(\textit{convergence rate}) = \mathrm{CR}_\varepsilon^{(\mu,\lambda)}. \tag{11.1}$$

It is known (Auger, 2005) that some evolution strategies lead to

$$\lim_{\varepsilon\to 0} \mathrm{CR}_\varepsilon^{(\mu,\lambda)} = O(1/d)$$

as $d \to \infty$, for a fixed μ and λ, e.g., on the sphere function or OneMax[§]. Note that the convergence ratio increases when the algorithm converges faster. Hence, a lower bound on the number of iterations of an algorithm for a given precision and a given confidence corresponds to an upper bound on its convergence ratio.

The ratio $CR_\varepsilon^{(\mu,\lambda)}$ is relevant in the parallel setting, i.e., when working on a parallel computer, with parallel evaluation of the offspring. In the sequential setting (i.e., when individuals are evaluated sequentially), one should consider the *normalized convergence ratio* (normalized by the number of individuals generated per iteration) defined by $NCR_\varepsilon^{(\mu,\lambda)} = CR_\varepsilon^{(\mu,\lambda)}/\lambda$. In this chapter we will provide explicit statements on $NCR_\varepsilon^{(\mu,\lambda)}$. Under some approximation, it has been shown that $NCR_\varepsilon^{(\mu,\lambda)}$ is roughly proportional to $\log(\lambda)/\lambda$ for $\mu = 1$, and roughly independent of λ for μ freely chosen (Beyer, 1995). We will here show that the latter is true only in the case of λ small in front of the dimension and we will precisely quantify this and its relation with parallel speed-up, both in the continuous and the discrete cases. Theorem 11.1 can be rewritten in terms of the convergence ratio as follows. Consider a $(\mu \overset{+}{,} \lambda)$-ES satisfying the hypothesis of Theorem 11.1. Let $\alpha(\varepsilon) = (1 - 1/\log N(\varepsilon))^{-1}$ (we shall use this notation throughout the chapter.) Then

$$CR_\varepsilon^{(\mu,\lambda)} \leqslant \frac{\log K}{d} \cdot \alpha(\varepsilon) \tag{11.2}$$

and of course $NCR_\varepsilon^{(\mu,\lambda)} \leqslant \frac{\log K}{d\lambda} \cdot \alpha(\varepsilon)$. This type of equation will be helpful for expressing results in a unified manner.

11.5. Results in the General Case

In this section, directly inspired by Teytaud and Gelly (2006), we do not assume anything on the fitness function. The results are therefore very general.

Formalizing the complexity of optimization algorithms is not a trivial task. In some cases, optimization is termed *expensive*: this means that the computational cost lies essentially in the evaluations of the fitness functions. Then, the cost is the number of fitness evaluations, or possibly the number of iterations in the case of parallel optimization (when all the population is evaluated simultaneously). The other extremal case is when the fitness

[§] Some authors (but not all) consider evolution strategies only on continuous domains; however, our definition does not forbid discrete cases and our results include this case.

function is very cheap - then, the internal cost of the algorithm has to be taken into account. A particular lower bound on the internal cost of the algorithm is the number of comparisons performed; the following theorem therefore lower bounds the number of comparisons necessary for reaching a given precision with a given confidence; the next subsection will consider the case of lower bounds on the number of fitness evaluations.

11.5.1. *Lower Bound on the Number of Comparisons*

As pointed out above, the number of comparisons is always a lower bound on the computational cost; if the cost of the fitness function is low and if the internal cost of the optimization algorithm is low, then this becomes the main computational cost. There are other cases in which the number of comparisons is a very natural criterion: when it is the only available information, as, *e.g.*, when parametrizing a strategy for two-players games. The following result is an immediate consequence of Theorem 11.1.

Theorem 11.2 (Convergence rate w.r.t the number of comparisons). Assume the same hypothesis as in Theorem 11.1 with $K = 2$ corresponding to the result of a comparison between the fitnesses of two previously visited points: SB-(μ, λ)-ES *with* $\mu = 1$, $\lambda = 2$. Importantly, this is not restricted to algorithms usually formalized with $\mu = 1$, $\lambda = 2$: *all* algorithms based on comparisons can be rewritten under this form, with one iteration of the algorithm (with population size 2) for each comparison (the population is the couple of points to be compared).

Then, with $\log_2(x) = \log(x)/\log(2)$, the number of comparisons n_c required for ensuring with probability $1 - \delta$ a precision ε is $n_c \geqslant \log_2(1 -$

Algorithm 22 A simplified version of the Hooke&Jeeves algorithm, matching the bound in Theorem 11.2. It is a $(1+1)$-ES (the difference between SB and FR does not make sense in $(1+1)$-ES).

```
Initialize x = (x_1, ..., x_d) at any point.
for In lexicographic order on (j, i) ∈ ℕ × [[0, d − 1[[ do
    Try to replace the j^th bit b of x_i by 1 − b; let x' be the result.
    Evaluate the fitness at x'
    if x' is better than x then
        x = x'
    end if
end for
```

$\delta) + \log_2(N(\varepsilon))$. *I.e., formally,* $\Pr(\|x_{n_c} - x^*(f)\| < \varepsilon) \geqslant 1 - \delta \Rightarrow n_c \geqslant \log_2(1-\delta) + \log_2(N(\varepsilon))$.

This bound, leading to $\lim_{\varepsilon \to 0} \mathrm{CR}_\varepsilon{}^{(1,2)} = O(1/d)$ is tight: e.g., for $D = [0,1]^d$, $\mathcal{F} = \{x \mapsto \|x - x^*(f)\|_1; x^*(f) \in [0,1]^d\}$ is solved with convergence ratio $\Omega(1/d)$ by Algorithm 22 (close to the Hooke&Jeeves algorithm by Hooke and Jeeves (1961)).

11.5.2. *Lower Bound on the Number of Function Evaluations*

We already mentionned that our approach covers almost all existing evolutionary algorithms. We can now check the value of K, depending on the algorithm, and consider convergence rates with respect to the number of fitness-evaluations instead of the number of comparisons. The convergence ratio will verify Equation (11.2).

- (μ, λ)**-ES (or SA-ES)** : at each step, then, we only know which are the selected points. Then, $K = \binom{\lambda}{\mu} \leqslant \binom{\lambda}{\lfloor \lambda/2 \rfloor} \leqslant (2^\lambda/\sqrt{2\pi\lambda})$ (see (Devroye *et al.*, 1997, p587) or (Feller, 1968) for classical properties of $\binom{\lambda}{\mu}$).
- Consider **more generally, any selection based algorithm**, i.e., any algorithm in which v_n encodes only a subset of μ points among λ. Then, the algorithm provides only a subset of $[[1,\lambda]]$, i.e., $K \leqslant 2^\lambda$ (even if μ is not fixed and chosen by the algorithm at each iteration). Therefore, $\mathrm{CR}_\varepsilon^{(\mu,\lambda)} \leqslant \frac{\lambda}{d} \cdot \alpha(\varepsilon)$. This bound can be outperformed by full-ranking algorithms as shown by Teytaud and Gelly (2006); this shows that in some cases, full-ranking can significantly (more than of a constant factor only) outperform selection. However, we will see that this does not hold anymore under some mild assumptions, at least for (μ, λ)-ES.
- $(\mu + \lambda)$**-ES:** then, we only know which are the selected points. Then, $K = (\lambda + \mu)!/(\mu!\lambda!)$. This does not allow a proof of $\mathrm{CR}_\varepsilon^{(\mu,\lambda)} = \Theta(\lambda/d)$ if μ increases as a function of d. This point will be further analyzed using VC-dimension in the next section; essentially the ultimate rates for elitist strategies (+) are an open problem.

Teytaud and Gelly (2006) also show that superlinear convergence rates are possible, in some very specific cases, when using the full ranking information. Next section will show that this is not possible for more reasonnable fitness functions (as shown in Table 11.1).

11.6. Bounds on the Convergence Speed using the VC-dimension

Lower bounds on the convergence speed of evolutions strategies can be obtained from Equation (11.2) (Section 11.4) as soon as an upper bound on the branching factor K is known. In the results above, we used essentially elementary combinatorial properties: in particular, the fact that the number of subsets of size μ of a set of λ points (and thus the branching factor K) is at most $\binom{\lambda}{\mu} \leqslant \binom{\lambda}{\lfloor \lambda/2 \rfloor}$. This surely holds, but it is a worst case on all possible selections. If we have the additional hypothesis that the fitness functions are *reasonnable*, many subsets of size μ of a set of size λ cannot be realized by a selection step. This is precisely quantified by the theory of VC-dimension and the shatter function lemma (also known as Sauer's lemma). In this section, we show how a VC-dimension hypothesis allows to obtain improved lower bounds on the convergence speed of $(\mu \overset{+}{,} \lambda)$-ES. VC-dimension is the central tool of many works, in particular for learning theory (Vapnik and Chervonenkis, 1968).

Given a function f defined over D and $r \in \mathbb{R}$, let $B_{f,r} = \{x \in D : f(x) < r\}$. We define the *sublevel sets* $L_\mathcal{F}$ of a set $\mathcal{F}$ of functions defined over the domain D as

$$L_\mathcal{F} = \{B_{f,r} \ : \ f \in \mathcal{F},\ r \in \mathbb{R}\}.$$

This section presents lower bounds on the convergence speed of algorithms optimizing $\mathcal{F}$ in terms of the VC-dimension of the sublevel sets $L_\mathcal{F}$. The results were originally proved by Fournier and Teytaud (2008).

We now briefly recall the definition of VC-dimension and the shatter function lemma (Vapnik and Chervonenkis, 1971; Sauer, 1972) – our presentation is based on the work by Matoušek (2002). A set system on a set A is a family $\mathcal{S}$ of subsets of A. For $B \subseteq A$, we define the restriction of $\mathcal{S}$ to B as $\mathcal{S}|_B = \{S \cap B \ : \ S \in \mathcal{S}\}$. The VC-dimension of the set system $\mathcal{S}$ defined over A is defined as $\sup\{|B| \ : \ \mathcal{S}|_B = 2^B\}$ where 2^B denotes the power set of B; in other words, it is the size of the largest subset B of A such that any subset of B can be obtained by intersecting B with an element of $\mathcal{S}$. Given a set system $\mathcal{S}$ over A, the *shatter function* $\pi_\mathcal{S}$ is defined by $\pi_\mathcal{S}(m) = \max\{|\mathcal{S}|_B| \ : \ B \subseteq A,\ |B| = m\}$; thus $\pi_\mathcal{S}(m)$ is the maximum number of different subsets of A which can be obtained by intersecting a single subset of size m of A with all elements of $\mathcal{S}$. We next recall the following lemma which gives an upper bound on $\pi_\mathcal{S}$ in terms of the VC-dimension of $\mathcal{S}$.

Lemma 11.3 (Shatter function lemma). *For any set system $\mathcal{S}$ of VC-dimension d, then for all integer m, it holds that $\pi_{\mathcal{S}}(m) \leqslant \sum_{i=0}^{d} \binom{m}{i}$.*

At last, let us recall the following classical bound (Devroye *et al.*, 1997) which is valid whenever $d \geqslant 3$:

$$\sum_{i=0}^{d} \binom{m}{i} \leqslant m^d. \tag{11.3}$$

Note that the trivial bound $\sum_{i=0}^{d} \binom{m}{i} \leqslant 2^m$ is tight when $m \leqslant d$. The interesting case happens when m is large with respect to the VC-dimension d: the bound stated in Equation (11.3) becomes polynomial in m in this case.

We now give a couple of examples of bounded VC-dimension for some classical set systems in $\mathbb{R}^d$ (Devroye *et al.*, 1997). Axis-parallel hyper-rectangles have a VC-dimension bounded by $2d$. Sphere functions in $\mathbb{R}^d$ (for the Euclidean norm) have a VC-dimension equal to $d+1$. The VC-dimension of ellipsoids in $\mathbb{R}^d$ is bounded by $d(d+1)/2+d$. More generally, subsets of $\mathbb{R}^d$ defined by polynomial inequalities involving a finite number of real parameters, and Boolean combinations of these, have a finite VC-dimension (Matoušek, 2002, chapter 10.3). Also, any set of functions which is the set of translations of a given function has finite VC-dimension.

In the rest of this section, we shall state bounds for set systems of VC-dimension at least 3. However, the case of VC-dimension smaller than 3 can be handled in a similar way: the bound above has to be replaced with $\sum_{i=0}^{d} \binom{m}{i} \leqslant m^d + 1$.

Upper bounds on the convergence ratio of evolution strategies obtained in this section are summarized and compared to results of the previous section in Table 11.1 (we recall that $\alpha(\varepsilon) = (1 - 1/\log N(\varepsilon))^{-1}$).

Hence, any algorithm of type $(\mu \overset{+}{,} \lambda)$-ES optimizing a set of fitness functions with sublevel sets of VC-dimension $\leqslant V$ on a domain $D \subset \mathbb{R}^d$ has a convergence ratio which satisfies the bounds given in Table 11.1. Let us remark that the sphere functions, i.e., the set of fitness functions $\mathcal{F} = \{f_c : c \in \mathbb{R}^d\}$ where $f_c(x) = \|x - c\|_2$, are a special case of functions with spheres as sublevel sets, but are not the only ones. (The same remark applies to hyper-rectangles and ellipsoids.)

We now derive specific results for various families of ES: non-elitist strategies (μ, λ)-ES, non-elitist full ranking ES, and elitist $(\mu + \lambda)$-ES.

11.6.1. *Selection-Based Non-Elitist Evolution Strategies*

Theorem 11.4 (Selection based (μ, λ)-ES). *Consider an SB-(μ, λ)-ES (Algorithm 20) in a domain $D \subset \mathbb{R}^d$, such that $D = \{x^*(f) : f \in \mathcal{F}\}$. Let $V \geqslant 3$ be the VC-dimension of the sublevel sets of $\mathcal{F}$. The convergence ratio of this algorithm satisfies*

$$\mathrm{CR}_{\varepsilon}^{(\mu,\lambda)} \leqslant \frac{V \log \lambda}{d} \cdot \alpha(\varepsilon),$$

or equivalently

$$n_{\varepsilon,\frac{1}{2}} \geqslant \frac{\log N(\varepsilon)\alpha(\varepsilon)}{V \log(\lambda)}.$$

The detailed proof is given by Fournier and Teytaud (2008) and combines a VC-type bound and Equation (11.2) from Theorem 11.1.

Remark. If $V = 1$ or $V = 2$, then the bound obtained in Theorem 11.4 becomes $\mathrm{CR}_{\varepsilon}^{(\mu,\lambda)} \leqslant \frac{V \log(\lambda+1)}{d} \cdot \alpha(\varepsilon)$. The case $V < 3$ can be handled in a similar way throughout the chapter: $V \log \lambda$ is replaced with $V \log(\lambda + 1)$ in this case.

The bound obtained in Theorem 11.4 is interesting when λ is large, and μ not too close to 1 or λ. Indeed, the trivial bound $K \leqslant \binom{\lambda}{\mu}$ combined with Equation (11.2) as above, yields better bounds on the convergence ratio in these cases.

We now briefly consider Full Ranking (μ, λ) evolution strategies. That is, we move from Algorithm 20 to the more general setting of Algorithm 21.

11.6.2. *Full Ranking (μ, λ)-ES with μ Not Too Large*

For evolutions strategies of type Full Ranking (μ, λ)-ES, an upper bound on the number of possible outcomes of the selection step (including the ranking of children) is obtained by multiplying by $\mu!$ the number of possible outcomes in the case of selection only. This multiplies the branching factor K accordingly and leads to

$$\mathrm{CR}_{\varepsilon}^{(\mu,\lambda)} \leqslant \frac{1}{d}(V \log(\lambda) + \mu \log \mu) \cdot \alpha(\varepsilon).$$

However, we can say better in the case where μ is large with respect to the VC-dimension V of the sublevel sets of the fitness functions; this will be the subject of the next section.

Application to optimization of the OneMax function

We end this section with an application by Fournier and Teytaud (2008) to optimization in the discrete domain $D = \{0,1\}^d$. For ε sufficiently small, the balls are singletons; it follows that $N(\varepsilon) = N(0) = 2^d$ and $\alpha(\varepsilon) = \alpha(0) = 1/(1-1/d)$ when ε is small enough and $d \geqslant 2$. Let us consider the ONEMAX function defined by $x \longmapsto \sum_{i=1}^{d} x_i$, and all its symmetries; the set of fitness functions is

$$\mathcal{F} = \{f_\eta : x \longmapsto \sum_{i \in [[1,d]]} |x_i - \eta_i|,\ \eta \in \{0,1\}^d\}.$$

Classical results around ONEMAX are as follows:

- A (1+1)-ES finds the optimum after $\Theta(d\log(d))$ iterations on average[¶];
- All black-box algorithms need average computation time at least $\Theta(d\log(d))$.

Let us discuss the case of Selection based (μ,λ)-ES for this family of functions. The main strength of results below is that we improve these lower bounds as a function of λ for the whole family of selection-based evolutionary algorithms. The aim is to find the point where this function is maximum (hence, the selection step of an $(\mu \overset{+}{,} \lambda)$-ES keeps μ points with *highest* fitness values). Notice that no $(\mu \overset{+}{,} \lambda)$-ES can avoid to generate points with equal fitness values in a single step as soon as $\lambda \geqslant 3$. Indeed, given three different points of $\{0,1\}^d$, there must be two of them x and y such that the Hamming weight of the symmetric difference of x and y is even; then there exists η such that $f_\eta(x) = f_\eta(y)$.

The VC-dimension of sublevel sets of $\mathcal{F}$ satisfies $V \leqslant d+1$ (Devroye *et al.*, 1997, chapter 13). Therefore, the convergence ratio $\mathrm{CR}_\varepsilon^{(\mu,\lambda)}$ is at best, for ε sufficiently small, $2(1-1/d^2)\log\lambda$ in the case of Selection based (μ,λ)-ES. Bounds on the convergence ratio of other kinds of strategies are obtained in the same way. As well as in many cases for the continuous domain, we get $O(\log(\lambda))$ independently of the dimension; this is a consequence of the definition of $\mathrm{CR}_\varepsilon^{(\mu,\lambda)}$.

The bound on the convergence ratio obtained above yields the following runtime for solving ONEMAX in dimension d with probability 1/2:

$$\begin{cases} n_{0,\frac{1}{2}} = \Omega\left(1/\log(\lambda)\right), & \text{if } \lambda/\log\lambda \geqslant d, \text{ (degenerate bound)} \\ n_{0,\frac{1}{2}} = \Omega\left(d/\lambda\right), & \text{if } \lambda/\log\lambda < d. \end{cases} \tag{11.4}$$

[¶] This is realized by a mutation in which each bit is flipped independently with probability $1/d$.

The case where $\lambda \leqslant 2d$ above is obtained by Equation (11.2), using the trivial bound $K \leqslant \binom{\lambda}{\lambda/2} \leqslant \binom{2d}{d}$.

We can give some elements of tightness in the (quite strong) model where the algorithm has access to the whole set of points in $\{x \in E_n; \forall y \in E_n, \text{ONEMAX}(x) \leqslant \text{ONEMAX}(y)\}$; this algorithm has access to the set of points with optimal fitness in E_n, of size possibly larger than μ in case of equality. Indeed, Equation (11.4) is tight in this model in the special cases $\lambda = d+1$ and $\lambda = O(1)$ (independent of d); the proof is omitted for short.

For the sake of parallelization of ONEMAX solving with ES, this suggests the use of λ linear in the dimension d; at least for optimally parallel comparison-based algorithms. Indeed, for a parallel evaluation of the λ generated points, we get an almost linear speed-up for $\lambda \leqslant d$ – the convergence ratio is $\Omega(1)$ for $\lambda = d+1$ whereas it is $O(1/d)$ for $\lambda = O(1)$ – while the speed-up is asymptotically logarithmic in λ when $\lambda \to \infty$.

11.6.3. *Full Ranking (μ, λ)-ES with μ Large*

Using a lemma by (Fournier and Teytaud, 2008) around the number of possible rankings of λ points by functions with finite VC-dimension, the following theorem provides an upper bound on the number of possible orders of a fixed set of points with respect to fitness functions $f \in \mathcal{F}$, when the sublevel sets of $\mathcal{F}$ have a finite VC-dimension.

Theorem 11.5 (Full ranking (μ, λ)-ES). *Consider a (μ, λ)-ES (Algorithm 21) in a domain $D \subset \mathbb{R}^d$, such that $D = \{x^*(f) \ : \ f \in \mathcal{F}\}$. Let $V \geqslant 3$ be the VC-dimension of the sublevel sets of $\mathcal{F}$. The convergence ratio of this algorithm satisfies*

$$\text{CR}_{\varepsilon}^{(\mu,\lambda)} \leqslant \frac{V\,(4\mu + \log\lambda)}{d} \cdot \alpha(\varepsilon),$$

or equivalently

$$n_{\varepsilon,\frac{1}{2}} \geqslant \frac{\log(N(\varepsilon)\alpha(\varepsilon)}{V\,(4\mu + \log\lambda)}$$

An important point here is the μ out of the logarithm. We will see that in some cases (typically the sphere function) we can do better.

11.6.4. *Elitist Case: $(\mu + \lambda)$-ES*

The elitist case (full ranking or not) is a corollary of Theorem 11.4.

Corollary 11.6 (Selection based $(\mu+\lambda)$-ES and full ranking $(\mu+\lambda)$-ES). Let $\mathcal{F}$ be a set of fitness functions defined in a domain $D \subset \mathbb{R}^d$, such that $D = \{x^*(f) \ : \ f \in \mathcal{F}\}$. Let V be the VC-dimension of the sublevel sets of $\mathcal{F}$. The convergence ratio of an algorithm of type Selection Based $(\mu+\lambda)$-ES for $\mathcal{F}$ satisfies:

$$\mathrm{CR}_{\varepsilon}^{(\mu,\lambda)} \leqslant \frac{V \log(\mu+\lambda)}{d} \cdot \alpha(\varepsilon),$$

or equivalently

$$n_{\varepsilon,\frac{1}{2}} \geqslant \frac{\log(N(\varepsilon))\alpha(\varepsilon)}{V \log(\mu+\lambda)}.$$

For an algorithm of type Full Ranking $(\mu+\lambda)$-ES, the following holds:

$$\mathrm{CR}_{\varepsilon}^{(\mu,\lambda)} \leqslant \frac{V\,(4\mu + \log(\mu+\lambda))}{d} \cdot \alpha(\varepsilon),$$

or equivalently

$$n_{\varepsilon,\frac{1}{2}} \geqslant \frac{\log(N(\varepsilon))\alpha(\varepsilon)}{V\,(4\mu + \log(\lambda))}.$$

11.7. Sign Patterns and the Case of the Sphere Function

This sections presents a technique proposed by Fournier and Teytaud (2008) based on the number of sign patterns to derive lower bounds on the convergence speed of full ranking algorithms. Applied to the special case of the sphere function, it shows that the speed-up is asymptotically at most logarithmic in λ (see Proposition 11.6). This improves on the bounds obtained by VC-dimension arguments in Theorem 11.5 which did not forbid a speed-up linear in μ.

Although it is applied to the sphere function, the technique used here applies to any system where the number of sign patterns can be efficiently bounded, such as quadratic functions with positive Hessian. In fact, polynomials of bounded degree are amenable to this technique – we refer the reader to the works by Rónyai *et al.* (2001) and Matoušek (2002) (chapter 6.2) for further details. However, as opposed to the previous section, the bound obtained by the sign pattern technique presented here does not apply to the general case of functions with spheres (or ellipsoids) as sublevel sets[‖]. On the other hand, as all results in this chapter, they can be applied to any monotonic transformations of sphere functions.

[‖]We recall that there are functions which have spheres as level sets (all level sets are spheres) without these sphere to be centered on the optimum.

For the sphere function, we point out in Section 11.7.2 that λ linear in the dimension provides an almost linear speed-up. Indeed, the straightforward parallelization performed by distributing the λ fitness evaluations over λ processors has an almost linear speed-up until at least λ equal to twice the dimension (while the speed-up is asymptotically at most logarithmic in λ by Proposition 11.7.)

11.7.1. *Lower Bounds on Full Ranking Strategies via the Number of Sign Patterns*

We present an alternative method to obtain improved lower bounds on the convergence speed of evolution strategies which use full ranking. For a real x, we define its sign to be $\text{sign}(x) = 0$ if $x = 0$, $\text{sign}(x) = 1$ if $x > 0$, and $\text{sign}(x) = -1$ if $x < 0$. Given a fitness function f and a finite set of points $x_1, \ldots, x_\lambda \in D$, we define the set of realizable sign conditions as

$$\text{Sign}_{f,(x_1,\ldots,x_\lambda)} = \{\text{sign}(f(x_i) - f(x_j)) : 1 \leqslant i < j \leqslant \lambda\}.$$

This method can be applied as soon as the number of sign conditions, i.e., the number of possible sign vectors, can be efficiently bounded. Indeed, the following inequality on the branching factor holds:

$$K \leqslant \max\left\{|\text{Sign}_{f,(x_1,\ldots,x_\lambda)}| : f \in \mathcal{F},\, x_1, \ldots, x_\lambda \in D\right\}. \tag{11.5}$$

We now apply the above remark to the sphere functions, in order to obtain an improved lower bound on the convergence speed of full ranking strategies for these functions. We recall that the sphere function is** in fact the the set of fitness functions $\mathcal{F} = \{f_c \ : \ c \in \mathbb{R}^d\}$ with $f_c(x) = \|x - c\|_2$ (where $\|\cdot\|_2$ denotes the Euclidean norm, i.e., $f_c(x) = ((x_1 - c_1)^2 + \ldots + (x_d - c_d)^2)^{1/2}$).

Proposition 11.7. *Let $d \geqslant 3$. Consider a Full Ranking (μ, λ)-ES, as in Algorithm 21, optimizing the sphere function in a domain $D \subset \mathbb{R}^d$. Then*

$$\text{CR}_\varepsilon^{(\mu,\lambda)} \leqslant 2\log(\lambda) \cdot \alpha(\varepsilon).$$

Remark. The case of equality (i.e., when it is possible for two generated points x and y to satisfy $f(x) = f(y)$) can be handled thanks to bounds on the number of faces of the hyperplanes arrangements, using the work by Matoušek (2002) (chapter 6.1).

**However, for comparison-based algorithms, replacing a function by its composition with an increasing function from $\mathbb{R}$ to $\mathbb{R}$ does not change the behavior of the algorithm; therefore, the sphere function is often extended to all its compositions with increasing functions.

11.7.2. *Fast Convergence Ratio with $\lambda = 2d$ for Optimizing the Sphere Function*

We point out here that for the specific case of the sphere function, a convergence ratio of $\Theta(1)$ can be reached with $\mu = \lambda = 2d$ in the domain $[0,1]^d$ by some algorithm of type Full Ranking (μ, λ)-ES.

This convergence ratio is easily obtained with the following algorithm in the spirit of the Hooke and Jeeves algorithm (Hooke and Jeeves, 1961). Let e_i denote the vector $(0,\ldots,0,1,0,\ldots,0)$ with a unique 1 in position i. First split $[0,1]^d$ into the 2^d cells delimited by the d hyperplanes of equations $x_i = 1/2$; the full ranking of the $2d$ points $\{(\frac{1}{2},\frac{1}{2},\ldots,\frac{1}{2}) + \frac{\eta}{2}e_i \ : \ 1 \leqslant i \leqslant n,\ \eta \in \{-1,1\}\}$ allows to decide in which of these cells the optimum lies; then the algorithm proceeds recursively. (See Algorithm 23.)

Algorithm 23 Fast algorithm of type Full Ranking (μ, λ)-ES for the sphere function in the domain $[0,1]^d$ in the special case $\mu = \lambda = 2d$.

$x \leftarrow (1/2,\ldots,1/2)$; $\sigma \leftarrow 1/2$
while true **do**
  Generate $\lambda = 2d$ distinct points equal to $x \pm \sigma e_i$
    (e_i denotes the vector $(0,\ldots,0,1,0,\ldots,0)$ with a unique 1 in position i)
  With the ranking information, decide in which octant Δ of $x + [-\sigma,\sigma]^d$ is the optimum
  Move x to the center of the octant Δ
  Set $\sigma \leftarrow \sigma/2$
end while

After n iterations, the point x_n proposed by this algorithm satisfies $\|x_n - x^*(f)\|_2 \leqslant \sqrt{d}/2^n$. Moreover, this distance is realized by some fitness function. It follows that $n_{\varepsilon,\frac{1}{2}} = \log\frac{1}{\varepsilon} + \frac{1}{2}\log d$. On the other hand $\log(N(\varepsilon)) = \Theta(d\log\frac{1}{\varepsilon})$. Thus, we have obtained an algorithm for the sphere function in dimension d which satisfies:

$$\mathrm{CR}_\varepsilon^{(2d,2d)} = \Theta(1). \tag{11.6}$$

Notice that for $\lambda = O(1)$, independent from d, the branching factor of any algorithm satisfies $K = O(1)$; it follows by Equation (11.2) that any algorithm optimizing the sphere function in dimension d satisfies $\mathrm{CR}_\varepsilon^{(\lambda,\lambda)} = O(1/d)$ in this case. Hence, Algorithm 23 achieves an almost linear speed-up $\Omega(d)$ when λ moves from $O(1)$ to $2d$. On the other hand, the asymptotic speed-up for λ large (and d fixed) is known to be at most logarithmic ($\mathrm{CR}_\varepsilon^{(\lambda,\lambda)} = O(\log(\lambda))$ by Proposition 11.7).

11.8. Conclusions and Further Work

The main strength of the approach presented above is its generality: it can be applied to evolution strategies, direct search methods, cross-entropy methods, estimation of distribution algorithms. It shows that

Table 11.1. Upper bound on the convergence ratio. The first column is the general case (no assumption on the set of fitness functions except that the optimum can be anywhere in the domain). The second column is when the sublevel sets of fitness functions have VC-dimension V in $\mathbb{R}^d$. The third column is just the application of the second column to the case of convex quadratic functions ($V = \Theta(d^2)$). The fourth column is the special case of the sphere function, or, equivalently, any set of functions in which the sublevel sets are homotheties of a fixed ellipsoid.

Framework	General case	Bounded VC-dimension	Quadratic case	Sphere function
SB-(μ,λ)-ES	$\frac{1}{d}(\lambda\log(2) - \frac{1}{2}\log(2\pi\lambda))$	$\frac{V}{d}\log(\lambda)$	$O(d)\log(\lambda)$	$(1+\frac{1}{d}) \times \log(\lambda)$
SB-$(\mu+\lambda)$-ES	$\frac{1}{d}\log(\frac{(\lambda+\mu)!}{\lambda!\mu!})$	$\frac{V}{d}\log(\mu+\lambda)$	$O(d)\log(\mu+\lambda)$	$(1+\frac{1}{d}) \times \log(\mu+\lambda)$
FR-(μ,λ)-ES		$\frac{V}{d}(4\mu+\log\lambda)$	$O(d)(4\mu+\log\lambda)$	$2\log(\lambda)$
FR-$(\mu+\lambda)$-ES		$\frac{V}{d}(4\mu+\log(\lambda+\mu))$	$O(d)(4\mu+\log(\lambda+\mu))$	$(1+\frac{1}{d}) \times (4\mu+\log(\lambda+\mu))$

- with respect to the number of comparisons, the convergence is at best linear with convergence rate $\exp(-O(1/d))$ where d is the dimension;
- with respect to the number of fitness evaluations and for sphere functions, the convergence is at best linear with convergence ratio $O((1+1/d)\log(\lambda))$ where d is the dimension; dividing by λ, this leads to $(1+1/d)\log(\lambda)/\lambda$ for λ large, and $\Omega(1)$ for $\lambda = d$. This suggests that choosing λ of the same order of the dimension does not significantly reduces the efficiency on a sequential computer. The classical trick consisting in increasing λ for avoiding local minima has therefore no significant cost from the point of view of the rate, even on a sequential computer.
- with respect to the number of iterations (typically the running time on a parallel computer when the internal cost of the algorithm is negligible and with one evaluation of individual per processing unit), the convergence, in most cases, is at best linear; the convergence ratio $O(p/d)$ for a number $p = \Theta(d)$ of computation units, and then linear with convergence rate $O(\log(p)/d)$ for $p \gg d$; this is proved for bounded VC-

dimension and SB non-elitist algorithms, and also for FR non-elitist strategies applied to the sphere fitness function.

The most important points, for the author, are: (i) the presence of μ out of the logarithm in some cases; (ii) the $O(d)$ for the quadratic case: it's likely that if the considered space of functions is translation invariant then this should be $O(1)$. This is already the case in the quadratic case, if the sublevel sets are homotheties of one and only one ellipsoid (becasue in this case the VC-dimension is the same as for the sphere).

We have illustrated the results mainly in the continuous domain (bounded open subsets of $\mathbb{R}^d$), a little on $\{0,1\}^d$; similar results have also been applied in multiobjective cases (Teytaud, 2007). However, the scope of the result is probably larger; the main strength of evolutionary algorithms is that they are robust and stable in structured spaces; it is likely that one can apply these bounds for, *e.g.*, sets of permutations, sets of trees, neural networks with their structures, etc.

A main implication of our results is the limit on the parallelization. In the case of evolution strategies based on selection only (algorithms of type SB-(μ, λ)-ES), the linear speed-up of selection based evolution strategies shown by (Beyer, 1995) cannot be obtained for λ large enough. Asymptotically, the speed-up obtained with such an algorithm is at most logarithmic as shown in Theorem 11.4. On the other hand, we show that the speed-up can be nearly linear for up to $2d$ processors on the sphere function in dimension d. A similar phenomenon is observed in the case of the discrete function OneMax.

The case of equality between fitness values of distinct points is not handled here; it is possible to show that, in many cases, the bound is just augmented by a constant factor (Fournier and Teytaud, 2010).

A somewhat puzzling point is that bounds on the convergence speed obtained by VC-dimension arguments are weaker when the function is more "complex" (i.e., when the VC-dimension of its sublevel sets is higher). This is not surprising after all: it is easier to have very fast algorithms on very strange fitness functions than on "natural" fitness functions; indeed, one can wonder if it is possible to build *ad hoc* fitness functions matching the bounds obtained by VC-dimension arguments. Such constructions were given by (Teytaud and Gelly, 2006) to match lower bounds on the convergence speed of algorithms obtained from the sole branching factor[††] - this

[††]More precisely, (Teytaud and Gelly, 2006) show that there is a FR algorithm a and a set of fitness function F so that a is "very" fast on F, and in particular much more than

provides very strange fitness functions, of low practical interest but showing that improving the results requires some more assumption. For example, the lower bounds for quadratic fitness functions are improved (they become equivalent to sphere fitness functions) if we consider only quadratic fitness functions which are translations of only one quadratic fitness function.

Also we know that the bounds obtained from VC-dimension arguments can be far from optimal for a specific family of fitness functions (this does not imply that the bound is not tight in the general case of a given VC-dimension): the bound obtained for Full Ranking (μ, λ)-ES is greatly improved in the special case of the sphere function in Section 11.7.

A related question is the parallelization of optimization of ill-conditioned functions. Can the speed-up be linear until roughly $\lambda = d^2$ computation units?

When moving from algorithms of type Selection Based (μ, λ)-ES to Full Ranking (μ, λ)-ES, and in the general case of any set of fitness functions, the speed-up that can be achieved moves from logarithmic to linear in λ (we have only presented here only lower bounds for the general case, but Teytaud and Gelly (2006) have shown that they can be reached.). However, we know from Proposition 11.7 that the speed-up is at most logarithmic for a Full Ranking (μ, λ)-ES in the special case of the sphere function and this linear speed-up is therefore essentially theoretical – see also the discussion following Proposition 11.7. An important question is then the existence of important families of functions for which a linear speed-up can be achieved. The answer is probably yes within a setting adapted to highly multimodal cases (with much bigger hitting times however), and no in the "easy" case as in this paper (we need a family of functions which is more parallelizable than the sphere functions!).

A related intriguing question is the convergence ratio that can be reached for selection based algorithms (i.e., without keeping the full ranking information) for the sphere function. In particular, is it possible to achieve a constant convergence ratio with λ linear in the dimension, as in Equation (11.6)? To the best of our knowledge, this is an open problem - preliminary investigation suggests that the answer is positive.

Finally, the results can be extended beyond the comparison-based case. Fitness-based or gradient-based or Hessian-based algorithms are concerned also when taking into account the finite number bits of the answers from the oracle. In particular, it is known that theoretically quadratically converging

rates achieved by any algorithm on the sphere; this is of theoretical interest only as the corresponding algorithm is useless except for these specific families of functions.

algorithms are in fact linear due to finite precision. The finite number of bits is for sure a limitation which can be handled by bounds as in this chapter.

Acknowledgments

We would like to thank Anne Auger and Fabien Teytaud for constructive talks. We point out also that Hervé Fournier contributed a lot to this work, and could almost be called a coauthor. This work was partially supported by the Pascal Network of Excellence. This work has been supported by French National Research Agency (ANR) through COSINUS program (project EXPLO-RA nANR-08-COSI-004).

References

Arnold, D. V. (2005). Optimal weighted recombination, in *Foundations of Genetic Algorithms 8, Lecture Notes in Computer Science*, Vol. 3469 (Springer-Verlag, Berlin Heidelberg), ISBN 3-540-27237-2, pp. 215–237.

Auger, A. (2005). Convergence results for (1,λ)-SA-ES using the theory of φ-irreducible Markov chains, *Theoretical Computer Science* **334**, 1-3, pp. 35–69.

Auger, A., Schoenauer, M. and Teytaud, O. (2005). Local and global order 3/2 convergence of a surrogate evolutionary algorithm, in *GECCO '05: Proceedings of the 2005 conference on Genetic and evolutionary computation* (ACM, New York, NY, USA), ISBN 1-59593-010-8, pp. 857–864, doi:http://doi.acm.org/10.1145/1068009.1068154.

Bäck, T., Hoffmeister, F. and Schwefel, H.-P. (1991). Extended selection mechanisms in genetic algorithms, in R. K. Belew and L. B. Booker (eds.), *Proceedings of the Fourth International Conference on Genetic Algorithms* (Morgan Kaufmann Publishers, San Mateo, CA), pp. 92–99.

Baker, J. E. (1987). Reducing bias and inefficiency in the selection algorithm, in *Proceedings of the Second International Conference on Genetic Algorithms on Genetic algorithms and their application* (Lawrence Erlbaum Associates, Inc., Mahwah, NJ, USA), ISBN 0-8058-0158-8, pp. 14–21.

Baluja, S. (1994). Population-based incremental learning: A method for integrating genetic search based function optimization and competitive learning,, Tech. Rep. CMU-CS-94-163, School of Computer Science, Carnegie Mellon University, Pittsburgh, PA.

Beyer, H.-G. (1995). Toward a theory of evolution strategies: On the benefit of sex – the $(\mu/\mu, \lambda)$-theory, *Evolutionary Computation* **3**, 1, pp. 81–111.

Beyer, H.-G. (2001). *The Theory of Evolution Strategies*, Natural Computing Series (Springer, Heidelberg).

Beyer, H.-G. and Sendhoff, B. (2008). Covariance matrix adaptation revisited - the CMSA evolution strategy, in G. Rudolph, T. Jansen, S. M. Lucas, C. Poloni and N. Beume (eds.), *Proceedings of PPSN*, pp. 123–132.

Conn, A., Scheinberg, K. and Toint, L. (1997). Recent progress in unconstrained nonlinear optimization without derivatives, URL `citeseer.ist.psu.edu/conn97recent.html`.

de Boer, P.-T., Kroese, D., Mannor, S. and Rubinstein, R. (2005). A tutorial on the cross-entropy method, *Annals of Operations Research* **134**, 1, pp. 19–67, doi:10.1007/s10479-005-5724-z, URL `http://dx.doi.org/10.1007/s10479-005-5724-z`.

Devroye, L., Györfi, L. and Lugosi, G. (1997). *A probabilistic Theory of Pattern Recognition* (Springer).

Droste, S. (2005). Not all linear functions are equally difficult for the compact genetic algorithm, in *Proc. of the Genetic and Evolutionary Computation Conference (GECCO 2005)*, pp. 679–686.

Feller, W. (1968). *An introduction to Probability Theory and its Applications* (Willey).

Fournier, H. and Teytaud, O. (2008). Lower bounds for evolution strategies using VC-dimension, in G. Rudolph, T. Jansen, S. M. Lucas, C. Poloni and N. Beume (eds.), *PPSN*, *Lecture Notes in Computer Science*, Vol. 5199 (Springer), ISBN 978-3-540-87699-1, pp. 102–111.

Fournier, H. and Teytaud, O. (2010). Lower bounds for comparison based evolution strategies using VC-dimension and sign patterns, *Algorithmica* , pp. 1–22.

Gelly, S., Ruette, S. and Teytaud, O. (2007). Comparison-based algorithms are robust and randomized algorithms are anytime, *Evolutionary Computation Journal (MIT Press), Special issue on bridging Theory and Practice* **15**, 4, pp. 411–434.

Hansen, N. and Ostermeier, A. (2001). Completely derandomized self-adaptation in evolution strategies, *Evolutionary Computation* **9**, 2, pp. 159–195.

Harik, G. R., Lobo, F. G. and Goldberg, D. E. (1999). The compact genetic algorithm, *IEEE Trans. on Evolutionary Computation* **3**, 4, p. 287.

Hooke, R. and Jeeves, T. A. (1961). "Direct search" solution of numerical and statistical problems, *Journal of the ACM* **8**, 2, pp. 212–229.

Kolmogorov, A.-N. and Tikhomirov, V.-M. (1961). ε-entropy and ε-capacity of sets in functional spaces, *Amer. Math. Soc. Transl. 17, pp 277-364* .

Larranaga, P. and Lozano, J. A. (2001). *Estimation of Distribution Algorithms. A New Tool for Evolutionary Computation* (Kluwer Academic Publishers).

Matoušek, J. (2002). *Lectures on Discrete Geometry*, *Graduate Texts in Mathematics*, Vol. 212 (Springer).

Mühlenbein, H. and Höns, R. (2005). The estimation of distributions and the minimum relative entropy principle, *Evolutionary Computation* **13**, 1, pp. 1–27.

Mühlenbein, H. and Mahnig, T. (2002). Evolutionary computation and Wright's equation. *Theoretical Computer Science* **287**, 1, pp. 145–165.

Nelder, J. and Mead, R. (1965). A simplex method for function minimization, *Computer Journal 7* , pp. 308–311.

Powell, M. J. D. (1974). Unconstrained minimization algorithms without computation of derivatives, *Bollettino della Unione Matematica Italiana* **9**, pp. 60–69.

Rechenberg, I. (1973). *Evolutionstrategie: Optimierung Technischer Systeme nach Prinzipien des Biologischen Evolution* (Fromman-Holzboog Verlag, Stuttgart).

Rónyai, L., Babai, L. and Ganapathy, M. K. (2001). On the number of zero-patterns of a sequence of polynomials, *Journal of the American Mathematical Society* **14**, 3, pp. 717–735.

Rudolph, G. (1997). Convergence rates of evolutionary algorithms for a class of convex objective functions, *Control and Cybernetics* **26**, 3, pp. 375–390.

Sauer, N. (1972). On the density of families of sets, *Journal of Combinatorial Theory, Ser. A* **13**, 1, pp. 145–147.

Schwefel, H.-P. (1974). Adaptive Mechanismen in der biologischen Evolution und ihr Einfluss auf die Evolutionsgeschwindigkeit, Interner Bericht der Arbeitsgruppe Bionik und Evolutionstechnik am Institut für Mess- und Regelungstechnik Re 215/3, Technische Universität Berlin.

Teytaud, O. (2007). On the hardness of offline multi-objective optimization, *Evolutionary Computation* **15**, 4, pp. 475–491.

Teytaud, O. and Gelly, S. (2006). General lower bounds for evolutionary algorithms, in *Proceedings of PPSN*, *Lecture Notes in Computer Science*, Vol. 4193 (Springer), ISBN 3-540-38990-3, pp. 21–31.

Vapnik, V. and Chervonenkis, A. (1968). On the uniform convergence of frequencies of occurence events to their probabilities, *Soviet Mathematics-Doklady 9, 915-918* .

Vapnik, V. and Chervonenkis, A. (1971). On the uniform convergence of relative frequencies of events to their probabilities, *Theory of Probability and its Applications* **16**, 2, pp. 264–280.

Whitley, D. (1989). The GENITOR algorithm and selection pressure: Why rank-based allocation of reproductive trials is best, in J. D. Schaffer (ed.), *Proceedings of the Third International Conference on Genetic Algorithms* (Morgan Kaufman, San Mateo, CA), pp. 116–121.

Witt, C. and Jägersküpper, J. (2005). Rigorous runtime analysis of a (mu+1) es for the sphere function, in *Proceedings of the 2005 Conference on Genetic and Evolutionary Computation* (ACM), pp. 849–856.

Subject Index